Measuring Vulnerability to Natural Hazards

The United Nations University created the Institute for Environment and Human Security (UNU-EHS) to address risks and vulnerabilities that are the consequence of complex environmental and human-induced hazards, both creeping and of sudden onset. UNU-EHS aims to improve the in-depth understanding of the cause-effect relationships to find possible ways to reduce risks and vulnerabilities. The Institute is conceived to support policy- and decision-makers with authoritative research and information. One core area of the research agenda of UNU-EHS is to develop and test tools to measure and assess vulnerability at different scales to various hazards of natural origin. This book emanates from discussions within the UNU-EHS Expert Working Group on Measuring Vulnerability held in Kobe in January 2005 and in Bonn in October of the same year. The Expert Working Group is an ongoing platform for exchange and policy-advice in this field.

**UNITED NATIONS
UNIVERSITY**

UNU-EHS

**Institute for Environment
and Human Security**

Measuring Vulnerability to Natural Hazards: Towards Disaster Resilient Societies

Edited by Jörn Birkmann

United Nations University Press

TOKYO · NEW YORK · PARIS

United Nations University Press
United Nations University, 53-70, Jingumae 5-chome,
Shibuya-ku, Tokyo, 150-8925, Japan
Tel: +81-3-3499-2811   Fax: +81-3-3406-7345
E-mail: sales@hq.unu.edu
general enquiries: press@hq.unu.edu
http://www.unu.edu

United Nations University Office at the United Nations, New York
2 United Nations Plaza, Room DC2-2062, New York, NY 10017, USA
Tel: +1-212-963-6387   Fax: +1-212-371-9454
E-mail: unuona@ony.unu.edu

United Nations University Press is the publishing division of the United Nations University.

Cover design by Mea Rhee

Printed in Hong Kong

ISBN 92-808-1135-5

Library of Congress Cataloging-in-Publication Data

Measuring vulnerability to natural hazards : towards disaster resilient societies /
edited by Jörn Birkmann.
    p. cm.
  Includes index.
  ISBN 92-808-1135-5 (pbk.)
  1. Natural disasters. 2. Natural disaster warning systems. I. Birkmann, Jörn.
GB5014.M4   2006
363.34—dc22   2006028268

Contents

Contents

List of tables and figures

List of colour figures

Materials in colour are indicated by (c), and are grouped together in the centre of the book.

List of acronyms

ADRC	Asian Disaster Reduction Center
ARMONIA	Applied Multi Risk Mapping of Natural Hazards for Impact Assessment
BBC	approach to vulnerability analysis that goes beyond assessment of deficiencies and impacts, and views vulnerability in the context of a dynamic process. See discussion by Birkmann, Chapter 1, section on "Reflection and introduction of the BBC conceptual framework".
BCPR	Bureau for Crisis Prevention and Recovery (of the United Nations)
BGR	German Federal Institute for Geosciences and Natural Resources
BMZ	German Federal Ministry for Economic Cooperation and Development
CATSIM	IIASA catastrophe simulator
CBDM	community-based disaster management
CBDRM	community-based disaster risk management
CCI	Coping Capacity Index
CE	capital expenditure
CFMS	community-based flood management strategy
CHS	Commission on Human Security
CI	critical infrastructures
CIESIN	Center for International Earth Science Information Network
CRA	community risk assessment
CRED	Centre for Research on Epidemiology of Disasters
DALYs	disability-adjusted life years (WHO risk factors)
DDI	Disaster Deficit Index

DGMAE	Directorate of Geology and Mining Area Environment
DIPECHO	regional disaster preparedness programme of the EU Commission's Humanitarian Aid department
DPSIR	driving forces – pressure – state – impact – response
DRI	Disaster Risk Index
DTM	digital terrain model
DVGHM	Directorate of Volcanology and Geological Hazard Mitigation
ECLAC	Economic Commission for Latin America and the Caribbean
EM-DAT	Emergency Disasters Data Base
ENSO	El Niño Southern Oscillation
ESPON	European Spatial Planning Observation Network
FEMA	Federal Emergency Management Agency
GCM	global climate models
GDI	Gender Development Index
GEO	Global Environmental Outlook (of UNEP)
GIS	geo-information systems
GLM	generalised linear models
GN	"Grama Niladari" subdivision of city, etc., smallest statistical unit in Sri Lanka
GNI	gross national income
GRIP	Global Risk Identification Programme
GTZ	German Technical Cooperation Agency (Deutsche Gesellschaft für Technische Zusammenarbeit)
GUI	Global Urban Indicator
HDI	Human Development Index
HIPC	Highly Indebted Poor Countries Initiative
HSI	Human Security Index
HPI	Human Poverty Index
IADB or IDB	Inter-American Development Bank
IDEA	Instituto De Estudios Ambientales
IDNDR	International Decade for Natural Disaster Reduction (1990–1999)
IDP	internally displaced persons
IFI	international financial institution
IFRC	International Federation of the Red Cross and Red Crescent societies
IHI	Index of Human Insecurity
IIASA	International Institute for Applied Systems Analysis
IOC	Intergovernmental Oceanographic Commission
IPCC	Inter-Governmental Panel on Climate Change
IRI	integrated risk index
IRPUD	Institute of Spatial Planning, University of Dortmund
ISDR	International Strategy for Disaster Reduction (of the United Nations)
ISFEREA	Information Support for Effective Rapid External Action
JRC	Joint Research Centre of the European Commission

LDI	Local Disaster Index
MA	Millennium Ecosystem Assessment
MARS	Major Accident Reporting System
MCE	maximum considered event
MDGs	Millennium Development Goals
NCCK	National Christian Council of Kenya
NEDIES	Natural and Environmental Disaster Information Exchange System
NGI	Norwegian Geotechnical Institute
NGO	non-governmental organisation
NOAA	National Oceanic and Atmospheric Administration
NUTS	nomenclature of territorial units for statistics
OCHA	Office for the Coordination of Humanitarian Affairs
PAL	probable annual losses
PAR	persons/populations at risk
PDF	probability density function
PLA	participatory learning activities
PML	probable maximum losses
PPEW	Platform for the Promotion of Early Warning
PREVIEW	Project for Risk Evaluation, Vulnerability, Information and Early Warning
PVI	Prevalent Vulnerability Index
RMI	Risk Management Index
RRA	rapid rural assessment
SoVI	Social Vulnerability Index
SRTM	Shuttle Radar Topography Mission
TWG	technical working group
UCLAS	University College of Lands and Architectural Studies (Tanzania)
UNCED	UN Conference on Environment and Development
UNDP	United Nations Development Programme
UNEP	United Nations Environment Programme
UNU	United Nations University
UNU-EHS	United Nations University, Institute for Environment and Human Security
WBGU	German Advisory Council on Global Change
WCDR	World Conference on Disaster Reduction
WHO	World Health Organization
WWDR	World Water Development Report

Foreword

Hans van Ginkel

The increase of large-scale disasters in recent years such as the devastating tsunami event in December 2004, the extreme floods in India, Germany and Switzerland in July and August 2005, the extensive bushfires due to severe droughts in Portugal and Spain in the same period, and Hurricane Katrina, which devastated the south-east coast of the United States in August 2005 have caused fatalities, disruptions of livelihood, and enormous economic loss. These events show dramatically how the ongoing global environmental change and also inadequate coastal defence, lack of early warning and unsustainable practices, and even neglect can affect people all over the world. The international community and organisations, national governments and local communities have to cope with the consequences of unprecedented extreme events. In this regard we have to acknowledge the "un-natural" dimension of the so-called natural disasters. We have to change our approach to disasters. Frequency analysis of hazard events, once the start of all considerations, becomes unreliable as non-stationary time series overthrow 100-year return period records every few years in a merciless pace. We do not only need to think the unthinkable, and prepare to face it should it occur, but we need to explore how to be better prepared. Saving people from the worst would require taking the assessment of human (in)security as the starting point of disaster preparedness and management.

It is important to understand that disasters deriving from hazards of natural origin are only partially determined by the physical event itself.

Growing economic losses, high numbers of casualties and the disruption of livelihoods in various places of the world, at an even higher rate than the increase of magnitude and frequency of extreme events, indicate forcefully that the other factor, often characterised as our vulnerability to hazards of natural origin, must have grown over proportionally. Reducing disaster risk implies therefore also taking into account the various vulnerabilities of the affected society; that of its economy, and that of its environment, including the built environment with all its complex mega-structures. The last decades have proven that our primarily engineering approach, controlling and conquering extreme events with infrastructural measures, is not the appropriate answer. Humanity is at the threshold of taking the step from an ill-perceived "security society" into "risk society", acknowledging the limit of how far we can master nature and learning to live with risks.

The World Conference on Disaster Reduction (WCDR) held in Kobe in January 2005 formulated the goal of creating societies more resilient to disasters. The development of a system of indicators of disaster risk and vulnerability that would enable the decision makers to assess the potential impact of disasters and to promote the formulation of appropriate policy responses – while identifying the most threatened areas and social groups – is viewed as a key activity to accomplish this goal (Hyogo Framework for Action 2005–2015).

The United Nations University Institute for Environment and Human Security (UNU-EHS) took the initiative to invite leading scholars and practitioners to discuss the state of the art of measuring vulnerability, to devise potential research initiatives on how to capture vulnerability at different aggregation levels of society. This publication, entitled *Measuring Vulnerability to Natural Hazards – Towards Disaster Resilient Societies*, edited by Jörn Birkmann, is the first summary of this work started just after the WCDR. It examines various methodologies from global indexing projects to local participatory self-assessment approaches. It reviews retrospective studies and takes stock of the efforts to "predict" vulnerability. A critical review of current methodologies of how to measure vulnerability is provided. The book leaves no doubt that there is still a long way to go from concepts and experiments to the full practical use of anticipative vulnerability measurement. In this context we may introduce this book as the first volume of a collaborative series.

I am very proud that the United Nations University is among the organisations that started immediately the implementation of the Hyogo Framework for Action. I wish to thank the authors for contributing to this book and want to invite every interested scientist, colleagues from the UN organisations and professionals from all over the world, to contribute to the work ahead of us.

Prof. Dr. Hans van Ginkel
Rector of the United Nations University, Japan
Under-Secretary-General of the United Nations

Preface

Sálvano Briceño

The tsunami tragedy of 26 December 2004, Hurricanes Katrina and Rita, the South Asia earthquake and other smaller disasters in 2005 are shattering reminders of how people's lives and property can be swept away in a matter of minutes. The recent earthquake left more than 78,000 people dead, with a colossal loss of livelihoods and infrastructure, which has again highlighted the long-term failure of rapidly developing countries to reduce disaster risk. In the last decade alone, disasters affected 3 billion people, killed over 750,000 people and cost around US\$ 600 billion.

In 2004 more than 240,000 people perished in 396 natural disasters that affected over 146 million people. Over 225,000 of these deaths were a result of the Indian Ocean tsunami that hit 12 countries on 26 December 2004. The South Asia earthquake and the tsunami and other disasters are a wake-up call to what should have been realised long ago. Disasters are undermining the world's development as never before. The current widespread disregard for disaster risks, hazards and their impacts presents an extraordinary challenge to communities and nations in their efforts to move closer to the Millennium Development Goals. Now is the time to realise that we are far from powerless: communities and nations can build their resilience to disasters by investing in proactive measures to reduce risk and vulnerability. Disaster risk reduction is essential to meet global challenges including sustainable development and the eradication of poverty. The case for disaster reduction is clear. Disaster risk concerns every person, every community, and every nation; indeed, disaster impacts are slowing down development, and their impact and actions in one region can have an impact on risks in another, and vice versa. Without taking into consideration the urgent need to reduce risk and vulnerability, the world simply cannot hope to move forward in its quest for sustainable development and reduction of poverty.

The tsunami disaster also gave additional relevance to the work of the World Conference on Disaster Reduction (WCDR, Kobe, Hyogo, Japan, 18–22 January 2005), and the blueprint agreed by Governments during the conference, the *Hyogo Framework for Action 2005–2015: Building the Resilience of Nations and Communities to Disasters*. The Hyogo Framework carries a strong commitment and ownership of Governments and regional, international and non-governmental organisations. We need to proceed to ensure effectiveness in translating the hopeful expectations of the Hyogo Framework for Action into the practical measures at international, regional, national and community levels, and into tangible activities by which progress in disaster reduction must be measured. The emphasis of the Hyogo Framework on the focus for national implementation and follow-up, with the primary responsibility of States, requires, as a corollary, the development of strong participatory and collaborative ties with civil society and authorities at national and local levels, involving all development sectors (health, education, agriculture, tourism, etc.), national disaster management systems, business sector, academic, scientific and technical support organisations.

The Hyogo Framework sets out specific priorities for action on early warning for all hazards and on associated risk assessment and preparedness. A primary lesson learnt from the tsunami was the importance of early warning systems for protecting people and property. Unlike the Pacific Ocean basin, countries located along the Indian Ocean do not have a regional early warning system. The ISDR secretariat through its Platform for the Promotion of Early Warning (UN/ISDR-PPEW) based in Bonn is supporting the Intergovernmental Oceanographic Commission (UNESCO/IOC) through the UN Tsunami Flash Appeal to build the capacities in the Indian Ocean. The ISDR secretariat is also promoting a coordinated UN-wide approach to early warning systems, involving not only technical organisations such as UNESCO/IOC and WMO, but also those involved in disaster management and development such as UN/ OCHA, UNDP, UNEP, UNESCAP, UNU, ADPC, ADRC and others to promote and coordinate related activities in warning response, preparedness and education, such as workshops for disaster managers, community leaders, media and the production of information materials, lessons learned and community based projects.

I am pleased to support the work of the UNU-EHS through its Expert Working Group on Measuring Vulnerability. The Hyogo Framework represents the most comprehensive action-oriented policy guidance in universal understanding of disasters induced by vulnerability to natural hazards and reflects a solid commitment to implementing an effective disaster reduction agenda. In this context, the UNU-EHS Expert Working Group is a valuable contribution to the implementation of the Hyogo Framework. I look forward to an increased collaboration between UNU-EHS and the ISDR Secretariat.

Sálvano Briceño
Director, Secretariat of the International Strategy for Disaster Reduction
Switzerland

Acknowledgements

Of course, putting this book together would not have been possible without the immense amount of behind-the-scenes and often unnoticed work done by the production team. Therefore, I am extremely grateful to the Rector of UNU, Hans van Ginkel, and the Director of UNU-EHS, Janos Bogardi, for their continued support since the beginning of our publishing endeavours.

Many thanks also to all the contributors from various countries, who often had to respond to many remarks and requests under the constraints of tight deadlines.

Moreover, I would like to express my deep gratitude to Janos Bogardi, Omar Cardona, Katharina Thywissen, Carlota Schneider, Fabrice Renaud and Juan Carlos Villagran for their comments on the manuscript and valuable suggestions.

My very special thanks and appreciation go to my research assistants Matthias Garschagen, Niklas Gebert and Benjamin Etzold for their enormous commitment and unflinching energy in bringing together relevant literature and in supporting the coordination of all the different papers and their revised versions.

My sincere words of thanks also go to the staff of UNU-PRESS in Tokyo, Rector Hans van Ginkel for his excellent foreword, Janos Bogardi for his compelling introduction, Sálvano Briceño, Director of UN/ISDR, for his preface, and Simon Horner of ECHO for his support.

Finally, I am honoured to mention the valuable role played by Alison McKelvey Clayson, Matthew Connolly and Ilona Roberts for helping to set up the editorial version and their careful copy-editing.

Introduction

Janos J. Bogardi

The beginning of a long road

The well-known statistical analysis of the MunichRe Georisk Research Group shows a close to threefold increase in the occurrence of extreme natural hazard events over the last three decades, an approximately six-fold increase in associated economic damages, and a constant number of casualties as a result of these disasters of natural origin. These trends underline the need for still more efforts, more focused disaster management. But they also reveal the necessity to recognise risk and make people aware of and prepared to live with risk, and to respond adequately should they face the occurrence of extreme events.

The World Conference on Disaster Reduction (WCDR) held in Kobe, Japan, in January 2005, was an excellent opportunity to take stock. The Hyogo Framework for Action agreed on during this conference gave the mandate and set the direction for professionals, scientists, individuals, and institutions alike. Among other priorities, it defines the development of indicator systems for disaster risk and vulnerability as one of the key activities enabling decision makers to assess the possible impacts of disasters. The subsequent Strategic Directions compiled by the United Nations International Strategy for Disaster Reduction (ISDR, 2005) should help to set the conference follow-up in motion. While the United Nations, State actors, non-governmental organisations, and many dedicated individuals are emphasising the disaster preparedness and management agenda, Mother Nature has dramatically confirmed this urgency. The

1

most recent mega-events, the 2004 Indian Ocean tsunami and Hurricane Katrina in 2005, will certainly strengthen the political momentum to act. At this juncture the scientific and professional community is expected to come up not only with concepts and strategies, but also with actions and capacity-building initiatives.

Do we know enough to advise parliaments and Governments how to find the best answers, and where to spend limited funds most efficiently? We have to ask ourselves whether and how fast we can come up with the required risk and vulnerability indicator system, one particular requirement of the Hyogo Framework for Action, with concepts and practical methods that are robust and ready to be used while sound enough to withstand critical scientific scrutiny. Unless the reply is a resounding "yes" we had better join forces to map the scientific issues and challenges involved, to debate, to develop, to test methods without losing sight of the mandate and requirements set by the World Conference on Disaster Reduction (WCDR).

There is plenty to debate. But are we well prepared for this process? We face even a terminological cacophony. Vulnerability and many other colloquial terms (risk, hazard, resilience, resistance) found in disaster management concepts are widely used irrespective of the fact that there are still no universally agreed definitions. An array of glossaries have been published to promote the use of a common terminology, or at least to serve as dictionaries for helping experts from different disciplines and schools to understand each other.

While this book also incorporates a comparative glossary, its main objective is to move the whole agenda forward. It documents the efforts being made by the scientific community to address issues well beyond these terminological concerns, by taking stock and summarising the state of the art of measuring vulnerability at the point where scientists and professionals have started the WCDR follow-up process.

Perspectives worth striving for

Vulnerability is broadly understood as the predisposition to be hurt should an event beyond a certain (though again ill-defined) threshold of magnitude occur and impact the society, its economic assets, the ecosystem, or its infrastructure. This general concept of vulnerability fits well into the ongoing scientific debate on security, and can be associated with the manifold dimensions of human security as defined by UNDP (1994) or represented and championed by the Commission on Human Security as "freedom from want" and by the Human Security Network as "freedom from fear" (Krause, 2004). As recently as 2005, Bogardi and Brauch suggested extending the human security concept by introducing a third pillar – "freedom from hazard impacts" – thus emphasising the environmental dimension of human security. In this context vulnerability would describe society's (in)security versus natural and human-induced hazards. This book deals with vulnerabilities to hazards of natural origin. We have to acknowledge however that human impact may influence both hazard magnitude and frequency.

Thus vulnerability, once it is properly assessed and preferably quantified, is the crucial feature that could serve to estimate the potential consequences of both rapid onset and/or creeping (natural) hazard events on the affected entities.

By following this line of thought, we can imagine that vulnerability assessment will become the crucial component of disaster preparedness. Monitoring vulnerability may be used to identify those target communities where proactive measures are needed, mostly to pre-empt the devastating consequences of extreme events should they occur. In a longer perspective, vulnerability assessment could become the core of a "political early-warning" system, at both national and international levels.

Our ability to assess a population's vulnerability and to use this information in the policy and decision-making sphere would be much easier if only we could develop indicators or indices to encapsulate the notion of vulnerability.

Some intriguing questions

How can we capture the idea of vulnerability or vulnerabilities? This is especially difficult in the human and social contexts, because vulnerabilities are hardly discernible without also looking at coping capacity, i.e. the ability of the potentially threatened group to overcome its vulnerabilities.

Thus there are a multitude of questions to answer.
- Can vulnerability be measured and quantified, and if yes, how?
- Can vulnerability be aggregated to characterise societies' overall susceptibility to several distinct hazards?
- Can vulnerability and coping capacity be conceived and assessed separately?
- At what aggregation level can vulnerability be measured?
- Could vulnerability assessment results be scaled up or down?
- What could be used as surrogate measures of vulnerability?
- How can vulnerability be assessed in advance of a devastating event?
- What lessons can be learned from retrospective assessment of vulnerability?

The above list is deliberately incomplete. Rather, it offers a sampling of questions meant to illustrate the great range of problems faced by the scientific community, practising professionals and decision makers alike. In the following chapters more than 40 authors from all corners of the world present the state of the art. They discuss potential developments, attempt to answer some of these questions, and seek to formulate yet more questions.

The book includes five parts, with 24 chapters, which address various aspects and approaches of measuring vulnerability.

Following the introduction, the first part deals with the concept of vulnerability and especially vulnerability indicators. Birkmann introduces different definitions and conceptual frameworks to systematise vulnerability developed and used by different schools of thought, such as the disaster risk community, development research and global change research. The second chapter gives an overview of theoretical aspects and requirements of vulnerability indicators. Both chapters include various links to approaches presented in the book, thus providing an important framework for the chapters that follow. Schneiderbauer and Ehrlich introduce a framework for determining vulnerability at different levels. They also address the question of whether vulnerability should be measured for a specific hazard or whether it should be hazard-independent. Thereafter Queste and Lauwe tackle the crucial question of what indicators are needed from a practitioner's perspective.

The second part gives insight into the relationship between vulnerability and environmental change. The environmental dimension of vulnerability is analysed and outlined by Renaud, then Kok, Narain, Wonink, and Jaeger examine the linkages between human vulnerability and environmental change.

The third part encompasses various approaches to measuring vulnerability and risk at global, national and sub-national scale. In the seventh chapter Pelling reviews the major global disaster risk index projects. Additional information regarding these approaches is presented by authors who were involved in the development of each approach. Thus, the intention and methodology of the Disaster Risk Index is shown by Peduzzi, the hotspots methodology by Dilley and the System of Indicators for Disaster Risk Management in the Americas are described by Cardona. On the basis of the global index projects a European approach of multi-risk assessment is presented by Greiving, followed by a study regarding the measurement of disaster vulnerability at national scale in Tanzania by Kiunsi and Meshack. Finally, Plate proposes a methodology to capture both vulnerability and coping capacity within a single human security index.

The fourth part focuses on approaches at the local level. It encompasses a community-based disaster risk assessment tested in Indonesia and presented by Bollin and Hidajat, as well as an overview of different methods to measure risk and vulnerability based on the experiences of the Asian Disaster Reduction Centre (ADRC) as explained by Arakida. Villagrán de León outlines a methodology to measure the vulnerability of different sectors illustrated by examples from Latin America. In contrast to quantitative approaches Wisner introduces more qualitative and participatory approaches to assess vulnerability and coping capacity using self-assessment tools. The first results of a study of United Nations University and Institute for Environment and Human Security (UNU-EHS), which uses different methods to measure vulnerability of communities to coastal hazards in Sri Lanka after the devastating tsunami event are presented by Birkmann, Fernando, and Hettige.

Part five deals with specific approaches to capturing and assessing institutional vulnerability, coping capacity and lessons learned. Lebel, Nikitina, Kotov, and Manuta underline the necessity of assessing institutional capacities to reduce risk using the example of flood disaster risk. The complexities of ensuring preparedness of institutions and the public sector for hazard events are also addressed by Mechler, Hochrainer, Linnerooth-Bayer, and Pflug who present a model to measure public sector financial vulnerability. The chapter by Billing and Madengruber focuses on the difficulties of measuring coping capacity, while Krausmann and Mushtaq introduce the approach of lessons learned as illustrated by examples drawn from European experience.

Chapter 23 summarises key aspects discussed in the preceding chapters and Birkmann, the author, draws important conclusions, which could also give some guidance for future research activities and research needs.

Finally, a comparative glossary of key terms in disaster risk reduction is presented by Thywissen, who illustrates the various definitions of the same terms by different institutions and experts.

Forums, platforms, networks: the UNU-EHS approach

Irrespective of the excellent contributions of so many co-authors to this book, it must be admitted that not all issues were captured, nor all concerns addressed. This book has focused mainly on vulnerability to rapid onset hazard events, whereas the scope and range of vulnerability research are much broader than this. Vulnerability to environmental change, capacity for adaptation, human-induced hazards, and many other areas are also being investigated. The UNU-EHS, which intends to move the scientific debate towards results that have practical applicability and are relevant to policy makers, expects to broaden its coverage in due course. But first it needs a firm conceptual basis.

The human security mandate of the Institute, which also reflects developments in the political arena worldwide, implies that any extension of the vulnerability debate should keep a strong social focus in mind. The recent establishment of a chair on social vulnerability at UNU-EHS, supported by the MunichRe Foundation, not only underlines the strong appeal of this approach to different stakeholder groups but also provides an excellent opportunity to broaden the interdisciplinary approach of the vulnerability debate.

I am very grateful to Professor Hans van Ginkel, Rector of UNU, for his encouragement to publish this book. It is based to a large extent on the contributions of participants at the first expert workshop on measuring vulnerability organised by UNU-EHS and co-organised and hosted by ADRC in Kobe, Japan, in January 2005. It is my pleasure to thank the many contributors to this book, and fellow scientists and practitioners who joined UNU-EHS in its quest to find answers to the question of how to measure the unmeasurable. My thanks are also due to Dr. Jörn Birkmann, whose enthusiasm and dedication as editor were instrumental in motivating the authors and bringing their contributions together.

We are at the beginning of a very long road. We know, both as scientists and concerned human beings, that we have an obligation to proceed towards better risk preparedness. Recognising our vulnerabilities is perhaps the first important step.

REFERENCES

Bogardi, J. and H.G. Brauch (2005) "Global Environmental Change: A Challenge for Human Security – Defining and Conceptualising the Environmental Dimension of Human Security", in A. Rechkemmer, ed., *UNEO – Towards an International Environmental Organization – Approaches to a Sustainable Reform of Global Environmental Governance*, Baden-Baden: Nomos, pp. 85–109.

Krause, K. (2004) "Is Human Security 'More than Just a Good Idea'?", in Bonn International Center for Conversion (BICC) ed., *Brief 30: Promoting Security: But How and for Whom?*, Bonn: BICC, pp. 43–46.

UN/ISDR (International Strategy for Disaster Reduction) (2005) *Strategic Directions for the ISDR System to Assist the Implementation of the Hyogo Framework for Action 2005–2015: Building the Resilience of Nations and Communities to Disasters*, Inter-Agency Task Force on Disaster Reduction, Geneva, available at http://www.unisdr.org/eng/task%20force/tf-meeting-11th-eng.htm.

Part I

Basic principles and theoretical basis

1

Measuring vulnerability to promote disaster-resilient societies: Conceptual frameworks and definitions

Jörn Birkmann

Introduction

This chapter stresses the need for a paradigm shift from quantification and analysis of the hazard to the identification, assessment and ranking of vulnerabilities. It underlines the importance of measuring vulnerability and developing indicators to reduce risk and the vulnerability of societies at risk, as mentioned in the final document of the 2005 World Conference on Disaster Reduction. Different conceptual frameworks of vulnerability in the context of disaster resilience are presented. The links between vulnerability and sustainable development are also discussed.

From hazard analysis to assessment of vulnerability

The ability to measure vulnerability is increasingly being seen as a key step towards effective risk reduction and the promotion of a culture of disaster resilience. In the light of increasing frequency of disasters and continuing environmental degradation, measuring vulnerability is a crucial task if science is to help support the transition to a more sustainable world (Kasperson et al., 2005).

UN Secretary-General Kofi Annan has underlined the fact that hazards only become disasters when people's lives and livelihoods are swept away (Annan, 2003). His view is in contrast to research and strategies in the past, which were often purely hazard-oriented (Lewis, 1999).

Instead of defining disasters primarily as physical occurrences, requiring largely technological solutions, disasters are better viewed as a result of the complex interaction between a potentially damaging physical event (e.g. floods, droughts, fire, earthquakes and storms) and the vulnerability of a society, its infrastructure, economy and environment, which are determined by human behaviour. Viewed in this light, natural disasters can and should be understood as "un-natural disasters" (Cardona, 1993; van Ginkel, 2005). Thus the promotion of disaster-resilient societies requires a paradigm shift away from the primary focus on natural hazards and their quantification towards the identification, assessment and ranking of various vulnerabilities (Maskrey, 1993; Lavell, 1996; Bogardi and Birkmann, 2004). It is part of UNU-EHS's mission to contribute to the identification of various vulnerabilities and the development and testing of relevant indicators and assessment tools (Birkmann, 2005) in order to expand the environmental dimension of human security further (Brauch, 2005).

In the final document of the World Conference on Disaster Reduction, "Hyogo Framework for Action 2005–2015", the international community underlined the need to promote strategic and systematic approaches to reducing vulnerabilities and risks to hazards (United Nations (UN), 2005, preamble). The declaration points out that:

The starting point for reducing disaster risk and for promoting a culture of disaster resilience lies in the knowledge of the hazards and the physical, social, economic and environmental vulnerabilities to disasters that most societies face, and of the ways in which hazards and vulnerabilities are changing in the short and long term, followed by action taken on the basis of that knowledge. (UN, 2005)

In this context the Hyogo Framework stresses the need to develop indicators of vulnerability as a "key activity":

Develop systems of indicators of disaster risk and vulnerability at national and sub-national scales that will enable decision-makers to assess the impact of disasters on social, economic and environmental conditions and disseminate the results to decision makers, the public and populations at risk. (UN, 2005)

Although the international community does not formulate guidelines on how to develop indicators or indicator systems to assess vulnerability, the Hyogo Framework for Action underlines the fact that impacts of disasters on (1) social, (2) economic, and (3) environmental conditions should be examined through such indicators. Since sustainable development is characterised by three pillars – social, economic and environmental (UN, 1993; WCED, 1987) – the formulation used in the Hyogo Framework for

Action can be interpreted as implying a link between vulnerability assessment and sustainable development. Moreover, the declaration underlines the necessity to develop methods and indicators which, based on those recommendations, can be used in policy and decision-making processes. Furthermore, it is evident that measuring vulnerability requires, first and foremost, a clear understanding and definition of the concept of vulnerability.

Definitions

The current literature encompasses more than 25 different definitions, concepts and methods to systematise vulnerability (for example, Chambers, 1989; Bohle, 2001; Wisner et al., 2004; Downing et al., 2006; UN/ISDR, 2004: 16; Pelling, 2003: 5; Luers, 2005: 215; Green, 2004: 323; UN-Habitat, 2003: 151; Schneiderbauer and Ehrlich, 2004; van Dillen, 2004: 9.; Turner et al., 2003: 8074; Cardona, 2004b: 37). The website of the ProVention Consortium includes about 20 manuals and different guidebooks on how to estimate vulnerability and risk (ProVention Consortium website). These manuals also include different definitions and various conceptual frameworks of vulnerability.

Although vulnerability has to be viewed in its multifaceted nature (Bohle, 2002a, 2002b), the different definitions and approaches show it is not clear just what "vulnerability" stands for as a scientific concept (Bogardi and Birkmann, 2004: 76). We are still dealing with a paradox: we aim to measure vulnerability, yet we cannot define it precisely. Although there is no universal definition of vulnerability, various disciplines have developed their own definitions and pre-analytic visions of what vulnerability means. An overview of different definitions is given by Thywissen in this book, and can also be studied for example in Schneiderbauer and Ehrlich (2004), Green (2004) and Cardona et al. (2003). Nevertheless, it is useful to give a brief introduction of the terms vulnerability, hazard, risk and coping capacity in order to discuss different concepts of how to systematise vulnerability.

Vulnerability

Vulnerability is a concept that evolved out of the social sciences and was introduced as a response to the purely hazard-oriented perception of disaster risk in the 1970s (Schneiderbauer and Ehrlich, 2004: 13). Since the 1980s, the dominance of hazard-oriented prediction strategies based on technical interventions has been increasingly challenged by the alternative paradigm of using vulnerability as the starting point for risk reduction. This approach combines the susceptibility of people and communities exposed with their social, economic and cultural abilities to cope with the damage that could occur (Hilhorst and Bankoff, 2004: 2). Additionally, some authors distinguish between social vulnerability on the one hand, which deals with the susceptibility of humans and the conditions necessary for their survival and adaptation, and biophysical vulnerability on the other (WBGU, 2005: 33). Biophysical vulnerability in this context is a concept developed from global environmental change research, where it is widely used to describe the extent to which a system is vulnerable to adverse effects of climate change and to what extent it is (un-)able to adapt to such impacts (see in detail WBGU, 2005: 33). Although there is still much uncertainty about what the term vulnerability covers, Cardona (2004b) underlines the fact that the concept of vulnerability helped to clarify the concepts of risk and disaster. He views vulnerability as an intrinsic predisposition to be affected by or to be susceptible to damage; that means vulnerability represents the system or the community's physical, economic, social or political susceptibility to damage as the result of a hazardous event of natural or anthropogenic origin (Cardona, 2004: 37–51).

One of the best-known definitions was formulated by the International Strategy for Disaster Reduction (UN/ISDR), which defines vulnerability as:

The conditions determined by physical, social, economic and environmental factors or processes, which increase the susceptibility of a community to the impact of hazards. (UN/ISDR, 2004)

In contrast, the United National Development Programme (UNDP) defines vulnerability as:

a human condition or process resulting from physical, social, economic and environmental factors, which determine the likelihood and scale of damage from the impact of a given hazard. (UNDP, 2004: 11)

While the definition of vulnerability used by the ISDR encompasses various conditions that have an impact on the susceptibility of a community, the UNDP definition understands vulnerability as a human condition or process. The human-centred definition used by UNDP affects the method used to calculate its Disaster Risk Index, especially with regard to the calculation of relative vulnerability (UNDP, 2004: 32). The Disaster Risk Index measures the relative vulnerability of a country to a given hazard by dividing the number of *people killed* by the number of people exposed (see Peduzzi, Chapter 8; Pelling, Chapter 7). Using people killed divided by people exposed as the indicator to measure relative vulnerability corresponds with the understanding that vulnerability is primarily a human condition. Furthermore, the lack of appropriate data at the global level has restricted UNDP's opportunities to establish a broader index. Although one has to take into account that human society is the main focus of concepts of vulnerability, a fundamental question has to be clarified: can human vulnerability be adequately characterised without considering simultaneously the vulnerability of the "surrounding" ecosphere? (e.g. Turner et al., 2003).

Furthermore, other authors, such as Vogel and O'Brien (2004: 4) stress the fact that vulnerability is:

- *multi-dimensional and differential* (varies across physical space and among and within social groups)
- *scale dependent* (with regard to time, space and units of analysis such as individual, household, region, system)
- *dynamic* (the characteristics and driving forces of vulnerability change over time).

Regarding the concept of social vulnerability, Cannon et al. (2003: 5) argue that social vulnerability is much more than the likelihood of buildings collapsing and infrastructure being damaged. They describe social vulnerability as a set of characteristics that includes a person's:

- initial well-being (nutritional status, physical and mental health)
- livelihood and resilience (assets and capitals, income and qualifications)
- self-protection (capability and willingness to build a safe home, use a safe site)
- social protection (preparedness and mitigation measures)
- social and political networks and institutions (social capital, institutional environment and the like).

The definition by Cannon et al. (2003) reflects the fact that vulnerability is only partially determined by the type of hazard; it is mainly driven by precarious livelihoods, the degree of self-protection or social protection, qualifications and institutional settings that define the overall context in which a person or a community experiences and responds to the negative impact of a hazardous event (Cannon et al., 2003: 5). However, the concept of social vulnerability also lacks a common definition, which means that different authors use it differently. Current literature reveals the fact that social vulnerability can encompass various aspects and features, which are linked to socially created vulnerabilities. Therefore, the concept of social vulnerability is not limited to social fragilities, but rather includes topics such as social inequalities regarding income, age or gender, as well as characteristics of communities and the built environment, such as the level of urbanisation, growth rates and economic vitality (Cutter et al., 2003: 243). Downing et al. (2006) define six attributes to characterise social vulnerability based on the experiences of over two decades of research on this topic. They emphasise that social vulnerability is:

- the differential exposure to stresses experienced or anticipated by the different units exposed
- a dynamic process
- rooted in the actions and multiple attributes of human actors
- often determined by social networks in social, economic, political and environmental interactions
- manifested simultaneously on more than one scale
- influenced and driven by multiple stresses.

Consequently, the concept of social vulnerability refers to more than socio-economic impacts, since it can also encompass features of potential physical damage in the built environment (Cutter et al., 2003: 243). Other experts such as Carreño et al. (2005a and 2005b) clearly distinguish between socio-economic fragilities and lack of resilience as social context conditions (that favour the second order impacts) on the one hand, and the physical damage caused by exposure and physical susceptibility of the built environment on the other hand (related to first-order impacts) (Cardona, 1999 and 2001; Cardona and Hurtado, 2000a, 2000b, 2000c; Cardona and Barbat, 2000; Carreño et al., 2004, 2005a, 2005b).

Downing et al. (2006) underline the fact that the concept of social vulnerability encompasses various vulnerability features, which are driven by multiple stresses and differential exposure, and are often rooted in multiple attributes of human actors and social networks.

One has to conclude that the concept of social vulnerability is much more broadly used than just for the estimation of traditional social aspects of vulnerability (gender, age and income distribution). Seen from the perspective of the social vulnerability school of thinking, "social vulnerability" can also encompass economic and physical aspects, provided they are the expressions of a socially constructed vulnerability. Although the conceptual classification of vulnerability differs, for example, between Cutter et al. (2003) and Carreño et al. (2005a and 2005b), both schools of thinking underline the fact that vulnerability should not be limited to an estimation of the direct impacts of a hazardous event. Rather, it has to be seen as the estimation of the wider environment and social circumstances, thus enabling people and communities to cope with the impact of hazardous events or, conversely, limiting their ability to resist the negative impact of the hazardous event. This underlines the fact that vulnerability can also take into account the coping capacity and resilience of the potentially affected society. However, it is important to acknowledge that also the analysis of damage patterns can contribute to the identification of revealed vulnerabilities as well as to the estimation of current and potential vulnerabilities in the future. Therefore, the challenge lies in developing a balanced approach between the general context and the macro indicators, on one side, and more precise and specific indicators on the other, which can also be based on revealed vulnerabilities in the past.

Coping capacity

According to ISDR, coping capacity can be defined as:

a combination of all strengths and resources available within a community or organization that can reduce the level of risk, or the effects of a disaster. (UN/ISDR, 2002)

Vulnerability and coping capacity manifest themselves once a vulnerable community is exposed to a hazardous event. In this context hazard is understood as:

A potentially damaging physical event, phenomenon and/or human activity, which may cause the loss of life or injury, property damage, social and economic disruption or environmental degradation. (UN/ISDR, 2002)

Compared to the terms hazard and vulnerability, the term risk can be described as the product of the interaction between hazard and vulnerability.

In risk sciences the term risk encompasses the probability and the amount of harmful consequences or expected losses resulting from interactions between natural or human induced hazards and vulnerable conditions. (UN/ISDR, 2002)

Moreover, the term *resilience* gained high recognition in the Hyogo Framework and the debate thereafter. The current literature reveals different interpretations of the term, especially concerning the question of whether resilience is defined as the capacity to absorb disturbances or shocks, and is thus more linked to the understanding of resistance, or whether the term refers to the regenerative abilities of a social or an ecosystem, encompassing the ability to learn and adapt to incremental changes and sudden shocks while maintaining its major functions. This meaning relates more to the coping and adaptation phase (see e.g. Adger et al., 2005: 1036; Allenby and Fink, 2005: 1034). In some cases resilience is also understood as the opposite of vulnerability (Adger et al., 2005), while others view vulnerability as the opposite and lack of human security (Bogardi and Brauch, 2005). Generally, a common ground can be seen in the understanding that resilience describes the capability of a system to maintain its basic functions and structures in a time of shocks and perturbations (Adger et al., 2005; Allenby and Fink, 2005). This definition of resilience also implies that the respective system or unit is able to adapt and learn, meaning that the system – e.g. social system, ecosystem or coupled human–environmental system – can mobilise sufficient self-organisation to maintain essential structures and processes within a coping or adaptation process.

What have we learned so far? Preliminary observations

The overview of key-terms associated with vulnerability and risk has revealed that although the concept of vulnerability has achieved a high degree of recognition in different fields, such as disaster management, environmental change research and development studies, the concept is still somewhat fuzzy and often used with differing connotations. In this context it might be misleading to try to establish a universal definition. Therefore the author provides an overview of the different spheres of the concept of vulnerability (Figure 1.1), without intending to be comprehensive.

Nearly all concepts of vulnerability view it as an "internal side of risk", closely linked with the discussion of vulnerability as an intrinsic characteristic of a system or element at risk. That means the conditions of the exposed element or community (susceptibility) at risk are seen as core characteristics of vulnerability (UN/ISDR, 2004; Cardona, 2004a/b: 37; Wisner, 2002: 12/7; Thywissen, in this book) and this can be defined as a common ground (the inner circle in Figure 1.1). Interestingly, the understanding that vulnerability is seen as an internal side of risk and as an intrinsic characteristic of an element at risk can be applied for very different elements, such as communities and social groups (socio-economic conditions, institutional framework), structures and physical characteristics of buildings and lifelines (physical structure), as well as eco-systems and environmental functions and services (ecosystem, environmental capital).

An extension of this definition can be seen in definitions such as Wisner's (2002), which defines vulnerability as the likelihood of injury, death, loss and disruption of livelihood in an extreme event, and/or unusual difficulties in recovering from negative impacts of hazardous events – primarily related to people (Wisner, 2002: 12/7). This definition underlines the fact that the main elements of vulnerability are those conditions that increase and determine the likelihood of injury, death, loss and disruption of livelihood of human beings. Thus a second sphere can be associated with this human-centred definition of the likelihood of death, injury and loss (Figure 1.1).

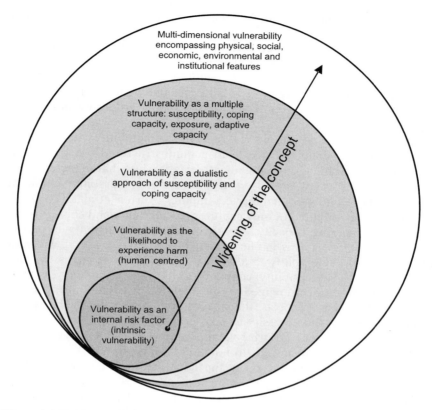

Figure 1.1 Key spheres of the concept of vulnerability.
Source: Birkmann 2005.

Furthermore, the "likelihood of injury" is extended by the focus of a dualistic structure of vulnerability, which can be observed in the definitions by Wisner (2002) and also partially by Chambers (1989) and Bohle (2001). Wisner clearly identifies the "likelihood of injury" and "unusual difficulties in recovering" from such events as the key features of vulnerability. This means the concept of vulnerability is widened by viewing vulnerability as implying a dualistic approach of susceptibility on the one hand and the unusual difficulties in coping and recovering on the other. However, Bohle's double structure of vulnerability (Figure 1.1) is not just "exposure" and "coping"; rather, it refers to vulnerability features which are external to an exposed element or unit at risk and those factors that are internal. The distinction between these two spheres "external exposure" and "internal coping" emphasises that vulnerability deals on the one hand with features and characteristics linked to capacities to anticipate and cope with the impact of a hazard, and on the other, with the exposure to risks and shocks (Bohle, 2001). In this context a third sphere can be associated with the "dualistic structure of vulnerability", which underlines the fact that vulnerability is shaped and determined by the likelihood of injury (susceptibility, negative definition) and by the ability and capacity to cope with (positive definition) and recover from these stresses and negative impacts of the hazardous event (Wisner, 2002: 12–17).

An additional extension of the concept of vulnerability can be seen in the shift from a double structure to a multi-structure. The conceptual framework of Bohle (2001) already stresses the fact that vulnerability is a multifaceted concept, and also the discourse of vulnerability within the climate change and sustainability community (Turner et al., 2003) highlights that vulnerability not only captures susceptibility and coping capacity, but also adaptive capacity, exposure and the interaction with perturbations and stresses. This implies a fourth sphere (Figure 1.1) widening the concept of vulnerability to a multi-structure that encompasses exposure, sensitivity, susceptibility, coping capacity, adaptation and response.

While the traditional engineering perspective of vulnerability focused primarily on physical aspects, the current debate regarding vulnerability clearly underlines the necessity to take into account various themes and parameters that shape and drive vulnerability (UN/ISDR, 2004), such as physical, economic, social, environmental and institutional characteristics. Some approaches also stress the necessity to integrate additional global drivers that have an impact on vulnerability, such as globalisation and climate change (Vogel and O'Brien 2004: 3; O'Brien and Leichenko, 2000). This implies that the focus of attention has shifted from a primarily physical structure analysis to a broad interdisciplinary analysis of the multidimensional concept of vulnerability (e.g. Cardona, 2004b: 39–49). The widening of the concept of vulnerability is illustrated in Figure 1.1. It shows that starting from a general basic understanding (first inner sphere), a process of broadening took place and this is shown through the arrow in the figure.

The different spheres of the concept of vulnerability are also reflected in the various conceptual frameworks to systematise vulnerability. Selected conceptual frameworks will be discussed in the following pages.

Conceptual frameworks of vulnerability

The different views on vulnerability are reflected in various analytical concepts and models of how to systematise it. Since these conceptual models are an essential step towards the development of methods mea-

suring vulnerability and the systematic identification of relevant indicators (Downing, 2004: 19), the following paragraphs give an insight into
different conceptual frameworks, such as the double structure of vulnerability as defined by Bohle, selected approaches of the disaster risk community, such as the UN/ISDR framework for disaster risk reduction, and
lastly the two conceptual frameworks developed by UNU-EHS.

The double structure of vulnerability

According to Bohle (2001), vulnerability can be seen as having an *external* and an *internal* side (see Figure 1.2). The *internal* side, coping, relates
to the capacity to anticipate, cope with, resist and recover from the impact of a hazard; in contrast, the *external* side involves exposure to risks
and shocks. In social sciences the distinction between the exposure to external threats and the ability to cope with them is often used to underline
the double structure of vulnerability (van Dillen, 2004). Based on the
perspective of social geography and the intensive famine research carried
out by Bohle (2001: 119), the pre-analytic vision of the double structure
underlines the fact that vulnerability is the result of interaction between
exposure to external stressors and the coping capacity of the affected
household, group or society. Thus the definition clearly identifies vulnerability as a potentially detrimental social response to external events and
changes such as environmental change. Interestingly, Bohle's conceptual
framework describes exposure to hazards and shocks as a key component
of vulnerability itself.

Viewed in this way, the term exposure goes beyond mere spatial exposure since it also encompasses features related to the entitlement theory
and human ecology perspective. Within the debate of social vulnerability
the term *exposure* also deals with social and institutional features, meaning processes that increase defencelessness and lead to greater danger,
such as exclusion from social networks. These alter the exposure of a person or a household to risk (Cannon et al., 2003). Moreover, the conceptual framework of the double structure indicates that vulnerability cannot
adequately be characterised without simultaneously considering coping
and response capacity, defined here as the internal side of vulnerability.

The sustainable livelihood framework

The 'sustainable livelihood framework' can also be seen as a framework
or vade-mecum for vulnerability assessment. Key elements of this
approach are the five livelihood assets or capitals (human, natural, financial, social and physical capital), the 'vulnerability context' viewed as
shocks, trends and seasonality, and the influence of transforming struc-

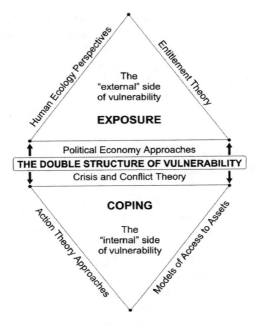

Figure 1.2 Bohle's conceptual framework for vulnerability analysis.
Source: Bohle, 2001.

tures for the livelihood strategies and their outcomes (see in detail DFID (1999) and Figure 1.3).

The sustainable livelihood framework encompasses two major terms, sustainability and livelihoods. The original concept developed by Chambers and Conway (1992) viewed livelihoods as the means of gaining a living, encompassing livelihood capabilities, and tangible and intangible assets. Within the livelihood framework, the term sustainability is often linked to the ability to cope with and recover from stresses and shocks as well as to maintain the natural resource base (DFID, 1999; Chambers and Conway, 1992). The framework emphasises that especially the transforming structures in the governmental system or private sector and respective processes (laws, culture) influence the vulnerability context, and determine both the access to and major influences on livelihood assets of people. The approach underlines the necessity of empowering local marginalised groups in order to reduce vulnerability effectively (see in detail DFID, 1999; Schmidt, 2005). A central objective of the approach was to provide a method that views people and communities on the basis of their daily needs, instead of implementing ready-made, general interven-

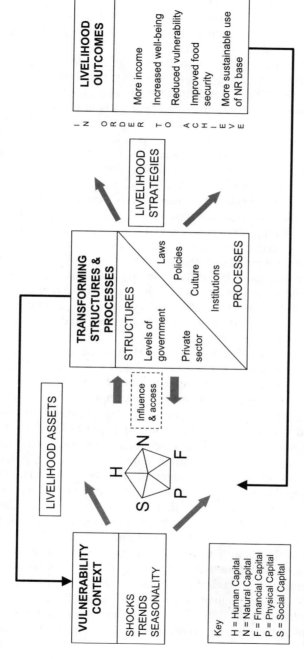

Figure 1.3 The sustainable livelihood framework.
Source: DFID, 1999.

tions and solutions, without acknowledging the various capabilities poor people offer (de Haan and Zoomers, 2005). The approach views vulnerability as a broad concept, encompassing livelihood assets and their access, and vulnerable context elements such as shocks, seasonality and trends, as well as institutional structures and processes.

Although the sustainable livelihood approach underlines the multiple interactions that determine the ability of a person, social group or household to cope with and recover from stresses and shocks, it remains abstract. The transforming structures and processes in particular, including influences and access aspects, remain very general. In this context, de Haan and Zoomers (2005: 33 and 45) emphasise that access and the role of transforming structures are key issues which have not been sufficiently examined so far. In particular, the flexibility of the interchanges of different capitals and assets (human capital, financial capital, social capital) has to be more closely considered, which means that the configuration of power around these assets and capitals as well as the power and processes of transforming structures need to be explored in more depth. They argue that access as a key element in the sustainable livelihood framework heavily depends on the performance of social relations, and therefore more emphasis in sustainable livelihood research should be given to the role of power relations. De Haan and Zoomers conclude that the current concept has the tendency to focus on relatively static capitals and activities within different livelihoods and livelihood strategies (de Haan and Zoomers, 2005).

Furthermore, it is interesting to note that the concept of livelihoods accounts solely for positive outcomes (livelihood outcomes). Additionally, some of the feedback processes underestimate the role of livelihood outcomes on the environmental sphere; for example, a "more sustainable use of natural resources" can be seen as an important tool to reduce the magnitude and frequency of some natural hazards such as droughts, floods or landslides. These linkages between the human–environmental system play a major role in the resilience discourse (see e.g. Allenby and Fink, 2005; Folke et al., 2002; Adger et al., 2005). Nevertheless, this approach, especially the five livelihood assets, can also serve as an important source and checklist for other approaches aimed at identifying susceptibility and coping capacity of people to hazards of natural origin. The framework can also be linked to categories used in the disaster risk community such as hazard, exposed and susceptible elements, driving forces/root causes, and potential outcomes and responses. While the various shocks encompass hazard components, the five livelihood assets could represent elements that are exposed and susceptible, while the transforming structures and processes in other frameworks are viewed as root causes, dynamic pressures or driving forces (see e.g. PAR framework).

The livelihood strategies and outcomes can be viewed as a mixture of intervention and response elements. However, the understanding of vulnerability in the sustainable livelihood approach is very broad, also encompassing the hazard sphere.

Vulnerability within the framework of hazard and risk

A second school, the disaster risk community, defines vulnerability as a component within the context of hazard and risk. This school usually views vulnerability, coping capacity and exposure as separate features. To illustrate this school of thinking three approaches will be presented: the definition of risk within the disaster risk framework by Davidson (1997), adopted by Bollin et al. (2003), the triangle of risk of Villagrán de León (2004), which reflects the "risk triangle" developed by Crichton (1999), and the UN/ISDR framework for disaster risk reduction (2004). Davidson's (1997) conceptual framework, adopted by Bollin et al. (2003), is shown in Figure 1.4. It views vulnerability as one component of disaster risk. The conceptual framework distinguishes four categories of disaster risk: hazard, exposure, vulnerability and capacity measures (Figure 1.4).

This conceptual framework views risk as the sum of hazard, exposure, vulnerability and capacity measures. While hazard is defined through its probability and severity, exposure is characterised by structures, population and economy. In contrast, vulnerability has a physical, social, economic and environmental dimension. Capacity and measures – which seem to be closely related to the subject of coping capacity – encompass physical planning, social capacity, economic capacity and management.

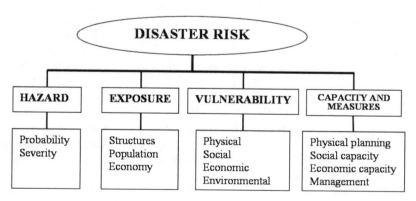

Figure 1.4 The conceptual framework to identify disaster risk.
Source: Davidson, 1997: 5; Bollin et al., 2003: 67.

Figure 1.5 Risk as a result of vulnerability, hazard and deficiencies in preparedness.
Source: Villagrán de León, 2001/2004.

In contrast to the framework of the double structure of vulnerability developed by Bohle (2001), this approach defines vulnerability as one component of disaster risk and differentiates between exposure, vulnerability and coping capacity (Davidson 1997; Bollin et al., 2003). Villagrán de León also explains vulnerability in the hazard and risk context. He defines a triangle of risk, which consists of the three components of vulnerability, hazard and deficiencies in preparedness (Villagrán de León, 2004: 10). His figure reflects the "risk triangle" developed earlier by Crichton (1999).

However, he defines vulnerability as the pre-existing conditions that make infrastructure, processes, services and productivity more prone to be affected by an external hazard. In contrast to the positive definition of coping capacities, he uses the term "deficiencies in preparedness" to capture the lack of coping capacities of a society or a specific element at risk (Villagrán de León, 2001, 2004). Although the term exposure is not directly mentioned, he views exposure primarily as a component of the hazard (Villagrán de León, Chapter 16).

The ISDR framework for disaster risk reduction

A different conceptual framework was developed by the UN/ISDR. The UN/ISDR framework views vulnerability as a key factor determining risk. According to UN/ISDR, vulnerability can be classified into social, economic, physical and environmental components (see Figure 1.6).

Vulnerability assessment is understood as a tool and a pre-condition for effective risk assessment (UN/ISDR, 2004: 14–15). Although the framework provides an important overview of different phases to be

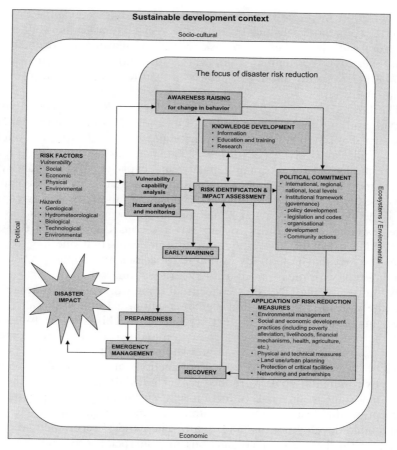

Figure 1.6 The ISDR framework for disaster risk reduction.
Source: UN/ISDR, 2004.

taken into account in disaster risk reduction, such as vulnerability analysis, hazard analysis, risk assessment, early warning and response, the framework does not indicate how reducing vulnerability can also reduce risk. Vulnerability is placed outside the risk response and preparedness framework. This makes it difficult to understand the necessity of also reducing risk through vulnerability reduction and hazard mitigation. In fact, in this conceptual framework risk and vulnerability cannot be reduced directly. The arrows from vulnerability and hazards only point out into the direction of the risk identification; the opportunity to reduce the vulnerabilities themselves is not explicitly shown. The figure underlines the fact that early warning, preparedness and response could reduce the disaster impact, even though a link between the risk factors (vulnerability and hazards) and the application of risk reduction measures is not included. Moreover, the conceptual framework does not give an answer as to whether exposure should be seen as a feature of the hazard or of the vulnerabilities.

The UN/ISDR report *Living with Risk* (UN/ISDR, 2004) views physical vulnerability as the susceptibility of location. This may be interpreted as a sign that physical vulnerability encompasses spatial exposure, but no precise answer is given (UN/ISDR, 2004: 42). Furthermore, the report differentiates between coping capacity and capacity. While capacity is understood as all the strengths and resources available within a community, society or organisation that can reduce risk, the term coping capacity is defined as the way in which people or organisations use available resources and abilities to face adverse consequences of a disaster (UN/ISDR 2004: 16). This differentiation indicates that one has to consider the fact that potentially available capacities and applied capacities are different with regard to disaster risk reduction.

Additionally, the UN/ISDR conceptual framework places vulnerability and the disaster risk reduction elements within a framework called the "sustainable development context" (Figure 1.6). This is meant to underline the necessity of linking risk reduction and sustainable development, which means risk reduction strategies should promote sustainable development by making the best use of connections among social, economic and environmental goals to reduce risk (UN/ISDR, 2004: 18). Although it is important to link risk reduction with sustainable development, the perception that risk reduction is similar to and always compatible with sustainable development is inadequate. The general recommendation of "making the best use of connections among social, economic and environmental goals" is a sort of ill-defined "balancing exercise" between social, economic and environmental goals. In practice, vulnerability reduction and sustainable development are confronted with deeply rooted social, economic and environmental conflicts, which cannot be wished away through a simple balancing exercise. There is therefore a need to define more precisely what sustainable development and risk reduction have in common as well as where the differences are (see section Vulnerability and sustainable development).

Vulnerability in the global environmental change community

The conceptual framework developed by Turner et al. (2003), considered here as being a representative of the global environmental change community, defines vulnerability in a broader sense. Their definition and

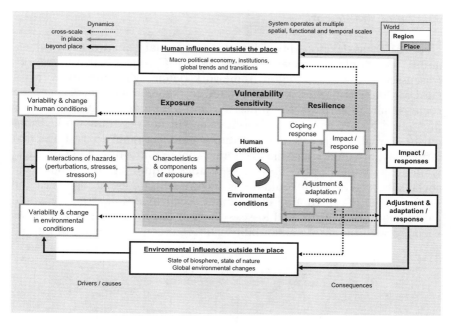

Figure 1.7 Turner et al.'s Vulnerability Framework.
Source: Turner et al., 2003: 8076.

analytical framework of vulnerability encompasses exposure, sensitivity and resilience. Moreover, vulnerability is viewed in the context of a joint or coupled human–environmental system (Turner et al., 2003: 8075; Kasperson, 2005). In contrast to the disaster risk community, this conceptual framework of Turner et al. (2003) defines exposure, coping response, impact response and adaptation response explicitly as parts of vulnerability (Figure 1.7). The framework also takes into account the interaction of the multiple interacting perturbations, stressors and stresses. Another important difference between the frameworks discussed earlier and this one lies in the fact that the conceptual framework of Turner et al. examines vulnerability within the broader and closely linked human–environment context (Turner et al., 2003: 8076; Kasperson, 2005).

The conceptual framework also takes into account the concept of adaptation, which is viewed as an element that increases resilience. This framework constitutes an interesting alternative to the conceptual frameworks discussed earlier. However, some questions remain, such as whether the distinction between drivers and consequences in this feedback-loop system is appropriate.

The onion framework

UNU-EHS has developed two different conceptual frameworks of vulner-ability, the "onion framework" and the "BBC conceptual framework" (discussed below). The onion framework defines vulnerability with re-gard to different hazard impacts related to the economic sphere and the social sphere. The impact of a disaster and the vulnerability it reveals is illustrated by the example of floods. Analytically the framework distin-guishes a reality axis and an opportunity axis. The reality axis shows that a flood event could affect the economic sphere and cause flood damage, while if the impact of the flood caused huge additional disruption in the social sphere, a disaster would occur (Figure 1.8). Economic assets can be replaced, but the disruption of the inner social sphere of a society would cause long-term injuries and losses, which in this model are primarily as-sociated with the term vulnerability. Different capacities exist within the

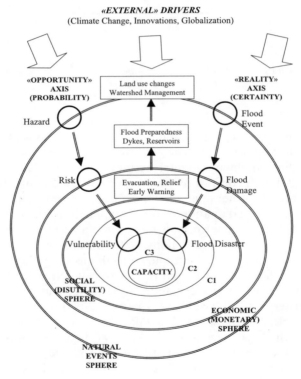

Figure 1.8 The onion framework.
Source: Bogardi/Birkmann, 2004.

centre of the social sphere (C1–C3), which means that whether a flood event becomes a disaster or not depends almost as much on the preparedness and coping capacity of the affected society as on the nature of the flood event itself (Bogardi and Birkmann, 2004). While C1 shows the fact that although the social sphere is affected, adequate coping capacities still exist; an impact of the flood event on the inner circle of the social sphere C3, however, would imply that social capacities are entirely insufficient to deal with the flood event, thus precipitating the occurrence of a disaster (Bogardi and Birkmann, 2004).

The "onion framework" relates the terms risk and vulnerability to potential losses and damages caused in the three different spheres. The framework emphasises that vulnerability deals with different "loss categories", such as economic and social losses. This means it stresses the fact that if a community's or a person's losses go beyond economic losses, for example, extending to loss of confidence and trust, the flood event has reached the "intangible" assets. This implies a serious disruption of the functioning of the society to the point that vulnerability becomes evident. According to this framework, the more comprehensive concept of social vulnerability should incorporate the monetary dimension (likelihood of economic harm) as well as "intangibles" like confidence, trust and fear as potential consequences of the flood. Furthermore, the onion framework shows potential response activities related to the different spheres. Finally, one has to remark that the onion framework does not account for environmental vulnerability. It defines the environment primarily as the event sphere. The aspect of exposure is also not specifically incorporated.

The pressure and release model (PAR model)

The pressure and release model (PAR model) views disaster as the intersection of two major forces: those processes generating vulnerability, on the one hand, and on the other, the natural hazard event. The PAR approach underlines how disasters occur when natural hazards affect vulnerable people (Blaikie et al., 1994; Wisner et al., 2004: 49–86). The conceptual framework stresses the fact that vulnerability and the development of a potential disaster can be viewed as a process involving increasing pressure on the one hand and the opportunities to relieve the pressure on the other. The PAR approach is based on the commonly used equation:

$$\text{Risk} = \text{Hazard} \times \text{Vulnerability}.$$

In this context vulnerability is defined within three progressive levels: root causes, dynamic pressures and unsafe conditions (Figure 1.9). Root

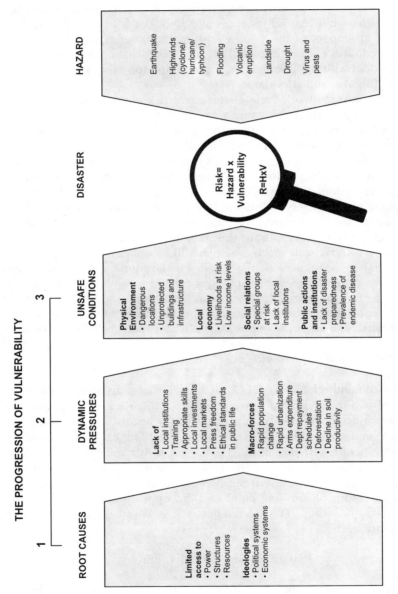

THE PROGRESSION OF VULNERABILITY

1   2   3

ROOT CAUSES

Limited access to
• Power
• Structures
• Resources

Ideologies
• Political systems
• Economic systems

DYNAMIC PRESSURES

Lack of
• Local institutions
• Training
• Appropriate skills
• Local investments
• Local markets
• Press freedom
• Ethical standards in public life

Macro-forces
• Rapid population change
• Rapid urbanization
• Arms expenditure
• Dept repayment schedules
• Deforestation
• Decline in soil productivity

UNSAFE CONDITIONS

Physical Environment
• Dangerous locations
• Unprotected buildings and infrastructure

Local economy
• Livelihoods at risk
• Low income levels

Social relations
• Special groups at risk
• Lack of local institutions

Public actions and institutions
• Lack of disaster preparedness
• Prevalence of endemic disease

DISASTER

Risk= Hazard x Vulnerability

R=HxV

HAZARD

Earthquake

Highwinds (cyclone/ hurricane/ typhoon)

Flooding

Volcanic eruption

Landslide

Drought

Virus and pests

Figure 1.9 The Pressure and Release (PAR) model.
Source: According to Wisner et al., 2004: 51.

causes can be, for example, economic, demographic and political processes, which determine the access to and distribution of power and various resources. These root causes are also closely linked with the subject of good governance, such as the nature of the control exercised by the police and military and the distribution of power in a society. The category dynamic pressure encompasses all processes and activities that transform and channel the effects of root causes into unsafe conditions, such as epidemic diseases, rapid urbanisation and violent conflicts (Wisner et al., 2004: 54). Interestingly, the authors of the approach stress the fact that dynamic pressure should not be labelled as negative pressure per se. Root causes implying dynamic pressures lead to unsafe conditions, which are a third column of the PAR model approach. Unsafe conditions are specific forms in which human vulnerability is revealed and expressed in a temporal and spatial dimension. These conditions can encompass lack of effective protection against diseases, living in hazardous locations, or having entitlements that are prone to rapid and severe disruption (Wisner et al., 2004: 52–80). The approach also accounts for access to tangible and intangible resources.

The differentiation of root causes, dynamic pressures and unsafe conditions underline the author's opinion that measuring vulnerability should go beyond the identification of vulnerability; rather, it should address underlying driving forces and root causes in order to be able to explain why people are vulnerable. However, the different elements of the PAR framework are dynamic in that they are subject to constant change, and hence the task of identifying and verifying the causal links between root causes, dynamic pressures and unsafe conditions in a quantitative way might be very difficult. Also Wisner et al. (2004) stress that, in multicausal situations and a dynamic environment, it is hard to differentiate between the causal links of different dynamic pressures on unsafe conditions and the impact of root causes on dynamic pressures. For example, although urbanisation as a dynamic pressure leads to unsafe conditions in many developing regions, such as Latin America or Asia, the general assumption that urbanisation leads to unsafe conditions is inappropriate. For example in Western European countries and the United States the increasing sub-urbanisation and urban sprawl (de-urbanisation) might be an appropriate surrogate indicator to point at unsafe conditions.

Overall, the PAR model is an important approach and one of the best-known conceptual frameworks worldwide that focuses on vulnerability and its underlying driving forces. It is particularly useful in addressing the release phase and the root causes that contribute to disaster situations. On the other hand the approach underlines the fact that the real effort to reduce vulnerability and risk involves changing political and economic systems, since they are viewed as root causes of, for example, dynamic pressures such as rapid urbanisation or rapid population change. This conceptual framework puts a heavy emphasis on the national and global levels, although many dynamic pressures and unsafe conditions might also be determined by local conditions.

A holistic approach to risk and vulnerability assessment

The conceptual framework for a holistic approach to evaluating disaster risk goes back to the work of Cardona (1999, 2001) and his developments with Hurtado and Barbat in 2000. In their first concept, vulnerability consisted of exposed elements that took into account several dimensions or aspects of vulnerability (Wilches-Chaux, 1989), which are characterised by three categories or vulnerability factors:
- physical exposure and susceptibility, which is designated as hard risk and viewed as being hazard dependent
- fragility of the socio-economic system, which is viewed as soft risk and being non hazard dependent
- lack of resilience to cope and recover, which is also defined as soft risk and being non hazard dependent (Cardona and Barbat, 2000: 53).

According to this framework vulnerability conditions depend on the exposure and susceptibility of physical elements in hazard-prone areas on the one hand, and on the other, on socio-economic fragility as well as on a lack of social resilience and abilities to cope. These factors provide a measure of the direct as well as indirect and intangible impacts of hazard events. The approach emphasises the fact that indicators or indices should measure vulnerability from a comprehensive and multidisciplinary perspective. They intend to capture conditions for the direct physical impacts (exposure and susceptibility), as well as for indirect and at times intangible impacts (socio-economic fragility and lack of resilience), of potential hazard events. Therefore the approach defines exposure and susceptibility as necessary conditions for the existence of physical (hard) risk. On the other hand, the likelihood of experiencing negative impacts, as a result of the socio-economic fragilities, and inability to cope adequately are also vulnerability conditions, which are understood as "soft" risk.

Although the classification of vulnerability conditions into "hard" and "soft" risk is controversial, the conceptual framework suggests a broader understanding of vulnerability, encompassing exposure, susceptibility and lack of resilience. The consequences of the interaction of the hazardous events and vulnerabilities are defined as risks from which a feedback loop starts: it encompasses a control and an actuation system that represent risk management organisation and corrective and prospective interventions. The feedback loop starts after the risk has become evident (Cardona and Barbat, 2000).

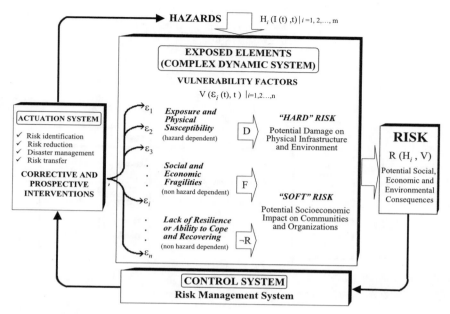

Figure 1.10 Theoretical framework and model for holistic approach to disaster risk assessment and management.
Source: Cardona and Barbat, 2000.

Carreño et al. (2004, 2005a, 2005b) have developed a revised version of the holistic model to evaluate risk that redefines the meanings of hard and soft risk in terms of "physical damage", obtained from exposure and physical susceptibility, and an "impact factor", obtained from the socio-economic fragilities and lack of resilience of the system to cope with disasters and recovery. The revised version of the holistic model of disaster risk views risk as a function of the potential physical damage and the impact factor (social and economic fragilities and lack of resilience). While the potential "physical damage" is determined by the susceptibility of the exposed elements (e.g. a house) to a hazard and its potential intensity and occurrence, the "impact factors" depend on the socio-economic context – particularly social fragilities and lack of resilience. Based on the theory of control and complex system dynamics, Carreño et al. (2004, 2005a, 2005b) also introduce a feedback loop encompassing corrective and prospective interventions, to underline the need to reduce both the vulnerabilities and the hazards. Thus risk management requires a system of control (institutional structure) and an actuation system (public policies and actions) to implement the changes needed.

The holistic approach to estimating vulnerability was also presented by

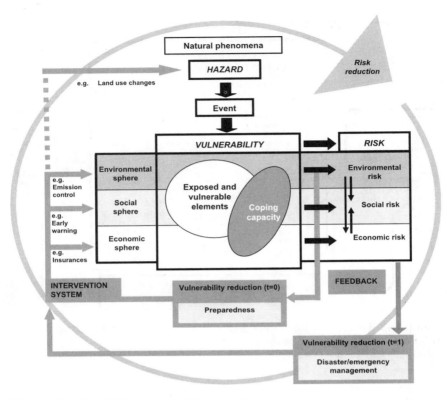

Figure 1.11 The BBC conceptual framework.
Source: Author, based on Bogardi/Birkmann (2004) and Cardona (1999/2001).

Cardona (2004) in Geneva. However, because his presentation outlined only some elements of the approach, we examine the original model here (Figure 1.10). At present, this model has been used to evaluate disaster risk at the national level in the Program of Indicators for Disaster Risk and Risk Management for the Americas (see Cardona, Chapter 10) as well as at the sub-national level and for cities, including Barcelona and Bogotá (Carreño et al., 2005a, 2005b).

Reflection and introduction of the BBC conceptual framework

The BBC conceptual framework combines different elements of the frameworks discussed earlier. Therefore, the presentation of this framework will also reflect on the frameworks analysed before and will stress some key aspects which are still controversial.

The term "BBC" is linked to conceptual work done by Bogardi and Birkmann (2004) and Cardona (1999 and 2001), which served as a basis for this approach. It grew from three discussions: how to link vulnerability, human security and sustainable development (Bogardi and Birkmann 2004; see also Birkmann section Vulnerability and sustainable development); the need for a holistic approach to disaster risk assessment (Cardona 1999, 2001; Cardona and Hurtado 2000a, 2000b, 2000c; Cardona and Barbat, 2000; Carreño et al., 2004, 2005a, 2005b, Cardona et al., 2005); and the broader debate on developing causal frameworks for measuring environmental degradation in the context of sustainable development (e.g. OECD, 1992: 6; Zieschnak et al., 1993: 144).

The BBC framework stresses the fact that vulnerability analysis goes beyond the estimation of deficiencies and assessment of disaster impacts in the past. It underlines the need to view vulnerability within a process (dynamic), which means focusing simultaneously on vulnerabilities, coping capacities and potential intervention tools to reduce vulnerabilities (a feedback-loop system). Furthermore, as shown in the BBC conceptual framework, vulnerability should not be viewed as an isolated feature. Rather, vulnerability assessment has also to take into account the specific hazard type(s) and potential event(s) that the vulnerable society, its economy and environment are exposed to, and the interactions of both that lead to risk. This means, the BBC framework underlines the necessity to focus on social, environmental and economic dimensions of vulnerability, clearly linking and integrating the concept of sustainable development into the vulnerability framework. Within the three sustainability dimensions (social, economic and environmental sphere), additional frameworks can be integrated, e.g. the sustainable livelihood framework within the social sphere.

In contrast to a risk analysis, the main focus of the BBC conceptual framework is on the different vulnerable or susceptible and exposed elements, the coping capacity and the potential intervention tools to reduce vulnerability.

In contrast to the model of holistic approach to estimate vulnerability and risk (Cardona and Barbat, 2000), the BBC conceptual framework does not account for hard and soft risk, but rather the three main thematic spheres of sustainable development define the inner thematic composition in which vulnerability should be measured: the economic, the social and the environmental dimensions. In this context the environmental dimension is not represented within the framework of the holistic approach to estimate vulnerability and risk developed by Cardona and Barbat (2000), but rather encompasses vulnerability regarding "exposure and physical susceptibility", "social and economic fragilities" and "lack of resilience or ability to cope and recovering". Another difference between the two frameworks refers to the response chains. The BBC framework distinguishes between the response before risk and disasters are manifested ($t = 0$) and the response needed when risk and disasters occur ($t = 1$). While during the disaster, emergency management and disaster response units play a crucial role, vulnerability reduction should give particular emphasis to responses, thus focusing on preparedness rather than on disaster response and emergency management.

Through the linkages between sustainable development and vulnerability reduction, the BBC conceptual framework emphasises the necessity to give due consideration to environmental considerations, on which human conditions depend (Turner et al., 2003). Organisational and institutional aspects are important, as are physical vulnerabilities, but they should be analysed within the three thematic spheres (economy, social and environmental) (Figure 1.11). Moreover, the BBC conceptual framework promotes a problem-solving perspective, by analysing the probable losses and deficiencies of the various elements at risk (e.g. social groups) and their coping capacities as well as the potential intervention measures, all within the three key thematic spheres. In this way it shows the importance of being proactive in order to reduce vulnerability before an event strikes the society, economy or environment ($t = 0$) (Figure 1.11). In this context, the framework is also open for links to other approaches, such as the sustainable livelihood approach. Especially within the social and economic spheres of vulnerability in the BBC framework, the five livelihood assets can serve as an important orientation and as a kind of vade mecum to select relevant sub-themes and indicators to assess susceptibility and coping within vulnerability to hazards of natural origin. Furthermore, potential intervention tools could also encompass measures and processes (e.g. planning processes) conducive to improve the access to important livelihood assets, e.g. to human, social and physical capital.

The various elements and links shown in the BBC conceptual framework – with a special emphasis on the key element vulnerability – also suggest a risk reduction strategy, since the intervention system encompasses measures to reduce vulnerability and also measures to reduce the frequency and magnitude of events, such as floods, droughts or landslides linked to a hazard of natural origin.

While some approaches view vulnerability primarily with regard to the degree of experienced loss of life and economic damage (e.g. DRI, Hotspots), the BBC conceptual framework addresses various vulnerabilities in the social, economic and environmental sphere. These three spheres have been defined as the three main pillars of sustainable development

(UN, 1993; WCED, 1987). Although the vulnerability of the society and the economy (anthroposphere) are seen as core areas, the BBC conceptual framework also takes into account the importance of the biophysical basis of human life: the environmental sphere. In this way the conceptual framework shows the close link between nature and society and does not limit the environment to the "hazard sphere".

In this regard, Oliver-Smith (2004, 12) points out that dominant Western constructions of the relationship between human beings and nature often place them in opposition to each other. This means that the understanding of dividing human and environmental issues is also culturally determined. In contrast to the pre-analytic vision of separating the human and environmental systems, the BBC conceptual framework views the environment on the one hand as the "event sphere" from which a hazard of natural origin starts, and on the other hand the environment itself is vulnerable to hazards of natural origin and to creeping processes, especially when it comes to natural-technological hazards.

According to Kraas (2003) and Cardona (2004b) vulnerability can also be directly related to environmental degradation in rural areas and to rapid urban growth patterns that bring about socio-economic fragmentation in urban agglomerations, particularly megacities (Kraas, 2003: 6; Cardona, 2004b: 49; MunichRe, 2004: 18). Interestingly, Cross (2001: 63) argues that, contrary to popular wisdom, small cities and rural communities are more vulnerable to disasters than megacities, since megacities are more likely to possess the resources needed to deal with the hazard and disasters, while in smaller cities and rural communities these capacities do not exist. In terms of the theoretical and conceptual development of vulnerability assessment, this debate is important because it underlines the fact that vulnerability estimation should also consider the capacities to cope with hazardous events.

The BBC conceptual framework stresses the fact that vulnerability assessment should take into account exposed, susceptible elements and coping capacities, which might have an important impact on the likelihood to suffer harm and injury due to a hazardous event. Although one should distinguish between vulnerable elements and coping capacity, there is a certain overlap (Figure 1.11), especially if one enters into the discussion of social capital; for example, whether to be part of a social network should be viewed as less vulnerable or whether the network itself can be associated with coping capacity. The role of social capital as 'social' or 'anti-social' capital is examined more in-depth in Bohle (2006). Also, the timescale for a natural disaster which is not defined per se is important: if a disaster is defined as ending whenever the community regains functionality, then coping capacities are crucial drivers.

Finally, the BBC conceptual framework shows that one has two options to reduce vulnerability ($t = 0$) and ($t = 1$) (see Figure 1.11). In this context it is important not to wait till the next disaster occurs, but rather to take into account the opportunities to reduce the various vulnerabilities before risk turns into catastrophe. Although disaster management capacities are important for limiting the impact of catastrophes and managing the crisis, the BBC conceptual framework points out the importance of anticipating risk and taking actions before it occurs ($t = 0$) (see Figure 1.11). Especially with regard to early warning at a political level, it is important to underline the necessity to promote vulnerability reduction as an integrated approach in daily decision-making processes. The improvement of disaster and emergency response capacity (t=1) is only one part of the picture and often occurs at the end of the chain. Instead, forward-looking and pro-active interventions are needed (preparedness, mitigation) in order to reduce vulnerability. For example, it is widely acknowledged that investments in mitigation and preparedness have a much higher return than investments needed to cover the costs of relief and recovery.

Regarding the controversial discussion of exposure, Cardona underlines the fact that an element or system is only at risk if the element or system is exposed and vulnerable to the potential phenomenon (Cardona, 2004b: 38). The BBC framework views exposure as being at least partially related to vulnerability. Although one can argue that exposure is often hazard-related, the total exclusion of exposure from vulnerability assessment could render this analysis politically irrelevant. If vulnerability is understood as those conditions that increase the susceptibility of a community to the impact of hazards, it also depends on the spatial dimension, by which the degree of exposure of the society or local community to the hazard or phenomena is referred to. The author views the location's general exposure primarily as a feature of the hazard, whereas, for example, the degree of exposure of a specific unit e.g. a critical infrastructure (schools) as well as the number of houses in the hazard-prone areas are a part of exposure that characterises the spatial dimension of vulnerability. Thus exposure is partially a characteristic of vulnerability.

Concerning vulnerability to climate change, O'Brien and Leichenko (2000) emphasise that extreme climate events can strike the wealthy and poor alike, particularly in high-risk zones. It follows, therefore, that all owners of coastal properties are susceptible to storm surges, even though their vulnerability will also depend on their capacity to recover from such impacts, meaning that the wealthy population will have less difficulties than the poor (O'Brien and Leichenko, 2000: 225).

The BBC conceptual framework considers the phenomenon of exposure, at least in part, since it recognises that the location of human settlements and infrastructure plays a crucial role in determining the susceptibility of a community. Yet it acknowledges the fact that within the given high risk zone there are other characteristics that will have a significant impact on whether or not people and infrastructure are likely to experience harm.

Besides the examination of the vulnerable elements within the society, the economy and the environment, the BBC conceptual framework shows the importance of reducing the risk by reducing vulnerability and mitigating hazard even before a risk can manifest itself. Vulnerability assessment should therefore also encompass the identification and analysis of potential intervention tools to reduce the various vulnerabilities and to increase the coping capacities of a society or system at risk (Figure 1.11).

Finally, the framework also stresses that the changes of vulnerability from one thematic dimension to another should be taken into account and viewed as a problem, since these shifts do not imply real vulnerability reduction. For example, if a company compensates for its economic vulnerability in a disaster situation by reducing loans to its employees, then it is the personnel who will have to deal with the negative financial impact of the event. Because the company did not have adequate disaster insurance, its vulnerability (economic vulnerability) is shifted to the employees (social sphere), but without achieving any real reduction in overall vulnerability.

First conclusions

The discussion of different conceptual and analytical frameworks on how to systematise vulnerability has revealed that at least six different schools can be distinguished:

- the school of the double structure of vulnerability (Bohle, 2001)
- the conceptual frameworks of the disaster risk community (Davidson, 1997; Bollin et al., 2003)
- the analytical framework for vulnerability assessment in the global environmental change community (Turner et al., 2003)
- the school of political economy, which addresses the root causes, dynamic pressures and unsafe conditions that determine vulnerability (Wisner et al., 2004)
- the holistic approach to risk and vulnerability assessment (Cardona, 1999 and 2001; Cardona and Barbat, 2000; Carreño et al., 2004, 2005a, 2005b)
- the BBC conceptual framework, which places vulnerability within a

feedback loop system and links it to the sustainable development discourse (based on work by Birkmann and Bogardi, 2004 and Cardona 1999 and 2001).

While the model of the double structure of vulnerability views vulnerability as the exposure to shocks and stressors and the ability to cope with these shocks (Bohle, 2001), the second approach widely used in the disaster risk community separates vulnerability from coping capacities and exposure (Davidson, 1997; Bollin et al., 2003; Villagrán de León, 2004). A third school, illustrated by the framework used by Turner et al. (2003), shows a broader definition of vulnerability, which also encompasses exposure, sensitivity and response capacity, including adaptation responses. The fourth school emphasises the root causes and dynamic pressures that determine vulnerability and unsafe conditions. This school of thinking is closely linked with the school of political economy. The fifth school, illustrated by the holistic approach to vulnerability and risk, considers exposure/susceptibility, socio-economic fragilities and lack of resilience, and uses complex system dynamics to represent risk management organisation and action (Cardona, 1999 and 2001; Cardona and Hurtado, 2000a, 2000b, 2000c; Cardona and Barbat, 2000; Carreño et al., 2004, 2005a, 2000b).

The sixth school, illustrated by the BBC conceptual framework and based on Bogardi and Birkmann (2004) and Cardona (1999 and 2001), includes elements of different schools and links – in particular – vulnerability assessment to the concept of sustainable development. The framework stresses the need to focus on exposed and susceptible elements and on coping capacities, at the same time. It includes an understanding of vulnerability, which goes beyond the estimation of damage and the probability of loss. Furthermore it stresses the fact that vulnerability should be viewed as a process. Vulnerability reduction also has to address both coping capacities and potential intervention tools at different levels. The BBC conceptual framework underlines the fact that the specific vulnerabilities and coping capacities at the different levels have to be examined with regard to the social, economic and environmental spheres that constitute the three dimensions of sustainable development.

Despite some similarities between the different schools of thinking, such as the understanding that vulnerability represents the inner conditions of a society or community that make it liable to experience harm and damage, as opposed to the estimation of the physical event (hazard), there remain many areas of uncertainty:

- Is coping capacity part of vulnerability or should it be viewed as a separate feature?
- Does vulnerability encompass exposure or should exposure be seen as a characteristic of the hazard or even a separate parameter?

- Which parts and characteristics of vulnerability are hazard dependent and which are hazard independent?
- What dimensions and themes should vulnerability assessment cover?
- How can the root causes of vulnerability be defined and measured?
- How far can one measure the interlinkages of the root causes at the national and global levels and the major driving forces and root causes at the local level that determine local vulnerability?
- Is resilience the opposite of vulnerability or a concept that covers coping and adaptation capacity as these relate to vulnerability?
- Should vulnerability focus primarily on human vulnerability alone or is it more appropriate to view vulnerability within a coupled human–environmental system?
- How far is environmental degradation a hazard or a revealed vulnerability of the environment?

Regarding the thematic focus of the different conceptual frameworks under review it is interesting to note that some concepts – such as the double structure of vulnerability (Bohle, 2001) – have no explicit thematic limits, while others define the precise thematic areas needed to be taken into account (UN/ISDR, 2004; Bogardi and Birkmann, 2004; Cardona, 1999 and 2001).

Before presenting fundamental principles and a theoretical basis for indicators to measure vulnerability, the links and the differences between sustainable development, vulnerability and disaster risk reduction will be discussed.

Vulnerability and sustainable development

Linking sustainable development, risk and vulnerability

International declarations and documents, such as the Hyogo Framework for Action 2005–2015, the UN/ISDR report "Living with risk" (UN/ISDR, 2004: 15) and the UNDP report "Reducing disaster risk" (UNDP, 2004: 19, 84), stress the necessity to integrate risk and vulnerability reduction into sustainable development. Therefore it is important to understand the links and also the differences between risk and vulnerability reduction, on the one hand, and sustainable development on the other. The Hyogo Framework for Action states:

There is now international acknowledgement that efforts to reduce disaster risks must be systematically integrated into policies, plans and programmes for sustainable development. Sustainable development, poverty reduction, good governance and disaster risk reduction are mutually supportive objectives. (UN, 2005: Chapter 1a)

It seems that international efforts to reduce disaster risk are increasingly being viewed within the context of sustainable development. On the other hand, the idea of integrating disaster risk reduction and vulnerability reduction into sustainable development does not appear in such important documents as AGENDA 21 or the Millennium Development Goals (MDGs).

Admittedly, some MDGs, especially MDG 1 ("eradicating extreme poverty and hunger"), MDG 3 ("promoting gender equality") and MDG 7 ("ensuring environmental sustainability") are indirectly linked to certain aspects of disaster risk and vulnerability reduction; for example, alleviating extreme poverty often also reduces vulnerability. However, the main focus of the MDGs is on socio-economic development and there is no reference to risk or vulnerability reduction as part of these development processes. This strong emphasis on issues of socio-economic development overlooks the fact that at a time of global environmental change (creeping environmental degradation processes),[1] traditional socio-economic development strategies are proving inappropriate to achieve a balance between socio-economic demands on the one hand and the environmental capacities of various ecosystems on the other. The MDGs pay very little attention to the new demands and challenges that global environmental change will make on the socio-economic development strategies that try to address sustainable development (Kempmann and Pilardeaux, 2005: 28). Although the MDGs can be linked to disaster risk reduction strategies and their goals, the current links and interrelations of global environmental change, socio-economic development and sustainable development remain abstract.

The UN/ISDR report "Living with risk" states the need to link sustainable development and risk reduction directly:

Promoting sustainability in disaster reduction means recognizing and making best use of connections among social, economic and environmental goals to reduce significant hazard risk. All countries require a healthy and diverse ecological system that is productive and life sustaining, a healthy and diverse economy that adapts to change and recognizes social and ecological limits. This cannot be achieved without the incorporating of disaster reduction strategies, one of the six principles of sustainability supported by strong political commitment. (UN/ISDR, 2004: 18–19)

The UN/ISDR is explicit about the need to integrate risk reduction into sustainable development. However, the question of how that can be

Figure 1.12 The six principles of sustainability.
Source: Natural Hazard Center 2006.

achieved and whether the two approaches are compatible remains open. In particular, the UN/ISDR definition of sustainable development is unclear. The formulation making the best use of connections among social, economic and environmental goals (UN/ISDR, 2004: 18) opens an enormous space for differing interpretations. The six principles of sustainable development given in the report by Monday (2002) could be considered as a first conceptual framework, even though these principles are very different and in same cases contradictory to each other (Figure 1.12).

The assumption, for example, that the concept of "quality of life" is part of community sustainability neglects important contradictions between the two concepts. The Brundtland Commission had already defined intra- and intergenerational justice as key principles of sustainable development, by pointing out that:

Sustainable development is development that meets the needs of the present without compromising the ability of future generations to meet their own needs. (WCED, 1987)

Although this root definition covers only a part of the current discourse of sustainability, the Brundtland Commission underlines the fact that intra- and intergenerational justice are constitutive elements of any development that might be described as sustainable. In contrast, "quality of life" approaches focus on the needs of the present and have very little in common with a strategy to reduce or balance the needs of the present to ensure the ability of future generations to meet their own needs (e.g. Ewringmann, 1999).

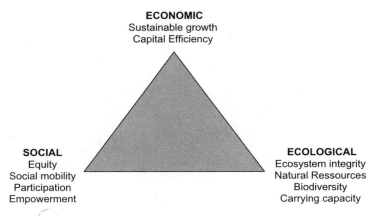

Figure 1.13 Serageldin's triangle of sustainability.
Source: Serageldin, 1995: 23.

From a scientific standpoint, the broadening of the concept of sustainable development and the accumulation of very different concepts under this label is misleading. The conflicts between current socio-economic development patterns and the limitations and changes of the surrounding environment cannot be whisked away through a simple balancing exercise. Implementing sustainable development means dealing with deeply rooted social, economic and environmental conflicts (Davoudi and Layard, 2001: 17).

Since reduction of disaster risks associated with hazards of natural origin was, until the 1970s, often viewed as a struggle against physical occurrences and environmental threats that required technological interventions and solutions (Hilhorst and Bankoff, 2004: 2), the contrast with an understanding of the vulnerability of the coupled human–environmental system, such as Turner et al. (2003) describe, is evident. Integrating sustainable development into risk and vulnerability reduction strategies (see e.g. Dikau and Weichselgartner, 2005) means recognising the fact that the social and the economic are closely linked with the environmental sphere. Thus, in the current discourse, two main analytical models can be distinguished: the triangle of sustainable development and the egg of sustainability.

Sustainability: the "triangle" versus the "egg"

The "triangle of sustainability" and the "egg of sustainable development" are two different schools within the discourse of sustainable development. While the "triangle of sustainable development" places the environment, the social system and the economy at the three different angles, the "egg" model defines a clear hierarchy between these dimensions. The model of the triangle was mainly developed by the World Bank (Serageldin, 1995: 3, 13), and it had broad repercussions, especially in Local Agenda 21 processes. According to Serageldin's conceptual framework:

This triangle recognizes that whatever we are talking about in terms of sustainability has to be economically and financially sustainable in terms of growth, capital maintenance and efficiency of use of resources and investments. But it also has to be ecologically sustainable, and here we mean ecosystem integrity, carrying capacity, and protection of species.... However, equally important is the social side, and here we mean equity, social mobility, social cohesion. (Serageldin, 1995: 17)

Although Serageldin points out that the economic, social and ecological spheres are interconnected, the conceptual model does not provide an integrative view. The three spheres are placed in relative isolation to each other. The questions of what sustainable growth means and whether the goal of sustainable economic growth is compatible with the goal of ecosystem integrity and carrying capacity remain open. In this regard, Daly argues:

The term sustainable growth when applied to the economy is a bad oxymoron.... When something grows it gets bigger. When something develops it gets different. The earth ecosystem develops (evolves), but does not grow. Its subsystem, the economy, must eventually stop growing, but can continue to develop. Politically it is very difficult to admit that growth, with its almost religious connotations of ultimate goodness, must be limited. But it is precisely the non-sustainability of growth that gives urgency to the concept of sustainable development. (Daly, 1993: 267–268)

Implicit in Daly's criticism is the notion that promoting traditional economic growth – also as a strategy of risk and vulnerability reduction – does not generally correspond with the concept of sustainable development. Within the international debate the criticism of the "triangle of sustainable development" focuses especially on the problematic isolation of the three dimensions. According to Prescott-Allen, the conceptual framework of the triangle of sustainability is misleading.

The common three-dimensional model of sustainability (economic sustainability + environmental sustainability + social sustainability) ... obliges people to balance economic, social and environmental concerns. It sets human and ecological needs against each other rather than accommodating both: sustainability cannot be achieved by compensating for reduced environmental goods with increased economic or social goods (or vice versa). (Prescott-Allen, 1995: 3)

The "triangle of sustainable development" does not show how the three main spheres of sustainability are interrelated (Birkmann, 2004; Fues, 1998). It implies an isolated goal definition for each of the three dimensions, neglecting the linkages between them (Bogardi and Birkmann, 2004: 77). When it comes to implementation, traditional conflicts between the social, the economic and the environmental spheres become apparent (Birkmann, 2004).

Contrasting with the triangle of sustainable development is an alternative conceptual model "the egg of sustainability", based on the science of the ecological economy. The alternative model defines a clear hierarchy and interdependency between the three dimensions (Figure 1.14).

The pre-analytic view of sustainable development as an "egg" should help to define goals that respect the linkages between the environmental sphere, the society and the economy and to put them into the right balance (Prescott-Allen, 1995; Busch-Lüty, 1995). Goals for sustainable economic development need to take into account goals of the social sphere as well as goals of the surrounding environmental sphere. The vulnerability and the sustainability of the human system both depend on conditions of the surrounding environmental sphere, as well as on the inner conditions of the socio-economic system. If vulnerability to disasters, as Wisner et al. (2004) argue, can also be seen as a function of the way in which humans interact with nature, the "egg of sustainability" is a good theo-

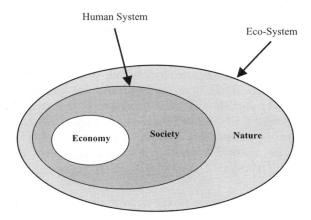

Figure 1.14 Egg of sustainable development.
Source: Busch-Lüty, 1995.

retical basis to start from. The degradation of the environmental sphere, especially through creeping processes like ongoing climate change and land degradation caused by unsustainable land use, production and consumption patterns, increases the risk of disasters for the inner human sphere. Unsustainable development due to higher risk and natural disasters can be interpreted in this regard as the loss of the ability of a (sub-)system (economic, social or environmental) to return to a state similar to the one prevailing prior to the disaster (Bogardi and Birkmann, 2004: 78). The ability to bounce back to a reference state after a negative hazardous event, as well as the capacity of a system to maintain certain structures and functions under stress conditions, is a key component of a broader vulnerability assessment and often captured by the term resilience (Turner et al., 2003: 8075). The concept of resilience is based on theories and experiences drawn from ecology.

If one accepts that sustainable development is based on the principles of intra- and intergenerational justice (WCED, 1987) that integrate social, environmental and economic aspects at the same time (UN, 1993, Agenda 21), as expressed by the "egg of sustainable development", it is evident that risk and vulnerability reduction have to promote strategies that increase the resilience of the inner spheres (human system) against the negative impacts of hazards of natural origin and at the same time ensure that socio-economic development acknowledges the limitations of the surrounding environmental sphere.

That does not mean that vulnerability assessment and risk reduction strategies have to focus only or primarily on the interaction between the human and natural systems, but it should be taken into account. Besides its importance for setting up stronger social protection of the inner spheres to deal with the impacts of a hazardous event by ensuring sustainable livelihoods, the conceptual framework of sustainable development also implies that such livelihood strategies are only sustainable if they take into account the surrounding environmental conditions and accepts intra- and intergenerational justice as a guiding principle. Viewing hazards of natural origin and vulnerability reduction strategies from the perspective of sustainable development means acknowledging the necessity of a dualistic approach: one that ensures a higher resilience of the inner human sphere while, in parallel, it promotes a more sustainable human–nature interaction by taking the limitations of the regional and local environmental capacity into account. Additionally, political ecologists and economic anthropologists emphasise the fact that human–environmental relationships are generated in social relations through the double nexus of production and consumption. In this regard, Oliver-Smith (2004) argues that environmental degradation and environmental resource limitations are not only a question of exceeding natural limits, but should be viewed as results of the socially contracted system of production and social exploitation (Oliver-Smith, 2004: 16). In terms of sustainable development and vulnerability reduction we therefore need to realise that we must take account not only of the biophysical resource budget, but also of the underlying patterns of production and consumption which define the other aspect of the limitations in which we constructed the relationship between human societies and the surrounding environment.

Overall, it can be concluded that the principles of intra- and intergenerational justice should be seen as key principles also for risk and vulnerability reduction. The exclusive focus on the needs of the present generation, along with the tendency to export risks and vulnerabilities from one dimension to another, is not sustainable. Moreover, it is important to acknowledge the fact that the environmental sphere is not only the event-and-hazard sphere (protection from environmental threats), but also the underlying biophysical basis of human activities. In this context a broader and long-term reduction of vulnerability would require also the analysis and reflection of how we construe our relationship with nature. This discussion was illustrated by two different pre-analytic visions of sustainable development and underlines the fact that the integrated perspective of the environmental sphere seems to be more appropriate for taking a holistic view of vulnerabilities to hazards of natural origin.

Note

1. Such degradation processes include for example land degradation, loss of biodiversity and climate change.

REFERENCES

Adger, W.N., T.P. Hughes, C. Folke, S.R. Carpenter and J. Rockström (2005) "Social-Ecological Resilience to Coastal Disasters", *Science* 309: 1036–1039.

Allenby, B. and J. Fink (2005) "Towards Inherently Secure and Resilient Societies", *Science* 309: 1034–1036.

Alexander, D. (2000): *Confronting Catastrophe. New Perspectives on Natural Disasters*, Oxford/New York: Oxford University Press.

Annan, K. (2003) "Message for the International Day for Disaster Reduction 8 October 2003", available at http://www.unisdr.org/eng/public_aware/world_camp/2003/pa-camp03-sg-eng.htm.

Birkmann, J. (2004) Monitoring und Controlling einer nachhaltigen Raumentwicklung, Indikatoren als Werkzeuge im Planungsprozess, Dortmund: Dortmunder Vertrieb für Bau- und Planungsliteratur.

Birkmann, J. (2005) "Danger Need Not Spell Disaster – But How Vulnerable Are We?", *Research Brief (1)*, Tokyo: United Nations University (ed).

Blaikie, P., T. Cannon, I. Davis, B. Wisner (1994) *At Risk: Natural Hazards, People's Vulnerability, and Disasters*, 1st edn, London: Routledge.

Bogardi, J. and J. Birkmann (2004) "Vulnerability Assessment: The First Step Towards Sustainable Risk Reduction", in Malzahn, D. and T. Plapp, eds, *Disaster and Society – From Hazard Assessment to Risk Reduction*, Berlin: Logos Verlag Berlin, pp. 75–82.

Bogardi, J. and H.-G. Brauch (2005) "Global Environmental Change: A Challenge for Human Security – Defining and conceptualising the environmental dimension of human security", in Rechkemmer, A., ed., *UNEO – Towards an International Environment Organization – Approaches to a sustainable reform of global environmental governance*, Baden-Baden: Nomos, pp. 85–109.

Bohle, H.-G. (2001) "Vulnerability and Criticality: Perspectives from Social Geography", *IHDP Update* 2/2001, Newsletter of the International Human Dimensions Programme on Global Environmental Change: 1–7.

Bohle, H.-G. (2002a) "Editorial: The Geography of Vulnerable Food Systems", *Die Erde* 133(4): 341–344.

Bohle, H.-G. (2002b) "Land Degradation and Human Security", in Plate, E.J., ed., *Environment and Human Security – Contributions to a Workshop in Bonn*, 23–25 October 2002, Bonn, pp. 3/1–3/6.

Bohle, H-G. (2006) "Soziales oder unsoziales Kapital? Das Sozialkapital-Konzept in der Geographischen Verwundbarkeitsforschung" ("Social or Anti-social Capital? The Concept of Social Capital in Geographical Vulnerability Research"), *Geographische Zeitschrift* (in press).

Bollin, C., C. Cárdenas, H. Hahn and K.S. Vatsa (2003) *Natural Disaster Network; Disaster Risk Management by Communities and Local Governments*, Washington, D.C.: Inter-American Development Bank, available at http://www.iadb.org/sds/doc/GTZ%2DStudyFinal.pdf.

Brauch, H.-G. (2005) "Threats, Challenges, Vulnerabilities and Risks in Environmental and Human Security", in Source No. 1/2005, Publication Series of UNU-EHS (United Nations University – Institute for Environment and Human Security), available at http://www.ehs.unu.edu/PDF/050818_Source1_final.pdf.

Busch-Lüty, C. (1995) "Nachhaltige Entwicklung als Leitmodell einer ökologischen Ökonomie", in Fritz, P., J. Huber and H. Levi, Hrsg., Nachhaltigkeit: in naturwissenschaftlicher und sozialwissenschaftlicher Perspektive, Stuttgart, S. 115–126.

Cannon, T., J. Twigg and J. Rowell (2003) Social Vulnerability. Sustainable Livelihoods and Disasters, Report to DFID Conflict and Humanitarian Assistance Department (CHAD) and Sustainable Livelihoods Support Office, available at http://www.benfieldhrc.org/disaster_studies/projects/soc_vuln_sust_live.pdf.

Cardona, O.D. (1993) "Evaluación de la Amenaza, la Vulnerabilidad y el Riesgo", in Maskrey, A. ed., Los Desastres No son Naturales, La Red, Bogotá: Tercer Mundo Editores.

Cardona, O.D. (1999) "Environmental Management and Disaster Prevention: Two Related Topics: A Holistic Risk Assessment and Management

Approach", in J. Ingleton, ed., *Natural Disaster Management*, London: Tudor Rose.

Cardona, O.D. (2001) *Estimación Holística del Riesgo Sísmico Utilizando Sistemas Dinámicos Complejos*, Barcelona: Technical University of Catalonia, available at http://www.desenredando.org/public/varios/2001/ehrisusd/index.html.

Cardona, O.D. (2004a) "Disasters, Risk and Sustainability", presentation regarding the Sasakawa Prize Ceremony in Geneva, Geneva (unpublished).

Cardona, O.D. (2004b) "The Need for Rethinking the Concepts of Vulnerability and Risk from a Holistic Perspective: A Necessary Review and Criticism for Effective Risk Management", in Bankoff, G., G. Frerks and D. Hilhorst, eds, *Mapping Vulnerability: Disasters, Development and People*, London: Earthscan, Chapter 3.

Cardona, O.D. and A.H. Barbat (2000) *El Riesgo Sísmico y su Prevención*, Cuaderno Técnico 5, Madrid: Calidad Siderúrgica.

Cardona, O.D. and J.E. Hurtado (2000a) "Holistic Approach for Urban Seismic Risk Evaluation and Management", Proceedings of Sixth International Conference on Seismic Zonation, EERI, November 2000, Palms Springs.

Cardona, O.D. and J.E. Hurtado (2000b) "Holistic Seismic Risk Estimation of a Metropolitan Center", Proceedings of 12th World Conference of Earthquake Engineering, January–February 2000, Auckland.

Cardona, O.D. and J.E. Hurtado (2000c) "Modelación Numérica para la Estimación Holística del Riesgo Sísmico Urbano, Considerando Variables Técnicas, Sociales y Económicas", in Oñate, E., F. García-Sicilia and L. Ramallo, eds, *Métodos Numéricos en Ciencias Sociales (MENCIS 2000)*, Barcelona: CIMNE-UPC.

Cardona, O.D., J.E. Hurtado, G, Duque, A. Moreno, A.C. Chardon, L.S. Velásquez and S.D. Prieto (2003) *The Notion of Disaster Risk. Conceptual Framework for Integrated Risk Management*, IADB/IDEA Program on Indicators for Disaster Risk Management, Universidad Nacional de Colombia, Manizales, available at http://idea.manizales.unal.edu.co/ProyectosEspeciales/adminIDEA/CentroDocumentacion/DocDigitales/documentos/01%20Conceptual%20Framework%20IADB-IDEA%20Phase%20I.pdf.

Cardona, O.D., J. E. Hurtado, A.C. Chardon, A. M. Moreno, S. D. Prieto, L. S. Velásquez, G. Duque (2005) Indicators of Disaster Risk and Risk Management. Program for Latin America and the Caribbean, Summary Report for World Conference on Disaster Reduction, IDB/IDEA Program of Indicators for Disaster Risk Management, National University of Colombia / Inter-American Development Bank, available at http://idea.manizales.unal.edu.co/ProyectosEspeciales/adminIDEA/CentroDocumentacion/DocDigitales/documentos/IADB-IDEA%20Indicators%20-%20Summary%20Report%20for%20WCDR.pdf.

Carreño, M.L, O.D. Cardona and A.H. Barbat (2004) *Metodología para la Evaluación del Desempeño de la Gestión del Riesgo*, Monografías CIMNE, Barcelona: Technical University of Catalonia.

Carreño, M.L, O.D. Cardona and A.H. Barbat (2005a) "Urban Seismic Risk Evaluation: A Holistic Approach", 250th Anniversary of Lisbon Earthquake, Lisbon.

Carreño, M.L, O.D. Cardona and A.H. Barbat (2005b) *Sistema de Indicadores para la Evaluación de Riesgos*, Monografía CIMNE IS-52, Barcelona: Technical University of Catalonia.

Chambers, R. (1989) "Editorial Introduction: Vulnerability, Coping and Policy", *IDS Bulletin* 20(2): 1–7.

Chambers, R. and G. Conway (1992) "Sustainable Rural Livelihoods: Practical Concepts for the 21st Century", *IDS Discussion Paper* 296, Brighton: Institute of Development Studies.

Crichton, D. (1999) "The Risk Triangle", in J. Ingleton, ed., *Natural Disaster Management*, London: Tudor Rose, pp. 102–103.

Cross, J.A. (2001) "Megacities and Small Towns: Different Perspectives on Hazard Vulnerability", *Environmental Hazards* 3(2): 63–80.

Cutter, S.L., B.J. Boruff and W.L. Shirley (2003) "Social Vulnerability to Environmental Hazards", *Social Sciences Quarterly* 84(2): 242–261.

Daly, H.E. (1993) "Sustainable Growth: An Impossible Theorem", in H.E. Daly and K.N. Townsend, eds, *Valuing the Earth: Economics, Ecology, Ethics*, Cambridge, Mass., MIT Press, pp. 267–273.

Davidson, R. (1997) *An Urban Earthquake Disaster Risk Index*, The John A. Blume Earthquake Engineering Center, Department of Civil Engineering, Report No. 121, Stanford: Stanford University.

Davoudi, S. and A. Layard (2001) "Sustainable Development and Planning: An Overview", in A. Layard, S. Davoudi and S. Batty, eds, *Planning for a Sustainable Future*, London/New York: Spon Press, pp. 7–17.

de Haan, L. and A. Zoomers (2005) "Exploring the Frontier of Livelihood Research", *Development and Change* 36(1): 27–47.

Department for International Development (DFID) (1999) *Sustainable Livelihood Guidance Sheets*, London: DFID, available at http://www.livelihoods.org/info/info_guidancesheets.html.

Dikau, R. and J. Weichselgartner (2005) *Der unruhige Planet: Der Mensch und die Naturgewalten*, Darmstadt, Primus Verlag.

Downing, T. (2004) "What Have We Learned Regarding a Vulnerability Science?" in *Science in Support of Adaptation to Climate Change*. Recommendations for an Adaptation Science Agenda and a Collection of Papers Presented at a Side Event of the 10th Session of the Conference of the Parties to the United Nations Framework Convention on Climate Change, Buenos Aires, 7 December 2004, pp. 18–21, available from www.aiaccproject.org/whats_new/Science_and_Adaptation.pdf.

Downing, T., J. Aerts, J. Soussan, S. Bharwani, C. Ionescu, J. Hinkel, R. Klein, L. Mata, N. Matin, S. Moss, D. Purkey and G. Ziervogel (2006) "Integrating Social Vulnerability into Water Management", *Climate Change* (in preparation).

Ewringmann, D. (1999) "Sustainability: Leerformel oder Forschungsprogramm?", Forschungsbericht Nr. 01–99, Sonderforschungsbereich 419, Teilprojekt C5, Universität zu Köln, Köln.

Fleischhauer, M. (2004) *Klimawandel, Naturgefahren und Raumplanung: Ziel- und Indikatorenkonzept zur Operationalisierung räumlicher Risiken*, Dortmund: Dortmunder Vertrieb für Bau- und Planungsliteratur.

Folke, C., S. Carpenter, T. Elmqvist, L. Gunderson, C.S. Holling and B. Walker

(2002) "Resilience and Sustainable Development: Building Adaptive Capacity in a World of Transformation", *Ambio* 31(5): 437–440.

Fues, T. (1998) *Das Indikatorenprogramm der UN-Kommission für nachhaltige Entwicklung: Stellenwert für den internationalen Rio-Prozess und Folgerungen für das Konzept von Global Governance*, Frankfurt a.M./Bern/New York/Paris: Peter Lang.

Green, C. (2004) "The Evaluation of Vulnerability to Flooding", *Disaster Prevention and Management* 13(4): 323–329.

Hilhorst, D. and G. Bankoff (2004) "Introduction: Mapping Vulnerability", in G. Bankoff, G. Frerks and D. Hilhorst, eds, *Mapping Vulnerability: Disasters, Development and People*, London: Earthscan.

Kasperson, J., R. Kasperson, B.L. Turner, W. Hsieh and A. Schiller (2005) "Vulnerability to Global Environmental Change", in J. Kasperson and R. Kasperson, eds, *The Social Contours of Risk. Volume II: Risk Analysis, Corporations & the Globalization of Risk*, London: Earthscan, pp. 245–285.

Kasperson, R. (2005) "Human Vulnerability to Global Environmental Change: The State of Research", Presentation at the Fifth Annual IIASA-DPRI Forum Integrated Disaster Risk Management, Innovations in Science & Policy, 14–18 September 2005, Beijing.

Kempmann, L. and B. Pilardeaux (2005) "Armutsbekämpfung durch globale Umweltpolitik", in *Entwicklungspolitik* 5: 25–28.

Kraas, F. (2003) "Megacities as Global Risk Areas", in *Petermanns Geographische Mitteilungen* 147(4): 6–15.

Lavell, A. (1996) "Degradación Ambiental, Riesgo y Desastre Urbano. Problemas y Conceptos: Hacia la Definición de una Agenda de Investigación", in M.A. Fernández, ed., *Ciudades en Riesgo*, La Red: USAID, pp. 12–42.

Lewis, J. (1999) *Development in Disaster-Prone Places: Studies of Vulnerability*, London: Intermediate Technology Publications.

Luers, A.L. (2005) "The Surface of Vulnerability: An Analytic Framework for Examining Environmental Change", *Global Environmental Change* 15: 214–223.

Maskrey, A. (1993) "Vulnerability Accumulation in Peripheral Regions in Latin America: The Challenge for Disaster Prevention and Management", in P.A. Merriman and C.W. Browitt, eds, *Natural Disasters: Protecting Vulnerable Communities*, IDNDR, London: Telford.

Monday, J. (2002) "Building Back Better: Creating a Sustainable Community After a Disaster", *Natural Hazards Informer* 3, January, available at http://www.colorado.edu/hazards/informer/infrmr3/informer3.pdf.

Müller-Mahn, D. (2005): "Von Naturkatastrophen zu Complex Emergencies – Die Entwicklung integrativer Forschungsansätze im Dialog mit der Praxis", in: D. Müller-Mahn, U. Wardenga, eds, *Möglichkeiten und Grenzen integrativer Forschungsansätze in Physischer Geographie und Humangeographie*, Leipzig: Institut für Länderkunde, pp. 69–78.

MunichRe (2004) *Megacities – Megarisks: Trends and Challenges for Insurance and Risk Management*, Munich: Munich Re Group.

Natural Hazard Center, University of Colorado (2006) *Holistic Disaster Recovery: Ideas for Building Local Sustainability after a Natural Disaster*, Fairfax, VA: Public Entity Risk Institute.

O'Brien, K.L. and R.M. Leichenko (2000) "Double Exposure: Assessing the Impacts of Climate Change within the Context of Economic Globalisation", in *Global Environmental Change* 10: 221–232.

OECD (1992) "Environmental Indicators: OECD Approach and Future Developments", Paper for the 21st Meeting of the Group of the State of Environment, OECD, Environmental Directorate, Environment Policy Committee, Paris.

Oliver-Smith, A. (2004) "Theorizing Vulnerability in a Globalized World: A Political Ecological Perspective", in G. Bankoff, G. Frerks and D. Hilhorst, eds, *Mapping Vulnerability: Disasters, Development and People*, London: Earthscan, pp. 10–24.

Pelling, M. (2003) *The Vulnerability of Cities: Social Resilience and Natural Disaster*, London: Earthscan.

Prescott-Allen, R. (1995) *Barometer of Sustainability: A Method of Assessing Progress toward Sustainable Societies*, Contribution to the IUCN/DRC Project on Monitoring and Assessing Progress Toward Sustainability, 2nd edn, British Columbia.

ProVention Consortium (2005) "Vulnerability and Risk Assessment", in *Tool Kit*, available at http://www.proventionconsortium.org/toolkit.htm.

Quarantelli, E.L. (no date) *Urban Vulnerability to Disasters in Developing Countries: Managing Risks*, available at http://www.worldbank.org/hazards/files/conference_papers/quarantelli.pdf.

Schmidt, A., L. Bloemertz and E. Macamo, eds (2005) *Linking Poverty Reduction and Disaster Risk Management, Eschborn: Bundesministerium für wirtschaftliche Zusammenarbeit und Entwicklung.*

Schneiderbauer, S. and D. Ehrlich (2004) *Risk, Hazard and People's Vulnerability to Natural Hazards: A Review of Definitions, Concepts and Data*, Brussels: European Commission–Joint Research Centre (EC-JRC).

Serageldin, I. (1995) "Promoting Sustainable Development: Toward a New Paradigm", in I. Serageldin and A. Steer, eds, *Valuing the Environment: Proceedings of the first Annual International Conference on Environmentally Sustainable Development*, Washington D.C.: World Bank, pp. 13–21.

Turner, B. L., R.E. Kasperson, P.A. Matson, J.J. McCarthy, R.W. Corell, L. Christensen, N. Eckley, J.X. Kasperson, A. Luers, M.L. Martello, C. Polsky, A. Pulsipher and A. Schiller (2003) "A framework for vulnerability analysis in sustainability science", *Proceedings of the National Academy of Sciences*, 100(14): 8074–8079.

United Nations (1993) *Agenda 21: Programme of Action for Sustainable Development*: The Final Text of Agreements Negotiated by Governments at the United Nations Conference on Environment and Development (UNCED), 3–14 June 1992, Rio de Janeiro, Brazil, New York: United Nations Publications.

United Nations (2000) *United Nations Millennium Declaration*, UN-Resolution 55/2, available at http://www.un.org/millennium/declaration/ares552e.pdf.

United Nations (2002) *Plan of Implementation of the World Summit on Sustainable Development in Johannesburg, South Africa*, available at http://www.un.org/esa/sustdev/documents/WSSD_POI_PD/English/WSSD_PlanImpl.pdf.

United Nations (2005) *Hyogo Framework for Action 2005–2015: Building the Re-*

silience of Nations and Communities to Disasters, World Conference on Disaster Reduction, 18–22 January 2005, Kobe, Hyogo, available at http://www.unisdr.org/wcdr/intergover/official-doc/L-docs/Hyogo-framework-for-action-english.pdf.

United Nations Development Programme (UNDP) (2004) *Reducing Disaster Risk: A Challenge for Development. A Global Report*, New York: UNDP – Bureau for Crisis Prevention and Recovery (BRCP), available at http://www.undp.org/bcpr/disred/rdr.htm.

UN-Habitat (2003) *The Challenge of Slums*, Global Report on Human Settlements 2003, London: Earthscan.

UN/ISDR (International Strategy for Disaster Reduction) (2002): *Living with Risk: A Global Review of Disaster Reduction Initiatives*, Geneva: UN Publications.

UN/ISDR (International Strategy for Disaster Reduction) (2004) *Living with Risk: A Global Review of Disaster Reduction Initiatives*, 2004 version, Geneva: UN Publications.

van Dillen, S. (2004) *Different Choices: Assessing Vulnerability in a South Indian Village*, Studien zur geographischen Entwicklungsforschung, Band 29, Saarbrücken: Verlag für Entwicklungspolitik.

van Ginkel, H. (2005) Introduction Speech regarding the Expert Workshop "Measuring Vulnerability", 23–24 January 2005, Kobe, in UNU-EHS *Working Paper No. 1*, Bonn: UNU-EHS.

Villagrán de León, J.C. (2001) "La Naturaleza de los Riesgos, un Enfoque Conceptual", in *Aportes para el Desarrollo Sostenible*, Guatemala: CIMDEN.

Villagrán de León, J.C. (2004) *Manual para la estimación cuantitativa de riesgos asociados a diversas amenazas*, Guatemala: Acción Contra el Hambre, ACH.

Vogel, C. and K. O'Brien (2004) "Vulnerability and Global Environmental Change: Rhetoric and Reality", *AVISO – Information Bulletin on Global Environmental Change and Human Security* 13, available at: http://www.gechs.org/publications/aviso/13/index.html.

WBGU (German Advisory Council on Global Change) (2005) *World in Transition: Fighting Poverty through Environmental Policy*, London: Earthscan.

World Commission on Environment and Development (WCED) (1987) *Our Common Future: Report of the World Commission on Environment and Development*, Brundtland Report, Oxford: Oxford University Press.

Wilches-Chaux, G. (1989) *Desastres, Ecologismo y Formación Profesional*, Popayán, SENA.

Wisner, B. (2002) "Who? What? Where? When? in an Emergency: Notes on Possible Indicators of Vulnerability and Resilience: By Phase of the Disaster Management Cycle and Social Actor", in: E. Plate, ed., *Environment and Human Security*, Contributions to a workshop in Bonn, 23–25 October 2002, Germany, pp. 12/7–12/14

Wisner, B, P. Blaikie, T. Cannon and I. Davis (2004) *At Risk: Natural hazards, People's Vulnerability, and Disasters*, 2nd edn, London: Routledge.

Zieschnak, R., J. van Nouhuys, T. Ranneberg and J.-J. Mulot (1993) *Vorstudie Indikatorensystem. Beiträge zur Umweltökonomischen Gesamtrechnung*, Wiesbaden: Statistisches Bundesamt Wiesbaden.

2

Indicators and criteria for measuring vulnerability: Theoretical bases and requirements

Jörn Birkmann

Introduction

Current approaches to measuring vulnerability often lack any systematic, transparent and understandable development procedures. In order to support a coherent and logical discussion of how different approaches have developed and generated their indicators and indices, it is important to explore theoretical foundations such as: quality criteria for indicator development; understanding of the relationship between indicators, goals and data; and the different phases of indicator development itself. Some of these are often taken into consideration either implicitly or explicitly when developing the quantitative and qualitative measurement tools for assessing vulnerability. This chapter will also address the differences between damage, impact and vulnerability assessment.

Rationale behind measuring vulnerability and vulnerability indicators

The ability to measure vulnerability is an essential prerequisite for reducing disaster risk, but it requires an ability to both identify and better understand exactly what are the various vulnerabilities to hazards of natural origin that largely determine risk. It is important to acknowledge that the term "measuring vulnerability" does not solely encompass quantitative approaches, but also seeks to discuss and develop all types of methods able to translate the abstract concept of vulnerability into practical tools to be applied in the field. This implies that approaches discussed under "measuring vulnerability" include quantitative indicators, qualitative criteria, as well as broader assessment approaches, such as trying to capture institutional aspects of vulnerability. The very complexity of the concept of vulnerability requires a reduction of potentially gatherable data to a set of important indicators and criteria that facilitate an estimation of vulnerability. Indicators and criteria are key tools for identifying and measuring vulnerability and related coping capacities. Just recently, in the final documents of the 2005 World Conference on Disaster Reduction (WCDR), held in Kobe, Japan, the international community stressed the need to develop vulnerability indicators. The final document of the WCDR, the Hyogo Framework for Action 2005–2015 (UN, 2005), stresses how important it is to:

develop systems of indicators of disaster risk and vulnerability at national and sub-national scales that will enable decision-makers to assess the impact of disasters on social, economic and environmental conditions and disseminate the results to decision-makers, the public and population at risk. (UN, 2005: 9)

Although the primary responsibility to reduce risk and vulnerability rests within individual countries and particularly local communities (e.g. Karl and Pohl 2003), there is a collective requirement throughout the international community to increase the knowledge and understanding of available methodologies for measuring risk and vulnerability (UN/ISDR, 2004: 397). This means the scientific community has the support of a number of international declarations in underlining the importance of developing indicators to measure the vulnerability and coping capacity of affected societies in order to improve disaster preparedness and to promote more disaster-resilient communities (Alexander, 2000: 37; UN, 2005; UNESCO, 2003; UN, 1993, 2002).

Because the concept of vulnerability is multidimensional and often ill-defined, it is difficult – and perhaps even impossible – to define a universal measurement methodology or to reduce the concept to a single equation (Downing, 2004: 18; Birkmann, Chapter 1; Schneiderbauer and Ehrlich, Chapter 3). This book aims to provide insights into the different techniques and methodologies for measuring vulnerability at different scales and with different thematic focuses – and even different tools, such as indicators, indices and criteria. Although considerable research has already been undertaken, we often know too little about the advantages of the different approaches and methodologies, their applicability in different areas and their limitations. Moreover, one has to acknowledge that we are also facing the problem that some areas of vulnerability, such as institutional vulnerability, are very complicated to measure and quantify, even though they are perceived as important. That means we have to bear in mind the limitations of measuring and simplifying the complex interactions that provide a context for and also shape the various vulnerabilities. Regarding the use of indicators and indices, Morse (2004) argues that we have been far too cavalier with indicators. In his book *Indices and Indicators in Development: An Unhealthy Obsession with Numbers?* he stresses the fact that indicators are necessary tools, but that one has to handle them with care.

Definition: indicator

Different authors define indicators differently and one can find many ambiguities and contradictions regarding the general concept of an indicator. Within the discourse of measuring sustainable development, Gallopín (1997) developed a general, but at the same time quite comprehensive, definition. He defined an indicator as a sign that summarises information relevant to a particular phenomenon (Gallopín, 1997: 14). A more precise definition views indicators as variables (not values), which are an operational representation of an attribute, such as a quality and/or a characteristic of a system (Gallopín 1997: 14). Consequently a vulnerability indicator for hazards of natural origin can be defined as:

- a variable which is an operational representation of a characteristic or quality of a system able to provide information regarding the susceptibility, coping capacity and resilience of a system to an impact of an albeit ill-defined event linked with a hazard of natural origin.

It is important to remember that any indicator – whether descriptive or normative – has significance besides its face value. This means the relevance of the indicator for estimating a certain quality or characteristic of a system arises from the interpretation made about the indicator and its relationship to the phenomena of interest. Therefore, assigning a meaning to a variable and defining the indicating function of the indicator makes an indicator out of the variable (Birkmann, 2004: 62; Gallopín, 1997: 16). In principle an indicator can be a qualitative variable (nominal), a rank variable (ordinal) and/or a quantitative variable (Gallopín, 1997: 17).

Current definitions of vulnerability indicators and the use of the terminology in this area are particularly confusing. Downing stresses the fact that the indiscriminate use of indicators for measuring vulnerability – pick any that seem to be relevant and/or available – must be avoided; rather, it is important to define and develop at least an implicit model to serve as a systematic basis for indicator development and selection (Downing, 2004: 19). Present concepts range from macroeconomic vulnerability estimation to individual assessment tools. The UN/ISDR emphasises the fact that measuring progress towards disaster risk and vulnerability reduction in a country or region requires different frameworks at different timescales (UN/ISDR, 2004: 396). This means that various concepts and methods of vulnerability assessment are not only shaped by the different scientific disciplines they derive from, but they are also necessary in order to capture the multifaceted nature of vulnerability to different phenomena and hazards, such as earthquakes, floods, droughts and various impacts of climate change (e.g. Leary et al., 2005: 1,4; Downing, 2004: 18–21).

Indicator development: a historical overview

Developing and using indicators is not a new phenomenon. Economic indicators had already emerged in the early 1940s (Hartmuth, 1998; Reich and Stahmer, 1983). Today, economic indicators such as "GDP" or "unemployment rate" are broadly used (and politically accepted) to estimate and communicate the state and evolution of the economy. In the 1960s and 1970s the development of social indicators was a hot topic in the social sciences (Cutter et al., 2003: 244), which crossed over into the political and social arena during the protest movements in the 1960s in the United States and Western European countries (e.g. Empacher and Wehling, 1999: 14). The development of environmental indicators followed in the 1970s, linked to the establishment of environmental policies. The last big impetus for indicator development emerged from the discussions about sustainable development. In this context various approaches to defining and operationalising sustainable development with indicators were undertaken (UNCSD, 1996; Birkmann, 2004: 61). However, no precise consensus emerged regarding the definition of indicators.

Indicators, goals and data

Some authors define indicators in relation to an aggregation process, starting with variables or basic data, followed by processed information and indicators, and finally ending up with highly aggregated indices (e.g. Adriaanse, 1995 (see Figure 2.1)). However, defining indicators in terms of the level of aggregation is not appropriate, since an indicator can be a single variable as well as a high aggregated measure, for instance the result of a complex computer model. The *UN World Water Development Report (WWDR)* makes it clear that an indicator can be a single piece of data (single variable) or an output value from a set of data (aggregation) that describes a system or process (UNESCO, 2003: 3/p.33). That means, the definition of indicators based on the level of aggregation (Figure 2.1) neglects an essential aspect: *goals*.

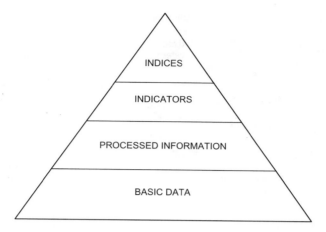

Figure 2.1 The data pyramid.
Source: Author, following Adriaanse, 1995: 5.

Every indicator-development process needs to be related (explicitly or implicitly) to goals, or at least to a vision which serves as a basis for defining the *indicandum* (state or characteristic of interest). To develop scientifically sound indicators it is necessary to formulate goals that serve as a starting point for the identification of relevant indicators. It is essential to acknowledge that the main interest is not in the indicator itself, but in the *indicandum*. The close link between the indicator and the *indicandum* should be recognised. The quality of the indicator is determined by its ability to indicate the characteristic of a system that is relevant to the underlying interest determined by the goal or guiding vision. The link between the indicator and the *indicandum* should be theoretically sound. The interrelation between indicators, data and goals can be illustrated as shown in Figure 2.2, which indicates that any indicator development must collect data as well as formulate goals that define the underlying interest.

The figure illustrates the fact that the assumptions and judgements made in selecting relevant issues and data for the indicator development, as well as the evaluation of the indicator's usefulness, require the existence of goals, whether implicit or explicit. In the case of vulnerability indicators, general goals would include, for example, reducing the vulnerability of potentially affected communities to hazards of natural origin, while more precise goals could be to reduce human fatalities or increase the financial capacity of households in order to reduce vulnerability and improve their potential for recovering from economic losses due to a hazardous event. In practice, one can distinguish two main types of indicator–goal relations (Weiland 1999: 252):

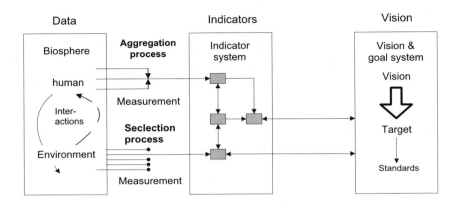

Figure 2.2 The model of the three pillars: indicators, data and goals.
Source: Birkmann, 1999: 122.

- In the first case the indicator focuses on the direction a development is taking. The assessment of the development trend allows one to evaluate its vulnerability; in other words, an increasing or decreasing development trend indicates a higher or lower vulnerability.
- In the second case the indicator focuses on a specific target that shows whether the state or the development has reached a defined value. This requires precise goals for the indicator. One has to be able to define whether a specific value indicates vulnerability or not.

The insurance industry, for example, is able to estimate precisely a value and target of potential economic losses of a firm or a household due to a specific event, such as a flood event of the magnitude HQ 200, and so to calculate the insurance risk (e.g. Kron, 2005: 66). In contrast, the definition of a single value to estimate social vulnerabilities, for example, is often problematic and needs additional interpretations, such as the Social Vulnerability Index (SoVI) developed by Cutter et al. (2003: 249) for the United States. Furthermore, regarding the environmental dimension, one should note that it is nearly impossible to derive values for environmental vulnerability (e.g. groundwater, air, soil conditions) solely on the basis of natural sciences. Goals for environmental services and qualities depend also on the definition of the specific ecological function assigned to an area by the local community or the State, which means quality standards have also to be based on specific sub-national and local contexts (Finke, 2003: 158; Kuehling, 2003: 2). However, there is often a lack of established environmental standards.

Since vulnerability assessment and the judgement of whether the value shows a high or low vulnerability are complex tasks, many approaches define a *relative* vulnerability that views, compares and interprets vulnerability between different groups, entities and geographic areas in order to assess it. Furthermore, due to a lack of precisely defined targets of vulnerability in many cases, trend evaluation and comparison is a useful surrogate or proxy for estimating vulnerabilities. Representatives of this approach include the Disaster Risk Index, the Hotspot project approach and the Americas project (see Chapters 7–10). In contrast, Plate's Human Security Index determines the vulnerability of a household and its potential to recover from a negative impact of a hazardous event through a single value (see Chapter 13). All these approaches focus mainly on a single or composite indicator to measure and estimate vulnerability and risk. As an alternative, Downing et al. (2006) propose viewing the vulnerability of socio-economic groups as a profile rather than merely as a single, composite number (Downing et al., 2006: 7).

Functions of vulnerability indicators

Within the general framework of functions of indicators and information, Wisner and Walter (2005) emphasise that gathering data and information has to serve the aggregation of knowledge (i.e. the understanding of how things work or are supposed to work), which in turn is essential for being able to make the right choices. Wisner stresses the fact that the entire process always has to aim at the final level of what he calls "wisdom" which allows for sound decision-making, which means making value judgement based on experience, understanding and principle (Wisner and Walter, 2005: 14).

In terms of a more specific focus on vulnerability indicators, one can say that their usefulness is determined by their success in achieving their objective and function, such as identification and visualisation of different characteristics of vulnerability, or evaluation of political strategies and monitoring of their implementation. In this context, Benson (2004) points out that measuring vulnerability across countries, within countries and between different events linked to different hazards helps to create an understanding of factors contributing to vulnerability, although the often-used ex-post focus cannot be directly equated with future vulnerability (Benson, 2004: 159). According to Benson the identification and the understanding of vulnerability and its underlying factors are important goals and functions of measuring vulnerability.

Queste and Lauwe stress how vulnerability measurement is needed for practical decision-making processes, such as to provide disaster managers with appropriate information about where the most vulnerable infrastructures are (Chapter 4). They argue that the indicators should enable the administration at different levels to integrate vulnerability reduction strategies into preventive planning. While Benson (2004) views the primary function of vulnerability measurement as serving the desire for knowledge of understanding, Queste and Lauwe stress the fact that indicators, from a practical point of view, should guide decision-making (knowledge for action). In this context Green underlines the fact that the primary interest in defining and measuring vulnerability lies in the goal of reducing it (Green, 2004: 324). If we cannot reduce vulnerability, the development of indicators and assessment tools would be of minor interest for political decision makers and practitioners. Billing and Madengruber underline the interest of aid agencies in developing and using tools for measuring vulnerability and coping capacity in order to design appropriate disaster reduction strategies and to achieve a better targeting of external assistance to countries most needing outside help to overcome disasters (Billing and Madengruber, Chapter 21). Additional functions of indicators and assessment tools for measuring vulnerability were derived through a "straw poll" at the first meeting of the Expert Working Group of UNU-EHS in Kobe (Birkmann, 2005). The experts present at the workshop in Kobe defined the following functions (Birkmann, 2005: 13) as most important for vulnerability indicators:

- setting priorities
- background for action
- awareness raising
- trend analysis
- empowerment.

These functions primarily address practical aspects, such as "awareness raising" and "promoting background for action". The literature on functions of indicators often encompasses more traditional features like simplification (reduction of complexity), comparison of places and situations, anticipation of future conditions and trends, and assessment of conditions and trends in relation to goals and targets (Gallopín, 1997; Tunstall, 1994).

Besides the discussion of major functions of indicators, and of vulnerability indicators in particular, the analysis of the process of indicator development is important in order to understand the different phases and judgements that the construction of indicators and criteria are based on.

The ideal phases of indicator development

In general, one can distinguish nine different phases in the development of indicators, which also apply in the development of vulnerability indicators. According to Maclaren (1996), indicator development starts with the definition or selection of relevant *goals*. Then it is necessary to carry out a *scoping process*, which implies clarifying the scope of the indicator by identifying the target group and the associated purpose for which the indicators will be used (goals and functions). It is also important to define the temporal and spatial bounds, which means identifying the timeframe over which indicators are to be measured and determining the spatial bounds of the reporting unit (community, sub-region or socio-economic regions, bio-geographical zones or administrative units). The third phase involves the identification of an appropriate *conceptual framework*, which means structuring the potential themes and indicators. The different approaches to measuring vulnerability presented in this book encompass various conceptual frameworks, like those that focus on sectors (Villagrán de León), issue-based frameworks (Pelling, Peduzzi, Dilley in this book) or causal frameworks, such as the one proposed by Turner et al. (2003) (see also Chapter 1). Which framework may be most appropriate for structuring vulnerability indicators depends on the purpose for which the indicators will be used, as well as on the target group and, finally, on the availability of data.

The fourth phase implies the definition of *selection criteria* for the potential indicators. Although the scientific debate about indicators has led to a set of general criteria for "good quality" indicators, like "scientifically valid", "responsive to change" and "based on accurate and accessible data", it is necessary to link these to the theme, function and goal of the specific approach. These criteria have to be interpreted, for example, in terms of such aspects as data accuracy and data accessibility. According to Benson (2004: 169), most disaster risk data is incomplete and based on historical disasters (past events), while the factors determining the outcome of future events are highly complex and can differ from those already experienced. Hence, it is a crucial task for all approaches aiming at measuring vulnerability to find the right balance between the accuracy of data and the limited data available.

The *identification of a set of potential indicators* (phase 5) is a key step in indicator development. An example of potential indicators for measuring vulnerability at different levels and scales is presented by Schneiderbauer and Ehrlich (Chapter 3). Finally, there is the evaluation and selection of each indicator (phase 6) with reference to the criteria developed at an earlier stage, which results in a set of indicators.

The *collection of data* for the indicator has to be followed in order to

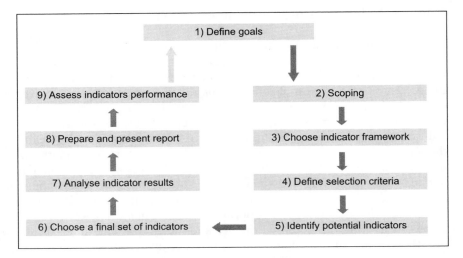

Figure 2.3 Development process of vulnerability indicators.
Source: Based on the general figure according to Maclaren, 1996: 189.

prove the applicability of the approach. This phase can often be the most difficult one, especially since vulnerability is characterised by many intangible factors and aspects, which are difficult to quantify or which can be measured only indirectly, such as social networks, confidence, trust and apathy, and institutional aspects such as good governance, appropriate early warning, appropriate legislation (Bogardi and Birkmann, 2004: 79; Cutter et al., 2003: 243). An interesting example of the challenges and limitations of assessing institutionalised capacities and practices in order to reduce flood disasters is given by Lebel et al. (Chapter 19). The final phases of the indicator development can be seen in the preparation of a report and the assessment of the indicator performance (Maclaren, 1996: 184). According to Maclaren, the whole development process is an "ideal process", which in practice is characterised through an iterative procedure of going backwards and forwards. Nevertheless, the distinction between different phases of indicator development is helpful in order to analyse current approaches and their development processes, including the underlying assumptions.

Quality criteria for indicator development

The development of indicators to measure vulnerability has to be based on quality criteria that support the selection of sound indicators. Stan-

Box 2.1 Standard criteria for indicator development

- measurable
- relevant, represent an issue that is important to the relevant topic
- policy-relevant
- only measure important key-elements instead of trying to indicate all aspects
- analytically and statistically sound
- understandable
- easy to interpret
- sensitivity; be sensitive and specific to the underlying phenomenon
- validity/accuracy
- reproducible
- based on available data
- data comparability
- appropriate scope
- cost effective.

(see EEA, 2004; Birkmann, 2004; NZOSA, 2004 (Internet); Berry et al., 1997; Parris, 2000)

dard criteria for indicator development prescribe, for example, that these indicators should be "relevant", "analytically and statistically sound", "reproducible" and "appropriate in scope", while participatory indicator development often focuses on criteria such as "understandable", "easy to interpret" and "policy-relevant". In contrast practitioners often emphasise that indicators that have to be applicable in practice should be "based on available data" as well as "cost effective" (Birkmann, 2004: 80; Gallopín, 1997: 25; Hardi, 1997: 29).

The different priorities and weightings of these criteria can also be viewed in current approaches to measuring vulnerability presented in this book. While, for example, international indexing projects define the availability of already existing data as a key criterion for providing useful global information to allow comparison of different countries, the method of self-assessment of vulnerability and coping capacity presented by Wisner in this book does not account for available data; rather it focuses on people's knowledge and policy-relevant recommendations (see Chapter 17). In this context also the methodology for measuring and monitoring the effectiveness of the lessons learned within risk management and presented by Krausmann and Mushtaq (Chapter 22) provide an insight into how one can gather new data while keeping to a reasonable timeframe and budget. Birkmann et al. present different methodologies on how to combine and use tools to measure revealed and emergent vulnerability based on existing data as well as methodologies that require and capture new and additional data for calculating the vulnerability of coastal communities to tsunami and coastal hazards in Sri Lanka (see Birkmann, Fernando and Hettige, Chapter 18).

Overall, one has to conclude that the general quality criteria for indicator development presented here are general guidelines. Specific approaches might have to define their own priorities and have to weigh the different criteria according to their specific needs and functions. However, one of the most difficult points in measuring vulnerability is the gathering of appropriate data.

Data for measuring vulnerability

The gathering of accurate, reliable and accessible data to estimate and measure vulnerability is a major problem when dealing with vulnerability assessment at various levels. Although one can find international databases on disasters, such as the database of the Centre for Research on Epidemiology of Disasters (CRED), the application of globally available data sources for sub-national and local vulnerability measurement approaches is often limited by the fact that this data does not give sufficient information to assess the various vulnerabilities at different spatial levels and units. Some of the most precise databases on potential and revealed economic losses are maintained by large reinsurance companies, such as MunichRe or SwissRe. The MunichRe database for natural catastrophes "MRNatCatSERVICE" (NatCat) includes over 20,000 entries on material and human loss events worldwide (MunichRe, 2003). Besides the classification of the type of disaster and its location, the database also contains information about the economic losses due to the specific hazardous event and the insured losses. NatCat accounts for effects on people with regard to such criteria as being injured, homeless, missing, affected and evacuated (see Figure 2.4), which can also be relevant for the estimation of revealed vulnerability. The damage to houses is also registered as well as the general impact on agriculture, infrastructure and life lines (MunichRe, 2004).

Overall, MunichRe focuses primarily on damage and gives a precise picture of the economic losses caused by hazards of natural origin. However, this data is of only limited use for estimating social, economic and environmental vulnerability, since the reported economic damage does not equal future or present economic vulnerability. Nevertheless, indicators of economic and social vulnerability can be tested and compared with the revealed vulnerabilities and losses gathered in the data source of MRNatCat. Another problem in using of this data to estimate vulner-

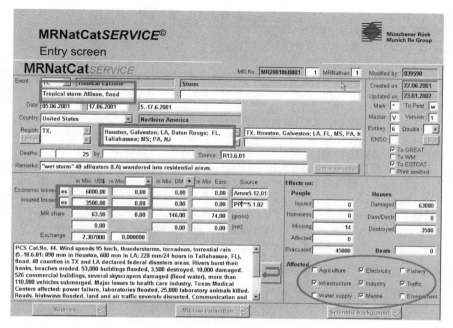

Figure 2.4 Example of the data gathered in the MRNatCatSERVICE.
Source: MunichRe, 2004.

ability is the fact that reported damages predominantly focus on direct
costs and losses, often excluding indirect losses as well as the long-term
socio-economic impacts of a disaster (Benson, 2004: 165). Additionally,
the different definitions of the categories used in these statistics, such as
"affected" or "injury" are a complication when trying to make comparisons and analysis (Wisner et al., 2004: 65). So-called soft assets and intangibles cannot be captured through loss assessment techniques. This requires the use of different methodologies, such as questionnaires or data
sampling through participatory approaches. Interestingly, in the current
literature one can find a debate about the appropriateness of social
science methodologies for estimating vulnerability. While for example
Benson (2004), Hilhorst and Bankoff (2004) and other authors (see Wisner in this book) underline the usefulness and benefits of using social
science methods to measure vulnerability, Tapsell et al. (2002) argue that
methodologies such as interviews and focus-group approaches are often
expensive and time consuming. However, Weichselgartner and Obersteiner (2002) point out that too little attention is being paid to the loss

of local knowledge and traditional adaptation strategies that have been valid and useful in the past.

In this publication, the various approaches to measuring vulnerability show the different opportunities and limitations that are implied in using methods that are either based on currently available data or that depend on gathering new data, for example through questionnaires and interviews.

Overall, it is evident that besides the need to improve the data on the impacts of past disasters, more comprehensive and holistic approaches are essential, that take into account the dynamic nature of vulnerability as well as trying to understand underlying causal factors of vulnerability at different levels for different hazards (Wisner et al., 2004; Vogel and O'Brien, 2004: 3). This underlines the importance of distinguishing – especially from a scientific point of view – between damage, impact and vulnerability assessment. These aspects will be discussed in greater depth in the following section.

Damage, impact and vulnerability assessment

In the literature and in practice, vulnerability, impact and damage assessment are often mixed or used synonymously. Although damage and impact assessments can overlap with vulnerability assessment, they generally imply different approaches. Damage assessment is based on the calculation of real losses, such as fatalities, economic losses and damage to the physical infrastructure. Impact assessment, however, is not necessarily limited to losses (one impact category); rather, it can also encompass positive impacts for social groups, specific economic sectors or the environment that result from an event linked to a hazard. For example, in the aftermath of the recent tsunami both the construction industry and local carpenters actually benefited from the overall need for enormous reconstruction efforts.

According to Vogel and O'Brien (2004), impact assessment has traditionally been used to document the potential consequences of a particular event (e.g. drought), while in contrast vulnerability assessment should focus also on the factors – of human or environmental origin – that, together or separately, drive and shape the vulnerability of the receptor, for example a community or ecosystem (Vogel and O'Brien, 2004: 2).

Furthermore, Wisner et al. (2004) and Benson (2004) point out that a major difference between damage assessment and vulnerability assessment is the time dimension. While damage assessment is often conducted as a rapid survey to provide information upon which to base appropriate and fast responses, such as direct humanitarian aid (Wisner et al., 2004:

13, 128; Benson, 2004: pp. 164), vulnerability assessment should focus on the likelihood of injury, loss, disruption of livelihood and other harm in an extreme event and/or the unusual difficulties in recovering (Wisner et al., 2004: pp. 13; Wisner, 2002: 12/7). This means the focus of vulnerability assessment should be on the identification of the variables that make people vulnerable and that show major differences in the (potential and revealed) susceptibility of people as well as those factors that drive and shape the vulnerability. While damage and impact assessments are based on an ex-post evaluation of revealed losses, vulnerability assessment can and should be a forward-looking concept (Wisner, Chapter 17; Vogel and O'Brien, 2004: p. 3; Benson, 2004: 167).

Vulnerability, then, cannot be estimated adequately solely through data based on past events. On the other hand, many indicators and criteria for measuring vulnerability often need to be linked, weighted and developed on the basis of analysing past events and their impacts (damage patterns and revealed vulnerabilities).

Interestingly, various approaches presented in this book have dealt with this complex problem, that on the one hand they aim to measure vulnerability through a forward-looking perspective, but on the other hand have often had to use data based on and linked to past events in order to derive sophisticated indicators and criteria to estimate vulnerability. Even the assessment of institutional capacities of risk reduction has to deal with this challenge, namely that the appropriateness and effectiveness of institutional capacities to mitigate risk and vulnerabilities only become evident under stress and disaster conditions (Lebel et al., Chapter 19). In order to illustrate the differences and the variety of current approaches to measuring vulnerability, a brief overview of selected approaches presented in this book is given.

Overview and classification of selected approaches

The following list encompassing a systematisation of selected approaches shows the main characteristics and differences of approaches presented in this book. It is not intended to be comprehensive. The systematisation and comparison is based on the following criteria:
- spatial level of the approach
- function of the approach
- thematic focus regarding vulnerability
- data basis
- target audience
- link to goals
- level of aggregation.

Table 2.1 Overview and systematisation of selected indicator approaches to measure vulnerability

Criteria	DRI (Chapter 8)	Multi risk assessment Europe (Chapter 11)	Sector approach (Chapter 16)	CATSIM model (Chapter 20)	Community-based disaster risk index (Chapter 14)	Self-assessment (Chapter 17)
Spatial level	Global (national resolution)	Europe (NUTS 3 regions, sub-national resolution)	Local (single building resolution)	National level	Municipal level	Local community (individual or group resolution)
Function of the approach	Identification of vulnerability Comparison of vulnerability between countries	Identification of vulnerability Comparison of vulnerability between NUTS-3 regions	Identification of vulnerability and options to reduce it	Identification and awareness rising	Identification and knowledge generation Empowerment of people Promoting gender equity	Identification of vulnerability Empowerment of people
Thematic focus on vulnerability	Mortality (average annual death) as the calculation of relative vulnerability, various socio-economic aspects (24 variables) are selected in order to explain the variation of the vulnerability between countries	Vulnerability encompasses hazard exposure and coping capacity Mainly expressed through GDP and population density	Sectors, such as housing, health, education, industry, agriculture, finance	Financial vulnerability of the public sector (subset of economic vulnerability)	Vulnerability regarding physical, demographic, social, economic and environmental assets	People, their assets and resources plus applications to address root causes

Table 2.1 (cont.)

Criteria	DRI (Chapter 8)	Multi risk assessment Europe (Chapter 11)	Sector approach (Chapter 16)	CATSIM model (Chapter 20)	Community-based disaster risk index (Chapter 14)	Self-assessment (Chapter 17)
Data basis	CRED (Center for Research on the Epidemiology of Disaster)	EUROSTAT, MunichRe	Field survey	National data	Questionnaire based data	Focus group discussions
Target group	International community and national states	European Commission	Institutions in charge of specific sector	Public authorities and private sector	Local population, local government	People at risk
Link to goals	No direct link to goals	Weighting of indicators through expert judgement No explicit link to goals	Classification of vulnerability (low, medium and high) No direct link to goals	Scenarios No direct link to goals	Classification of vulnerability (low, medium and high) No direct link to goals	No direct link to goals
Level of aggregation	Medium (the relative vulnerability measure shows a relatively low aggregation level, the exposure component is more complex)	High, medium (the multi-risk map involves several indicators and therefore classified High/medium aggregated)	High (aggregated values for each facility, sector or part of a municipality)	Medium (financial gap compared to other potential aid and income sources)	Medium, high indicators and index (47 single indicators, aggregation into 4 factor scores and 1 risk index)	Low No aggregation

Source: Author.

Final remarks and key questions

The following conclusions reflect the foregoing discussion and raise some important questions, which provide useful tools and guidelines for further investigating current approaches.

The formation of a theoretical basis of indicator development, the analysis of current data – especially of the insurance industry (MunichRe) – and the examination of differences between damage, impact and vulnerability assessment have revealed that vulnerability assessment must go beyond damage assessment. Vulnerability assessment should focus on the characteristics that determine the likelihood of injury, loss and other harm as well as the capacity to resist and recover from negative impacts. However, developing a more sophisticated approach to measuring – one that promotes a forward-looking perspective – is still a challenge. For practical reasons, and also for validation of vulnerability characteristics, vulnerability indication and assessment often needs also to take into account damage known to have occurred in the past. Therefore, the strict separation between damage, impact and vulnerability becomes less rigid in the course of practical application. Nevertheless, it is important to keep in mind the differences and also to underline how vulnerability is distinguished from a "pure" damage and impact assessment approach. Also important is a precise understanding of what the approach focuses on in terms of "vulnerability": vulnerability to what and vulnerability of what. In this context it is essential to focus on the definition and perception of what is meant by vulnerability in the specific approach, such as the Disaster Risk Index, the Human Security Index and the CATSIM model.

Furthermore, the presentation of theoretical bases has underlined the fact that there is a difference between the *indicator* and the *indicandum*. Thus, it is important to check whether the intention and conceptual framework of measuring vulnerability also correspond with the selected indicators. This also implies a need to examine how the perception of vulnerability in the approach corresponds with the revealed vulnerability.

One of the most important goals in developing tools for measuring vulnerability is to help bridge the gaps between the theoretical concepts of vulnerability and day-to-day decision-making. In this context it is interesting to learn about the target group of any particular approach, namely the group that is expected to use the results of the measurement tool. These questions are linked with the overall function of the approach. For example, is the approach mainly intended to provide knowledge for understanding or is it aimed at informing decision-making processes (knowledge for action)? Finally, the author also recommends keeping in mind that every approach to measuring vulnerability is based – explicitly or implicitly – on a vision or goals.

Based on these ideas, the following questions arise when examining more closely the various approaches presented in this book:
- What is meant by vulnerability in the specific approach?
- What is the pre-analytic view and conceptual framework of the approach?
- What criteria and indicators are used to measure vulnerability?
- How does the approach simplify the complex concept of vulnerability?
- What purpose and functions should the approach serve?
- How are vulnerability, coping capacity and exposure addressed?
- How do perceptions of vulnerability (hypotheses of main characteristics and driving forces) compare with the revealed vulnerability in disasters?
- Does the selection of indicators reflect the conceptual framework of vulnerability?
- How far is the approach applied in decision-making processes?

The variety of approaches presented in this book allows for a comparison between various conceptual frameworks, methodologies and tools to measure vulnerability.

REFERENCES

Adriaanse, A. (1995) "In Search of Balance: A Conceptual Framework for Sustainable Development Indicators", in A. MacGillivray, ed., *Accounting for Change. Papers from an International Seminar*, October 2004, London: The New Economic Foundation, pp. 3–10.

Alexander, D. (2000) *Confronting Catastrophe: New Perspectives on Natural Disasters*, Oxford/New York: Oxford University Press.

Benson, C. (2004) "Macro-economic Concepts of Vulnerability: Dynamics, Complexity and Public Policy", in G. Bankoff, G. Frerks and D. Hilhorst, eds, *Mapping Vulnerability: Disasters, Development and People*, London: Earthscan.

Berry, D. (1997) "Sustainable Development in the United States: An Experimental Set of Indicators", Interim Report, US Interagency Working Group on Sustainable Development Indicators, Washington D.C.

Birkmann, J., (1999) "Indikatoren für eine nachhaltige Entwicklung: Eckpunkte eines Indikatorensystems für räumliche Planungsfragen auf kommunaler Ebene", *Raumforschung und Raumordnung Heft 2/3 99*, Hannover: Akademie für Raumforschung und Landesplanung (ARL), pp. 120–131.

Birkmann, J., H. Koitka, V. Kreibich and R. Lienenkamp (1999) "Indikatoren für eine nachhaltige Raumentwicklung: Methoden und Konzepte der Indikatorenforschung", Institut für Raumplanung (ed.), Dortmunder Beiträge der Raumplanung, Band 96, Dortmund.

Birkmann, J. (2003) "Measuring Sustainable Spatial Planning in Germany: Indicator Based Monitoring at the Regional Level", *Built Environment* 29(4): 296–305.

Birkmann, J. (2004) *Monitoring und Controlling einer nachhaltigen Raument-wicklung, Indikatoren als Werkzeuge im Planungsprozess*, Dortmund: Dortm-under Vertrieb für Bau- und Planungsliteratur.

Birkmann, J. (2005) "Measuring Vulnerability – Expert Workshop in Kobe, UNU-EHS Working Paper, Bonn: UNU-EHS available at http://www.ehs.edu/file.php?id=60

Birkmann, J. and O. Frausto (2001) "Indicators for Sustainable Development for the Regional and Local Level – Objectives, Opportunities and Problems: Case Studies from Germany and Mexico", *European Journal of Regional Development* 1(9): 23–30.

Bogardi, J. and J. Birkmann (2004) "Vulnerability Assessment: The First Step to-wards Sustainable Risk Reduction", in D. Malzahn and T. Plapp, eds, *Disaster and Society: From Hazard Assessment to Risk Reduction*, Berlin: Logos Verlag Berlin.

Centre for Research on the Epidemiology of Disasters (CRED) Website, avail-able at http://www.cred.be/

Cutter, S.L., B.J. Boruff and L.W. Shirley (2003) "Social Vulnerability to Envi-ronmental Hazards", *Social Science Quarterly* 84(2): 242–261.

Downing, T. (2004) "What Have We Learned Regarding a Vulnerability Science?" in *Science in Support of Adaptation to Climate Change*. Recommen-dations for an Adaptation Science Agenda and a Collection of Papers Pre-sented at a Side Event of the 10th Session of the Conference of the Parties to the United Nations Framework Convention on Climate Change, Buenos Aires, 7 December 2004, pp. 18–21, available from www.aiaccproject.org/whats_new/Science_and_Adaptation.pdf.

Downing, T., J. Aerts, J. Soussan, S. Bharwani, C. Ionescu, J. Hinkel, R. Klein, L. Mata, N. Matin, S. Moss, D. Purkey and G. Ziervogel (2006) "Integrating Social Vulnerability into Water Management", *Climate Change* (in prepara-tion).

European Environment Agency (EEA) (2001) *Environmental Signals 2001: EEA Regular Indicator Report*, Copenhagen: EEA, available at http://reports.eea.eu.int/signals-2001/en/signals2001/.

EEA (2004) *Criteria for the Selection of the EEA Core Set of Indicators*, available at http://themes.eea.eu.int/IMS/About/CSI-criteria.pdf.

Empacher, C. and P. Wehling (1999) *Indikatoren sozialer Nachhaltigkeit: Grund-lagen und Konkretisierungen*, ISOE DiskussionsPapiere 13, Frankfurt a.M.: In-stitut für sozio-ökologische Forschung (ISOE).

Finke, L. (2003) "Der Umweltzielplan oder ökologische Funktionsplan als Vor-aussetzung einer nachhaltigen Raumentwicklung", in W. Kühling and C. Hild-mann, eds, *Der integrative Umweltplan: Chance für eine nachhaltige En-twicklung*, Dortmund: Dortmunder Vertrieb für Bau- und Planungsliteratur.

Fleischhauer, M. (2004) *Klimawandel, Naturgefahren und Raumplanung. Ziel-und Indikatorenkonzept zur Operationalisierung räumlicher Risiken*, Dort-mund: Dortmunder Vertrieb für Bau- und Planungsliteratur.

Gallopín, G.C. (1997) "Indicators and their Use: Information for Decision-Making. Part One: Introduction", in B. Moldan and S. Billharz, eds, *Sustain-ability Indicators: Report of the Project on Indicators of Sustainable Develop-

ment, SCOPE (Scientific Committee on Problems of the Environment), New York: John Wiley, available at http://www.icsu-scope.org/downloadpubs/scope58/ch01-introd.html.

Green, C. (2004) "The evaluation of vulnerability to flooding", *Disaster Prevention and Management* 13(4): 323–329.

Habich, R. and H.-H. Noll (1994) *Soziale Indikatoren und Sozialberichterstattung, Internationale Erfahrungen und gegenwärtiger Forschungsstand*, Bern: Bundesamtes für Statistik der Schweiz.

Hardi, P. (1997) "Measurement and Indicator Program of the International Institute for Sustainable Development", in B. Moldan and S. Billharz, eds, *Sustainability Indicators: Report of the Project on Indicators of Sustainable Development*, SCOPE (Scientific Committee on Problems of the Environment), New York: John Wiley, available at http://www.icsu-scope.org/downloadpubs/scope58/box1a.html.

Hartmuth, G. (1998) "Ansätze und Konzepte eines umweltbezogenen gesellschaftlichen Monitorings", in L. Kruse-Graumann, ed., *Ziele, Möglichkeiten und Probleme eines gesellschaftlichen Monitorings*, Tagungsband zum MAB-Workshop, 13–15 June 1996, Bonn: Potsdam-Institut für Klimafolgenforschung (PIK)/Bundesamt für Naturschutz, pp. 9–33.

Hilhorst, D. and G. Bankhoff (2004) "Introduction: Mapping Vulnerability", in G. Bankoff, G. Frerks and D. Hilhorst, eds, *Mapping Vulnerability: Disasters, Development and People*, London: Earthscan, pp. 1–9.

Jänicke, M. (1995) "Tragfähige Entwicklung: Anforderungen an die Umweltberichterstattung aus der Sicht der Politikanalyse", *Zeitschrift für angewandte Umweltforschung* 8(1): 90–98.

Karl, H. and J. Pohl, eds, (2003): *Raumorientiertes Risikomanagement in Technik und Umwelt. Katastrophenvorsorge durch Raumplanung*. Forschungs- und Sitzungsberichte 220. Hannover: Academy for Spatial Research and Planning (ARL).

Kron, W. (2005) "Flood Risk = Hazard*Values*Vulnerability", *Water International* 30(1): 58–68.

Kühling, W. (2003) "Einführung", in Kühling, W., Hildmann, C., eds, *Der integrative Umweltplan: Chance für eine nachhaltige Entwicklung*, Dortmund: Dortmunder Vertrieb für Bau- und Planungsliteratur.

Lass, W., F. Reusswig and K.D. Kühn (1998) *"Katastrophenanfälligkeit und 'Nachhaltige Entwicklung': Ein Indikatorensystem für Deutschland"*, Bonn: Deutsche IDNDR (International Decade for Natural Disaster Reduction)-Reihe 14.

Leary, N., J. Adejuwon, W. Bailey, V. Barros, M. Caffera, S. Chinvanno, C. Conde, A. Decomarmond, A. de Sherbinin, T. Downing, H. Eakin, A. Nyong, M. Opondo, B. Osman, F. Pulhin, J. Pulhin, J. Ratnasiri, E. Sanjak, G. von Maltitz, M. Wehbe, Y. Yin and G. Ziervogel (2005) *Dimensions of Vulnerability in a Changing Climate*, Draft synthesis paper of research from case studies of Assessment of Impacts and Adaptations to Climate Change (AIACC), Washington D.C.: International START Secretariat.

Maclaren, V.W. (1996) "Urban Sustainability Reporting", *Journal of the American Planning Association*, 62: 184–203.

Morse, S. (2004) *Indices and Indicators in Development: An Unhealthy Obsession with Numbers?*, London: Earthscan.

Moss, R.H., A.L. Brenkert and E.L. Malone (2001) *Vulnerability to Climate Change: A Quantitative Approach*, available at http://www.pnl.gov/globalchange/pubs/vul/DOE%20VCC%20report.pdf.

MunichRe (2003) *NatCatservice, a Guide to the Munich Re Database for Natural Catastrophes*, Munich: MunichRe.

New Zealand's Official Statistics Agency (NZOSA) (2004) *Criteria for Indicator Selection*, available at http://www.stats.govt.nz/user-guides/indicator-guidelines/indicator-guidelines-indicator-selection.htm.

Organisation for Economic Co-operation and Development (OECD) (1993) *OECD Core Set of Indicators for Environmental Performance Reviews*, OECD Environment Monographs, No. 83, Paris: OECD.

Parris, K. (2000) "OECD AGRI-Environmental Indicators", in *Frameworks to Measure Sustainable Development*, OECD Expert Workshop, Paris: OECD, pp. 125–136.

Plate, E.J. (2002) "Towards Development of a Human Security Index", in: E.J. Plate, ed., *Environment and Human Security: Contributions to a Workshop in Bonn*, 23–25 October 2002, Germany, pp. 13/1–13/13.

Reich, U.-P. and C. Stahmer (1983) *Gesamtwirtschaftliche Wohlfahrtsmessung und Umweltqualität. Beiträge zur Weiterentwicklung der volkswirtschaftlichen Gesamtrechnungen*, Frankfurt a. M./New York.

Tapsell, S.M., E.C. Penning-Rowsell, S.M.Tunstall and T.L. Wilson (2002) "Vulnerability to Flooding: Health and Social Dimensions", *Philosophical Transactions of the Royal Society A: Mathematical, Physical and Engineering Sciences* 360: 1511–1525.

Tunstall, D. (1994) *Developing and Using Indicators of Sustainable Development in Africa: An Overview*, Network for Environment and Sustainable Development in Africa (NESDA), Thematic Workshop on Indicators of Sustainable Development, Banjul, the Gambia, 16–18 May 1994.

Turner, B. L., R.E. Kasperson, P.A. Matson, J.J. McCarthy, R.W. Corell, L. Christensen, N. Eckley, J.X. Kasperson, A. Luers, M.L. Martello, C. Polsky, A. Pulsipher and A. Schiller (2003) "A Framework for Vulnerability Analysis in Sustainability Science", *Proceedings of the National Academy of Sciences*, 100(14): 8074–8079.

United Nations (UN) (1993) *Agenda 21: Programme of Action for Sustainable Development*: The Final Text of Agreements Negotiated by Governments at the United Nations Conference on Environment and Development (UNCED), 3–14 June 1992, Rio de Janeiro, Brazil, New York: United Nations Publications.

UN (2002) *Plan of Implementation of the World Summit on Sustainable Development in Johannesburg, South Africa*, available at http://www.un.org/esa/sustdev/documents/WSSD_POI_PD/English/WSSD_PlanImpl.pdf.

United Nations Commission on Sustainable Development and Division for Sustainable Development, Department for Policy Coordination and Sustainable Development (UNCSD) (1996) *Indicators of Sustainable Development: Framework and Methodologies*, New York: UNCSD.

United Nations Development Programme (UNDP) (2004) *Reducing Disaster*

Risk: A Challenge for Development. A Global Report, New York: UNDP–Bureau for Crisis Prevention and Recovery (BRCP), available at http://www.undp.org/bcpr/disred/rdr.htm.

United Nations Educational, Scientific and Cultural Organization (UNESCO) (2003) *The UN World Water Development Report (WWDR): Water for People, Water for Life.* Paris/Oxford/New York, available at http://www.unesco.org/water/wwap/wwdr/table_contents.shtml.

United Nations/International Strategy for Disaster Reduction (UN/ISDR) (2004), *Living with Risk: A Global Review of Disaster Reduction Initatives*, 1, Geneva: United Nations.

Villagrán de León, J.C. (2004) *Manual para la Estimación Cuantitativa de Riesgos Asociados a Diversas Amenazas*, Guatemala: Acción Contra el Hambre, ACH.

Vogel, C. and K. O'Brien (2004) "Vulnerability and Global Environmental Change: Rhetoric and Reality", *AVISO – Information Bulletin on Global Environmental Change and Human Security* 13, available at: http://www.gechs.org/publications/aviso/13/index.html.

Weichselgartner, J. and M. Obersteiner (2002) "Knowing Sufficient and Applying More: Challenges in Hazard Management", *Environmental Hazards* 4: 73–77.

Weiland, U. (1999) "Indikatoren einer nachhaltigen Entwicklung: vom Monitoring zur politischen Steuerung?", in U. Weiland, ed., *Perspektiven der Raum- und Umweltplanung angesichts Globalisierung, Europäischer Integration und Nachhaltiger Entwicklung. Festschrift für Karl-Hermann Hübler*, Berlin, pp. 245–262.

Werner, G. (1977) "Zur Funktion von Umweltindikatoren und Umweltmodellen – historischer Abriß", in Institut für Umweltschutz der Universität Dortmund, Hrsg., *Umweltindikatoren als Planungsinstrumente*, Dortmund: Beiträge zur Umweltgestaltung Heft B 11, pp. 9–15.

Wisner, B. (2002) "Who? What? Where? When? in an Emergency: Notes on Possible Indicators of Vulnerability and Resilience: By Phase of the Disaster Management Cycle and Social Actor", in E. Plate, ed., *Environment and Human Security: Contributions to a Workshop in Bonn*, 23–25 October 2002, Germany, pp. 12/7–12/14.

Wisner, B., P. Blaikie, T. Cannon and I. Davis (2004) *At Risk: Natural Hazards, People's Vulnerability, and Disasters*, 2nd edn, London: Routledge.

Wisner, B. and J. Walter (2005) "Data or Dialog? The Role of Information in Disasters", in: Walter, J., ed., *World Disasters Report: Focus on Information in Disasters*, Geneva: International Federation of Red Cross and Red Crescent Societies.

3

Social levels and hazard (in)dependence in determining vulnerability

Stefan Schneiderbauer and Daniele Ehrlich

Abstract

This chapter aims to contribute to the ability to determine people's vulnerability for large areas at sub-national scale. It is based on the assumption that disaster risk is determined as a function of three components: hazard, exposure and vulnerability. It elaborates on classifications of hazards and scrutinises basic principles of the concept of vulnerability. It introduces discriminative "social levels" of vulnerability and discusses the complex matter of distinguishing between a hazard-dependent and a hazard-independent vulnerability. A matrix has been developed that is composed of characteristics of vulnerability at different "social levels" and the corresponding proxy-indicators chosen according to their availability. This matrix also reveals the potential role played by spatial datasets and remote sensing technology in quantifying vulnerability. Finally, the chapter makes several recommendations for future work.

Introduction

Disasters are of major concern and reducing disaster risk is an urgent priority for the humanitarian/development community worldwide (Birkmann, Chapter 1). Between 1971 and 1995 natural disasters caused, on average, more than 128,000 deaths per year and affected the lives of 136 million people (Twigg, 2004). Nearly every country of the world is affected by natural hazards, but natural disasters cause most damage and fatalities in developing countries. Between 1971 and 1995 about 97 per cent of deaths and 99 per cent of people affected by disasters were in developing countries (Twigg, 2004).

The United Nations (UN) declared the last decade of the twentieth century the "International Decade for Natural Disaster Reduction (IDNDR)". This promoted research in the following areas:

- conceptual approaches to disaster management
- methodological development for risk evaluations and vulnerability assessments.

The results of the IDNDR have stressed the importance of socio-economic parameters when scrutinising the underlying reasons for the extent of natural disasters. Definitions of the most common terms such as "hazard", "vulnerability", "resilience" and "coping capacity" continue to be debated (see Birkmann, Chapter 1), and there is still very limited agreement on what the terminology means. Though the focus of disaster management research shifted from "hazard assessment" to "vulnerability analysis" during the last decades, the determination of vulnerability and/or coping capacity remains one of the weakest links in the chain of risk assessment.

This chapter scrutinises some of the basic principles of the concept of vulnerability that should be taken into account when trying to measure vulnerability. It does not consider the temporal aspects of vulnerability, which are described in detail in Schneiderbauer and Ehrlich (2004). The work aims to close the gap between local, very fine-resolution vulnerability assessments and global vulnerability estimations based on national data. The final objective is to contribute to the development of a methodology for determining vulnerability at sub-national grid cells. The focus is on natural disasters. The authors are aware that so-called "natural disasters" are not triggered solely by natural events, but are also strongly linked to the political, social, economic and ecological context (Blaikie et al., 1994; Smith, 2000).

The chapter is based on work conducted within the Information Support for Effective Rapid External Action (ISFEREA) project of the Joint Research Centre. One of ISFEREA's tasks is to support the European Commission's (EC) External Relations services (DG External Relations, DG Development, DG AIDCO, DG Enlargement and DG ECHO). The results of this work should contribute to the establishment of scientific methods, datasets and procedures aimed at increasing the speed, quality and efficiency of disaster response, improving humanitarian aid allocation and providing improved information for long-term development project design and planning. Our emphasis is on assessing the risk of loss of human lives, which is of the greatest importance for the EC services.

Figure 3.1 The risk triangle.
Source: Crichton, 1999.

Our ultimate goal, therefore, is to be able to specify *where* and *how many* people are living at risk of natural disasters and to *what* disasters they are most vulnerable. The assessment of spatial distribution and the socio-economic characteristics of the population at risk is key to the work.

Hazard and vulnerability as components of disaster risk

One important outcome of the research carried out within the scope of the IDNDR is the recognition that risk depends on three components: hazard, exposure and vulnerability, as visualised by Crichton (1999) in the "risk triangle" (see Figure 3.1). This concept has been widely applied to research on natural disasters (e.g. Peduzzi et al., 2002; Granger, 2003). In the following section we will elaborate on the first two key components of "hazard" and "vulnerability", while leaving out the less critical "exposure" part.

In this chapter *risk* is understood as the probability of harmful consequences or expected losses resulting from exposure to a given hazard (a given element of danger or peril), over a specified time period (UN/ISDR, 2004; Coburn et al., 1994b).

Disasters are triggered by *hazards*, a term that needs to be differentiated from the expression *risk*. We understand hazard as signifying a potentially damaging physical event, phenomenon and/or human activity which may cause loss of life or injury, property damage, social and economic disruption or environmental degradation. Hazards can be single, sequential or combined in their origin and effects (UN/ISDR, 2004). Garatwa and Bollin (2002) distinguish between truly natural hazards (such as earthquakes) and socio-natural hazards (such as forest fires, floods and landslides), which are triggered or aggravated by a combination of extreme natural events and human intervention in nature. Most authors agree that hazards have the potential to cause harm to people, property or the environment.

The problem of defining "vulnerability" has been discussed by Birkmann in Chapter 1. Within our work we describe vulnerability as the characteristics of individuals or groups in terms of their capacity to anticipate, cope with, resist and recover from the impact of a natural or anthropogenic disaster – noting that vulnerability is made up of many political-institutional, economic and socio-cultural factors (IFRC, 1999; Garatwa and Bollin, 2002). According to Chambers (1989) and Bohle (2001), the *internal* dimension of vulnerability refers to defencelessness and insecurity, or conversely to the capacity to anticipate, cope with, resist and recover from the impacts of a hazard. The *external* dimension involves exposure to risks and shocks. Since the latter is mainly dependent on the geo-location (or exposure) of the population, we explicitly separate it from the former dimension, which concerns the person or group's vulnerability (see also equation 1 below).

Having clarified the basic terms and identified people as the element at risk, we can therefore express the risk to a disaster of natural origin as follows:

$$R_{ah} = H_{ah} \times E_a \times V_{ah} \tag{1}$$

Subscript "h" relates to the type of hazard (determined in its severity and its temporal extent) and subscript "a" is the geographical region affected by hazard "h". Exposure is, for example, the number of people located in area "a". The resulting risk refers to the potential lives lost regarding hazard "h" in area "a". Vulnerability is people's ability to cope with hazard "h" in area "a". Since the degree of vulnerability of the people living in the affected area may vary, the vulnerability in eq. 1 has to represent the average vulnerability of a single individual within area "a".

The risk equals 0 if one of the three components of hazard, exposure or vulnerability is 0. In the case of earthquakes there is no risk if (1) there is no likelihood of an earthquake occurring and/or (2) the region affected is not populated and/or (3) the population is not vulnerable (for example, if all houses are built to a high level of earthquake security).

We note that:

- Vulnerability changes with the severity and type of hazard. For example the houses might be built earthquake proof, but only up to a certain standard, or they might be earthquake resistant but vulnerable to floods.
- Determining risk requires knowledge of the spatial distribution of hazardous events and the elements at risk.

- We consider people as the only *element at risk*. Other possible elements could be physical assets such as built-up areas, transport lines or similar types of infrastructure.

"Hazard" and "exposure" can be determined by using, respectively, physical parameters and demographic datasets. The concept of vulnerability is more complex and more difficult to describe. It is necessary to rely on approximating methods such as proxy indicators when attempting to quantitatively estimate a population's vulnerability.

The variety of negative outcomes and sources of risk raises the question of the necessity to specify causes for and effects of disasters in order to describe vulnerability. The World Food Programme (WFP, 2004: 2) states that in order "to be useful, vulnerability has to be defined in terms of what it is that a population is considered to be vulnerable to and its definition therefore requires specificity." In the context of disaster management this means specifying exogenous events and shocks (Table 3.1).

Table 3.1 Classification of groups and types of hazards

Hazard group	Hazard type	Examples
Natural	Geological	Earthquake, volcanic eruption, *landslides, subsidence*
Potentially socio-natural	*Meteorological*	*Cyclones, lightning and fires, drought, avalanche, hail storm, cold spell*
	Oceanographic	*Tsunami, sea storm*
	Hydrological	*Flood, flashflood*
	Biological	*Epidemics, crop blight, insect infestation*
Technological	Explosion	
	Release of toxic materials	
	Severe contamination	
	Structural collapse	
	Transportation, construction or manufacturing accident	
Social/anthropogenic hazards	Crowd-related	Riot, crowd crush
	Terrorist activity	Bombing, shooting, hijacking
	Political conflict	International and civil war, revolution and 'coup d'état'

Hazards

When describing a hazard in order to explain a certain risk it is important to know the basic characteristics of hazardous events such as location, time, intensity and frequency (Gravley, 2001). Hazards are often grouped into three main classes according to their causes: natural, technological and anthropogenic or social disasters (see Table 3.1). As mentioned above and pointed out in the definition, hazards may have interrelated causes and the allocation of a hazard to one class is often difficult. For example, a landslide might be triggered by heavy rainfall but its severity might be determined by deforestation. Often one hazard is triggered by another. For example, volcanoes may cause movements of rock masses, which in turn cause tsunamis. Or an earthquake may provoke the destruction of buildings and infrastructures such as dams, which will result in other hazards such as floods. Additionally, it is highly likely that in the near future the number of hazards triggered by disputes about access to limited natural resources such as water will increase significantly.

The expression "environmental hazard" is used with increasing frequency for events that are caused by a mix of natural and anthropogenic

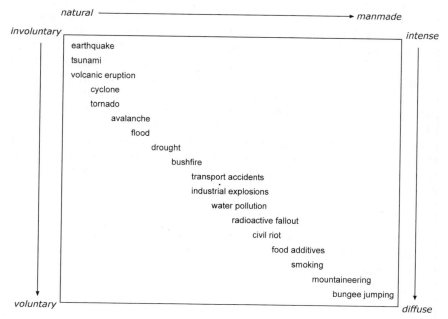

Figure 3.2 Spectrum of hazards
Source: After Smith, 2000.

incidents and circumstances. This is particularly the case for hazards associated with global climate change. According to Garatwa and Bollin (2002) these should be defined as "socio-natural" events. Those "natural" hazards that are potentially triggered by an environmental change have been listed in Table 3.1 in italics. Smith (2000: 16) has developed a

spectrum of hazards from geophysical events to human activities. Hazards that are increasingly man-made tend to be more voluntary in terms of their acceptance and more diffuse in terms of their impact.

Vulnerability

Vulnerability is related to poverty. The poorest societies have the fewest resources and opportunities to significantly reduce vulnerability. However, while poverty is generally linked to income or availability of goods and degree of well-being based on wealth, the concept of vulnerability has a broader remit that also embraces cultural and social components (see e.g. Chambers, 1989). Not being poor does not necessarily mean not being vulnerable, and vice versa.

It should be noted that the development process of a society might exclude certain social or cultural groups. This is particularly the case where rapid national economic development, measured by indicators such as GNP, can hide the fact that part of a population may remain disadvantaged, with a low development status. These groups are also most likely to be found in high-risk areas. Affiliation with a specific social or cultural group might therefore have certain implications for an individual's vulnerability. Indicators created particularly for measuring development, such as the widely used HDI (Human Development Index) and HPI (Human Poverty Index), are available globally but only rarely at sub-national scale, and hence are not adequate for vulnerability assessments at a finer resolution. In the context of measuring vulnerability at household or individual level, one has to take into account the linkages of vulnerabilities between the different social levels.

The social levels of vulnerability

The average vulnerability of an individual is made up of a set of vulnerabilities connected to different social levels that each individual belongs to. The social levels we have identified are:
- individual
- household
- administrative community
- cultural community
- national
- regional.

The individual, household, administrative community and national levels follow a hierarchical spatial order and the administrative partition of a country. The regional and cultural community levels may intersect the other social levels confined by administrative limits, as shown in Figure 3.3. The total vulnerability $V_{ahd\ tot}$ of an individual to a hazard "h" within an area "a" and for a day "d" can be computed by compounding vulnerability of the six defined social levels:

$$V_{ahd\ tot} = f(V_{ahd\ in}, V_{ahd\ hs}, V_{ahd\ ca}, V_{ahd\ cc}, V_{ahd\ cn}, V_{ahd\ rg}) \qquad (2)$$

With the levels "in" = individual, "hs" = household, "ca" = administrative community, "cc" = cultural community, "cn" = country and "rg" = region.

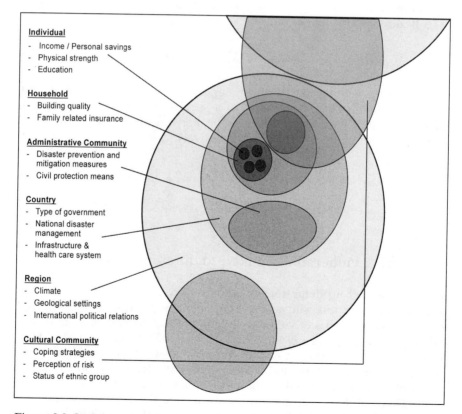

Figure 3.3 Social levels and relevant characteristics of vulnerability.
Source: Authors.

Hazard-dependent and hazard-independent vulnerability

The vulnerability of an individual or a group of individuals of a certain social level can be better quantified if one distinguishes between hazard-independent and hazard-dependent parameters.

- Hazard-independent parameters describe the strength or weakness of an individual or a group of people to withstand stresses derived from their exposure to any natural hazard. Typically, hazard-independent parameters describe general aspects of development including income, health and education, but also access to information or the existence of national disaster plans.
- Hazard-dependent parameters describe people's vulnerability to the impact of a given hazard. They are largely of a physical nature, such as the quality of building or the construction of dams, but also include social and cultural aspects, such as drought preparedness or the percentage of the population vaccinated.

Vulnerability can thus be broken up into (1) a general vulnerability and (2) a hazard specific vulnerability.

$$V_{adh\ tot} = f(V_{ad\ gen}, V_{adh}) \qquad (3)$$

With $V_{ad\ gen}$ describing the hazard-independent part of vulnerability and V_{adh} determining the vulnerability to a specific hazard "h". Both parts include variability measures that can be derived at the different social levels as shown in eq. 4 and eq. 5, respectively.

$$V_{ad\ gen} = f(V_{ad\ in\ gen}, V_{ad\ hs\ gen}, V_{ad\ ca\ gen}, V_{ad\ cc\ gen}, V_{ad\ cn\ gen}, V_{ad\ rg\ gen}) \quad (4)$$

$$V_{adh} = f(V_{adh\ in}, V_{adh\ hs}, V_{adh\ ca}, V_{adh\ cc}, V_{adh\ cn}, V_{adh\ rg}). \qquad (5)$$

Quantifying vulnerability linked to different social levels

There are numerous definitions of vulnerability corresponding to numerous ways of conceptualising and quantifying it (see Birkmann, Chapters 1 and 2). Most measures of vulnerability are tailored to a small area or specific region (e.g. Coburn et al., 1994b; IFRC, 2002; Cannon et al., 2003). Voss and Hidajat (2001) pose the question of whether it is possible to model vulnerability at all, and if so at what scale. Recently global assessments of *environmental* vulnerability to climate change have been developed. Their methodological approaches include the creation of composite indices (e.g. Schellnhuber, 2001; SOPAC, 2003). Methodologies to address the socio-economic characteristics of vulnerability worldwide are still at an early stage of development. Recently, global vulnerability assessments tackling various hazards have been attempted (UNDP, 2004; Dilley, Chapter 9; Peduzzi, Chapter 8).

A major difficulty in assessing vulnerability is the lack of any external reference, which eliminates the possibility of testing the model's quality or accuracy. For example, when a model is developed to estimate the population of a given area, one can assess this model's quality by comparing it with the real population number within sample sites. But for a model that attempts to assess vulnerability no tangible values for quality assessment exist. Damage records of past hazardous events may be used as substitutes for the absent reference data (UNDP, 2004; Dilley et al., 2005). However, this approach entails two problems: (1) the varying quality of data relating to previous disasters, and (2) the difficulty of normalising past events according to their strength in order to allow comparison of damage impact. As a result of these limitations, a vulnerability assessment for the whole world is at best only possible through the use of general proxies and will always result in relative rather than absolute estimations (Dilley et al., 2005).

We derive vulnerability measures as defined in eq. 4 and eq. 5. This follows previous research (Cardona et al., 2003), which also stresses that at each social level – as identified and shown before – at least one characteristic and a corresponding measurable indicator covering the physical, economic, social, educational, political, institutional, cultural, environmental and ideological dimension should be defined. In practice it is difficult to allocate a characteristic clearly to a specific level. For this reason we attempt to identify the most significant characteristics and parameters for each "social level" for both the hazard-dependent and hazard-independent fraction of vulnerability. We then associate potentially available and measurable indicators with these parameters.

We are aware of the complexity of this methodology, the lack of accuracy in making estimations and the impossibility of including all aspects of vulnerability. However, the results of the implementation of the concept described above should outweigh these constraints.

Vulnerability parameters and indicators

Hazard-dependent and independent parameters, and potential indicators, are listed in Table 3.2 and Table 3.3, respectively. The parameters were selected with the aim of including a representative number of dimensions at the different "socio-administrative levels". The selection of the corresponding indicators was based on three criteria:

Table 3.2 Selected hazard-independent parameters and potential indicators for vulnerability at different 'social levels'

Social levels	Parameters	Indicators
Individual and household	Age	Average age
	Income	GDP per capita
	Health/disability	Malnutrition of children <5
	Education	Life expectancy
	Subsistence economy in primary sector	HIV/AIDS infection rate
	Savings	Illiteracy rate
	Individual and family related insurance	Productivity per capita (primary sector)
	Neighbourhood network	Number of mobile phones, TVs, radios/per capita
	Access to information	
Administrative community	Infrastructure/accessibility	Traffic infrastructure/road network
	Presence and quality of civil protection, incl. early warning/emergency plans/ disaster management capacities	Density of rural population
		Level of urbanisation
		Level of corruption
	Disaster preparedness	
	Degree of autonomy/participation in decision making procedures and access to resources	
Country	Regulatory environment	Type of government/number of signed international agreements
	Armed conflicts with involvement of national government	Number and intensity of conflicts
	Population structure	Number of IDPs (internally displaced people) and refugees
	Economic system	Fertility rate
	Economic dependency	Sex ratio
	Infrastructure/services	Age average
	National disaster planning	Trading activities – rate of GDP
	Forecast and early warning system	External aid as ratio of GNI
	Emergency management system and capacities	Contribution of primary sector to GDP

	Insurance services
	Remittances from abroad
	Urban population growth
	Transportation and communication network
	Number of missing values of important indicators
Region	Climate
	Regional political stability
Cultural community	Status of community
	Armed conflicts with involvement of the community
	Gender inequality
	Perception of risk and approach towards emergencies (cultural beliefs)
	Coping strategies (incl. farming methods and land tenure systems)
	Climate records and their long-term changes
	Number and intensity of regional conflicts
	Political discrimination of ethnic groups
	Economic disadvantages of ethnic groups
	Cultural restrictions of ethnic groups
	Intra- and intercommunal conflicts and their intensity
	GDI

Table 3.3 Selected hazard-dependent parameters and potential indicators for vulnerability at different "social levels"

Social levels	Parameters	Indicators	Ea	Vo	Cy	Fl	Dr	Ep	SD	RS
Individual and household	Quality and age of building	Building construction date combined with law enforcement considering earthquake safety								
		Main building material							X	X
		Urban growth							X	X
	Size/height of building	Number of floors								X
		Number of families/residential building								
	Location of dwelling	Terrain information (for example slope gradient)							X	X
		Altitude (relating to sea level or local watersheds)							X	
	Hygiene	Access to drinking water								
		Quality of sewage system								
Administrative community	Preparedness for floods	Dams							X	X
		Legal regulations relating to floods								
	Preparedness for earthquakes	Percentage of earthquake resistant built houses								
		Law considering earthquake resistant buildings								

	Indicator						SD	RS
Country	Local environmental degradation	Soil degradation/soil sealing					X	X
		Erosion						X
	Constraints for agricultural use	Soil, terrain, climate conditions regarding agricultural activities					X	X
	Countrywide and regional environmental degradation	Deforestation rate					X	X
	Vaccination	Number of people vaccinated						
		Legal regulations for vaccinations						
Region	Sufferance from climate change	Significant change of measurable climate characteristics					X	
	Land use	Land cover					X	X
	Relief	Slope/Elevation					X	X
	Preparedness for droughts	Adaptation of land use methods according to climate conditions (culture of certain crops, sustainable use of resources)						
Cultural community	Customs of sexual behaviour	Prevalence of protected sexual intercourse practices						
		Contraception methods						

Ea = earthquakes, Vo = volcanoes, Cy = cyclones, Fl = floods, Dr = droughts, Ep = epidemics
SD = Spatial data, RS = Remote sensing data

☐ No or very low importance ▨ Low to medium importance ■ High importance [x] Data available
Source: Authors.

- availability and coverage, that is, the number of countries/areas covered
- measurability and accuracy
- frequency of update.

The following compilation of parameters and indicators does not pretend to be complete. It intends to summarise what we believe is important for determining vulnerability. Due to the complexity of the concept of vulnerability, many more aspects could be added.

The selection of parameters and indicators was made with a focus on developing countries. A number of the indicators are only relevant to the specific economic, institutional and environmental situation in those countries (e.g. access to drinking water or children's malnutrition). In order to maintain the general applicability of our methodology, we do not consider aspects that are specific to a certain region or to a particular group of people. For practical and computational reasons, from now on we will consider the individual and the household level to be one entity.

Hazard-independent parameters and indicators

Hazard-independent parameters are relevant for assessing vulnerability to any type of natural hazard. Table 3.2 lists these parameters and corresponding indicators, emphasising the level of development, the efficiency of administrative and disaster management, and any involvement in armed conflicts.

Individual/household level

At the individual/household level the hazard-independent parameters cover demographic, social and economic topics. "Age", "income", "health/disability" and "education" are basic features affecting the physical and economic strength of an individual and his/her dependents. People's health and education contribute to an explanation of their general capacity to cope and deal with external impacts. These can be expressed through "classic" indicators for development measurement such as life expectancy, malnutrition and illiteracy.

All hazards may have a very strong direct or indirect impact on natural resources within the affected area. Hence, dependency on a subsistence economy in the primary sector (i.e. agriculture, pastoralism and fishing) is likely to be highly relevant to levels of resilience. The ability to recover can be determined by household savings and individual and family-related insurance, as well as the existence of social or neighbourhood networks. For these parameters we could not identify any indicators that are currently available and that cover developing countries.

Having access to information can be one way of decreasing vulnerability. The population's access to information is important for knowledge relating to early warning or post-disaster emergency and relief actions. An appropriate indicator is the average number of communication devices per capita (e.g. TVs, radios).

Another important parameter is the HIV/AIDS infection rate, as AIDS victims are often unable to continue as breadwinners. In addition, their illness places a burden on the household budget in terms of added costs for medicine, medical attention and funeral ceremonies. For most developing countries data at individual level are not available. Any measure has to be derived from the average values at available country level.

Administrative community

Parameters of physical and institutional infrastructure are predominantly determined at the sub-national administrative community level. The physical infrastructure is important for permitting access to potentially hazard-struck areas or communication systems, and can be measured by the network of roads or other traffic lines and mobile phone coverage or Internet access, respectively. The institutional infrastructure provides the framework for disaster mitigation, preparedness and response activities, which are usually implemented and managed at this administrative level.

Assessment of the efficiency or quality of an institutional setting can often only be approached by using indirect indicators, such as, for example, the level of corruption. The suggested use of indicators relating to urban and rural populations is based on the assumption that the institutional setting for disaster management is of high quality when the ratio of urbanisation is high and the rural population density is low. A more in-depth discussion of this aspect can be found in Lebel et al. (Chapter 19) and Krausmann and Mushtaq (Chapter 22). The procedures of decision-making in disaster management and the potential for local community participation in these procedures are crucial for successful pre- and post-disaster activities. These parameters are not quantifiable, although they may be explained and classified qualitatively. However, worldwide indicators are not available.

Country

National Governments define policies that affect people countrywide and influence their level of vulnerability directly or indirectly. The lack of legal obligations to implement certain aspects of disaster management and civil protection, as well as involvement in conflicts, can weaken a nation with regard to its preparedness and resilience to the impact of hazardous events. The regulatory environment and the number of conflicts the Government is involved in (within the country or internationally) are therefore relevant parameters for an assessment of vulnerability at national scale. Targeted indicators could be the type of Government (democratic, autocratic, military regime) and the number of signed international agreements, as well as the number of armed conflicts, refugees and internally displaced persons (IDPs).

The population structure of a society may serve as an indirect indicator of the country's development status. For example, high fertility rates and a disproportionately high proportion of young people are both indicators of low development status. Population structure such as a high dependency ratio – the proportion of the economically active population to the economically inactive population – may indicate societal vulnerability. War or out-migration may create an imbalance in sex ratios or age patterns and weaken a society.

Economic factors are also highly influential in relation to vulnerability at the national scale. Financial resources, a strong vital economy and participation in international trading activities all contribute to a reduction in vulnerability. Possessing these features usually results in the construction of high-quality physical and medical infrastructure, the installation and maintenance of early warning systems and modern civil protection or the compensation of costs for reconstruction in disaster-struck areas. There are numerous relevant indicators for the assessment of a country's economic system. We suggest using trading and primary sector activities as well as the percentage which external aid contributes to the gross national income (GNI). Additionally, remittances from abroad (that is, the money sent home by expatriates) assist in revealing economic weakness and thus indicating a lower level of resilience.

As with the parameters discussed at the administrative community level, we evaluate the existing infrastructure and capacity for disaster management and its underlying institutional setting at the national level. Relevant direct indicators are those describing the national transport and communication network. Rapid urban population growth can often result in a lack of infrastructure and therefore of disaster management capacity. In addition, we suggest using the lack of existing values of important and widely-used indicators as measurement in itself of weak and unreliable public administration and institutions.

Region

Cross-national parameters acting at regional level can be of either a physical or socio-economic nature. Extreme climate conditions, such as droughts, cross administrative borders and can weaken the resistance of the population to other types of hazard. Global data on various climate characteristics and their variations are now available as a result of research conducted on global climate change. The stability or instability of social systems within a region may be based on ethno-linguistic groupings of people that also cross administrative borders. We therefore suggest monitoring the number and intensity of international conflicts as one component of vulnerability assessment.

Cultural community

A "cultural community" may be defined by shared cultural, social and/or ethnic traits. The characteristics of external relations and the internal value system contribute to determining its level of vulnerability. For example, a functioning cultural community may provide strong social networks, which can be used to support disaster victims. On the other hand culturally influenced fatalism towards the occurrence of natural hazards may result in failure to implement any pre-event mitigation measures.

We propose assessing external relations by looking at the status quo of the community and conflicts the community is, and has been, involved in. Relevant indicators are political discrimination, economic disadvantage and any cultural restrictions inflicted on the community, as well as the number and intensity of armed conflicts caused by ethnic, religious or similar tensions.

The internal value system may be approached by looking at gender inequality levels and the perception of risk based on cultural beliefs, combined with developed coping strategies. Coping strategies could include adapted farming methods and land tenure systems. However, the only available indicator identified by us as being useful for assessing internal values is the Gender Development Index (GDI), which looks at differences in life expectancy and levels of education and income among men and women.

Hazard-dependent parameters and indicators

Hazard-dependent parameters are usually relevant for just one *specific* hazard, or at most a few, and are largely physical in nature. Table 3.3 lists these parameters and corresponding indicators and includes information on the relative importance of each indicator for the different hazards. Table 3.3 also shows whether indicator data are available as "spatial data" and which indicators can be acquired through remote sensing techniques. "Spatial data" refers to geo-located information and can thus be represented on a map or analysed by using GIS techniques. Remote sensing data refers to satellite images but may also include aerial photographs.

Individual/household

Hazard-dependent parameters at household level are for example related to the dwellings and other infrastructure people use in their everyday lives. For example the quality, age, size and height of buildings are important. A building's general stability varies depending on the building material used. This is relevant for determining vulnerability to cyclones, floods and – with some reservations – to volcanoes. Due to the possible impact on the dwellers' health, it also has some relevance for epidemics. Direct indicators for the buildings' size and height are the number of floors and inhabiting families (residential houses). Indirectly, the overall quality of buildings may be assessed by the speed of physical urban growth, with high growth rates often corresponding to low quality due to the lack of regulatory control. The construction of earthquake resistant housing may depend on whether or not relevant legislation is in place and its date of enforcement.

The location of dwellings also influences the inhabitants' susceptibility to a number of hazards. Depending on its location, a building might be in danger of basaltic flows or landslides and mudflows triggered by earthquakes or exceptional rains. Buildings at low altitude near the coast or in occasionally flooded areas might be vulnerable to floods and cyclones. Areas at risk due to their elevation or terrain might be identified with the help of digital terrain models (DTMs).

The fact that poor people tend to live in locations of higher risk, such as polluted areas or regions with severe climate, is also relevant in determining vulnerability to epidemics. Levels of hygiene play a role in all hazards but in particular for epidemics. In addition, the location and accessibility of drinking water is of great importance when determining vulnerability to droughts.

Administrative community

Mitigation measures are typically implemented at administrative community level. Examples are the construction of dams or earthquake-resistant infrastructures and housing. Such precautions are usually not taken unless an appropriate legal framework is in place; the presence of such a framework would indicate reduced vulnerability.

The existing level of environmental degradation is of particular relevance for evaluating vulnerability to floods, droughts and cyclones. The effects of environmental degradation might vary with climate conditions and affect areas differently according to their size. In the case of droughts, degradation of the environment may aggravate an already existing natural constraint on agricultural use caused by, for example, low soil quality, steep terrain and/or severe climate conditions. In the table we mention environmental degradation twice, once in relation to administrative community, and a second time with regard to country level, taking into account the available indicators. This is despite the fact that in reality local, countrywide and regional environmental degradation cannot be separated from one another. Soil degradation and sealing, deforestation and erosion have a largely negative impact on water balance and infiltration rates, which in turn leads to rapid runoff and water shortages across political or administrative borders.

Country

Environmental degradation is listed again at country level in order to emphasise the extent of areas that might be affected by human impacts on the environment and the negative influence this could have on vulnerability to floods and droughts. An available indicator at this level is the countrywide deforestation rate.

The number of people vaccinated in a country is a good indicator for assessing vulnerability to epidemics and, like earthquake mitigation and protection, is largely dependent on whether there is relevant national legislation or not. Vaccination rates are also important when considering the aftermath of hazards. The ensuing disruption to health care and deteriorated sanitary conditions increase vulnerability, particularly in countries where vaccination levels are low.

Region

Human-induced environmental degradation, discussed in the "administrative community" section that includes soil degradation, deforestation and erosion, may extend beyond the administrative community to affect entire regions. Also, changes in climate and in particular global warming may have an impact on the vulnerability of a whole region. The consequences of global climate change are most likely to mean having to cope with a change in temperature and water supply, which are both important to vulnerability from slow-onset hazards such as droughts and epidemics. It must be pointed out that we risk stepping into the exposure part of our risk function when looking at increased vulnerability due to a change in climate, since climate change as such may be seen as a hazard.

The type of land use, though ultimately dependent on human activities, is initially determined by physical features of the region, such as soil quality and climate. As with environmental degradation, these characteristics will determine vulnerability to floods, droughts and cyclones. Certain land use types may influence vulnerability to epidemics such as malaria in irrigated agricultural areas. The terrain, assessed by slope and elevation values, may have an effect on vulnerability to floods and cyclones. Again, we risk stepping into the exposure part of our risk function when considering these aspects.

Cultural community

The cultural values that communities hold always play an important role in determining vulnerability. Drought preparedness is largely influenced by the cultural setting and determined by whether there is sustainable use of natural resources and an adaptation of land use methods to climate conditions. Cultural values also determine sexual customs and behaviour that are strongly dependent on affiliation with a social and/or ethnic group, and crucial when looking at vulnerability to HIV/AIDS infection (which we include as a slow-onset epidemic even though it cannot be described as being triggered by a natural hazard). Due to its taboo status, it is difficult to find reasonable values for the proposed indicators relating to the prevalence of unprotected sex and/or methods of contraception.

Discussion

From an analysis of the data available one has to note that:

- There are a number of indicator sets linked to the economic, social and educational dimensions. There are a very limited number of available indicators for determining the political, physical and environmental dimensions, while there are insufficient indicators describing the institutional, cultural and ideological setting.
- Even though the number of indicators that can be used to quantify a dimension might be quite large, their accuracy and availability may prove to be inadequate.
- The countries that are vulnerable have the most significant data gaps. In fact, the lack of indicator values may be used as an indicator in itself.
- Though a number of indicators are supposed to describe individual characteristics such as GDP per capita, they are often only available as countrywide averages.
- Ideally, rapid changes in the vulnerability of a population relating to the time of day or the phenological season when a hazard occurs should also be taken into account. However, we do not take account of the time when an earthquake occurs or whether an agricultural area is flooded just before or after the harvest.
- Table 3.3 shows that there are quite a number of indicators available as geo-datasets or acquired through remote sensing techniques. In contrast to the structural indicators describing the socio-economic parameters, which are usually available at country level, these data cover the physical and environmental aspects of vulnerability. By being available as either vector or raster datasets of relatively fine resolution, they allow a vulnerability assessment at sub-national level.

Table 3.3 also shows up a slight correlation between the dimension of the area potentially affected by a specific hazard and the social level of the most relevant parameters and indicators: the smaller the affected area, the more local the social level. That is, the vulnerability to more local events such as earthquakes, volcanoes and cyclones is predominantly determined by parameters of individual and household scale. In the case of hazards that have an impact on larger areas, such as floods, droughts and epidemics, however, the parameters at national and regional level are more important.

Conclusions

The concept of vulnerability is complex and a realistic determination of population vulnerability worldwide is extremely difficult to make. In order to incorporate all important elements the authors recommend allocating vulnerability parameters and corresponding indicators to "social levels", as well as indicating whether they have a general importance (hazard-independent) or are hazard-specific. The compilation of these parameters and indicators for a worldwide vulnerability assessment of populations clearly reveals the need to tackle the following issues for future applications:

- Some indicators chosen for determining vulnerability show a high correlation with one another. Given these correlations, those indicators that are not adding any significant value to vulnerability information should be excluded. Once the set of selected indicators is compiled for all relevant countries and years, and after their values have been normalised, the most valuable variables can be identified based on a calculation of correlation matrices and factor analyses.
- In order to develop a useful composite indicator for vulnerability assessment it is necessary to define weighting factors for each indicator. For this procedure a variety of techniques have been developed (Saisana and Tarantola, 2002). However, in the case of vulnerability there is no reference data – except for expert knowledge – with which the quality of the resulting composite indicator could be evaluated.
- Most of the indicators represent structural characteristics of the observed population, group or society and therefore change slowly with time. However, people's vulnerability may change quickly, for example due to conflicts or as a result of hazardous events. The composite indicator developed should be able to represent – at least partly – these possible rapid developments.
- A number of more qualitative parameters that are highly relevant for vulnerability assessments, such as disaster management capaci-

ties or risk perception, are difficult to describe, and none of the frequently globally surveyed indicators could support their determination. The relevant information can only be compiled with local expert consultation.

Furthermore, risk assessments of natural hazards at sub-national level require datasets that are spatially disaggregated. Global datasets are available but do not always satisfy the requirements of accuracy. For example, physical exposure data needed for our purposes would be data on the density of a population potentially affected in a defined area by a specific hazard. Worldwide datasets such as LandScan (Oak Ridge National Laboratory) or CIESIN (Center for International Earth Science Information Network, Columbia University) are available. These data have proven to be invaluable and would be even more so if provided with information about their accuracy. Vulnerability data are the most difficult parameter to assess. This is due to the complexity of the issues involved, a few of which have been discussed in this chapter. Limitations include the frequency of updating and the quality of many of the potential indicators, and most importantly the unavailability of data at fine spatial resolution. Almost all structural indicators are surveyed only at the national level and their disaggregation to a sub-national administrative level requires intensive modelling work. One of the challenges to improving vulnerability measures will be to derive the appropriate information from available GIS layers, maps and satellite imagery.

REFERENCES

Blaikie, P., T. Cannon, I. Davis and B. Wisner (1994) *At Risk: Natural Hazards, People's Vulnerability, and Disasters*, London: Routledge.

Bohle, H.-G. (2001) "Vulnerability and Criticality", Vulnerability Article 1, in *Newsletter of the International Human Dimensions Programme on Global Environmental Change*, Nr. 2/2001, available at http://www.ihdp.uni-bonn.de/html/ publications/update/update01_02/IHDPUpdate01_02_bohle.html.

Cannon, T., J. Twigg and J. Rowell (2003) "Social Vulnerability, Sustainable Livelihoods and Disasters", Report to DFID Conflict and Humanitarian Assistance Department (CHAD) and Sustainable Livelihoods Support Office, available at http://www.livelihoods.org/static/tcannon_NN197.htm.

Cardona, O.D., J.E. Hurtado, G, Duque, A. Moreno, A.C. Chardon, L.S. Velásquez and S.D. Prieto (2003) *The Notion of Disaster Risk: Conceptual Framework for Integrated Risk Management*, IADB/IDEA Program on Indicators for Disaster Risk Management, Universidad Nacional de Colombia, Manizales, available at http://idea.manizales.unal.edu.co/ProyectosEspeciales/adminIDEA/ CentroDocumentacion/DocDigitales/documentos/01%20Conceptual%20 Framework%20IADB-IDEA%20Phase%20I.pdf.

Chambers, R. (1989) "Editorial Introduction: Vulnerability, Coping and Policy", in *Vulnerability: How the Poor Cope*, IDS Bulletin 20(2): 1–7.

Coburn, A.W., R.J.S. Spence and A. Pomonis (1994a) *Disaster Mitigation*, 2nd edn, UNDP Disaster Management Training Programme, available at http://www.proventionconsortium.org/files/undp/DisasterMitigation.pdf.

Coburn, A.W., R.J.S. Spence and A. Pomonis (1994b) *Vulnerability and Risk Assessment*, 2nd edn, UNDP Disaster Management Training Programme, available at http://www.proventionconsortium.org/files/undp/VulnerabilityAnd RiskAssessmentGuide.pdf.

Crichton, D. (1999) "The Risk Triangle", in J. Ingleton, ed., *Natural Disaster Management*, London: Tudor Rose, pp. 102–103.

Dilley, M., R.S. Chen, U. Deichmann, A.L. Lerner-Lam, M. Arnold with J. Agwe, P. Buys, O. Kjekstad, B. Lyon, and G. Yetman (2005) *Natural Disaster Hotspots: A Global Risk Analysis*, New York/Washington: Columbia University/World Bank.

Garatwa, W. and C. Bollin (2002) *Disaster Risk Management: Working Concept*, Eschborn: Deutsche Gesellschaft für Technische Zusammenarbeit (GTZ), available at http://www2.gtz.de/dokumente/bib/02-5001.pdf.

Granger, K. (2003) "Quantifying Storm Tide Risk in Cairns", *Natural Hazards* 30: 165–185.

Gravley, D. (2001) *Risk, Hazard, and Disaster*, University of Canterbury, available at http://homepages.uc.edu/~huffwd/Volcanic_HazardRisk/Gravley.pdf.

International Federation of Red Cross and Red Crescent Societies (IFRC) (1999) *Vulnerability and Capacity Assessment: An International Federation Guide*, Geneva: IFRC, available at http://www.proventionconsortium.org/files/vca.pdf.

IFRC (2001) *World Disaster Report: Focus on Recovery*, Geneva: IFRC, available at http://www.ifrc.org/publicat/wdr2001/index.asp.

IFRC (2002): World *Disasters Report: Focus on Reducing Risk*, Geneva: IFRC, available at http://www.ifrc.org/PUBLICAT/wdr2002/index.asp.

International Strategy for Disaster Reduction (UN/ISDR) (2004) *Terminology: Basic Terms of Disaster Risk Reduction*, available at http://www.unisdr.org/unisdr/eng/library/lib-terminology-eng%20home.htm.

Peduzzi, P., H. Dao and C. Herold (2002) *Global Risk and Vulnerability Index Trends per Year (GRAVITY). Phase II: Development, analysis and results*, UNDP/BCPR, Geneva, available at http://www.grid.unep.ch/product/publication/download/ew_gravity2.pdf.

Saisana, M. and S. Tarantola (2002) *State-of-the-art Report on Current Methodologies and Practices for Composite Indicator Development*, European Commission-Joint Research Centre, available at http://farmweb.jrc.cec.eu.int/ci/Document/state-of-the-art_EUR20408.pdf.

Schellnhuber, H.-J. (2001) "Brainstorming: What is Vulnerability and How Do We Measure It?", Contribution to the Potsdam Sustainability Days, 28–30 September 2001, Potsdam Institute for Climatic Impact Research.

Schneiderbauer, S. and D. Ehrlich (2004) *Risk, Hazard and People's Vulnerability to Natural Hazards: A Review of Definitions, Concepts and Data*, Brussels: European Commission–Joint Research Centre (EC-JRC).

Smith, K. (2000): *Environmental Hazards: Assessing Risk and Reducing Disaster*, 3rd edn, London: Routledge.

South Pacific Applied Geoscience Commission (SOPAC) (2003) *The Demonstration Environmental Vulnerability Index (EVI)*, SPOAC Technical Report 356, available at http://list.sopac.org.fj/Projects/Evi/Files/Demo%20EVI%20Report%202003.pdf.

Twigg, J. (2004) "Disasters, Development and Vulnerability", in Benfield Hazard Research Centre (HRC), ed., *Development at Risk? Natural Disasters and the Third World*, available at http://www.benfieldhrc.org/activities/misc_papers/DEVRISK/TWIGG.HTM.

United Nations Development Programme (UNDP) (2004) *Reducing Disaster Risk. A Challenge for Development. A Global Report*, New York: UNDP – Bureau for Crisis Prevention and Recovery (BRCP), available at http://www.undp.org/bcpr/disred/rdr.htm.

Voss, H. and R. Hidajat (2001) "Vulnerabilität als Komponente zur Bewertung des Naturrisikos", 2. Forum Katastrophenvorsorge: "Extreme Naturereignisse – Folgen, Vorsorge, Werkzeuge". 24–26 September 2001, Leipzig, available at http://www.dkkv.org/forum2001/Datei23.pdf.

World Food Programme (WFP) (2004) *Vulnerability Analysis and Mapping: A Tentative Methodology (Annex III)*, Rome: WFP, available at http://www.proventionconsortium.org/files/wfp_vulnerability.pdf.

Figure 4.1(c) Germany: Elbe Flood 2002.
Source: German Remote Sensing Data Centre of DLR.

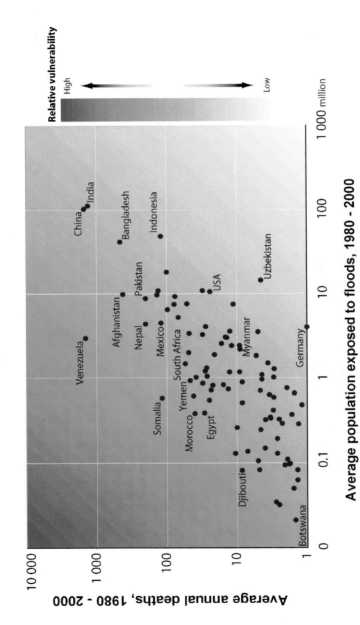

Figure 7.1(c) Relative vulnerability for flooding, 1980–2000.
Source: EM-DAT OFDA/CRED and UNEP/GRID-Geneva (in UNDP, 2004).

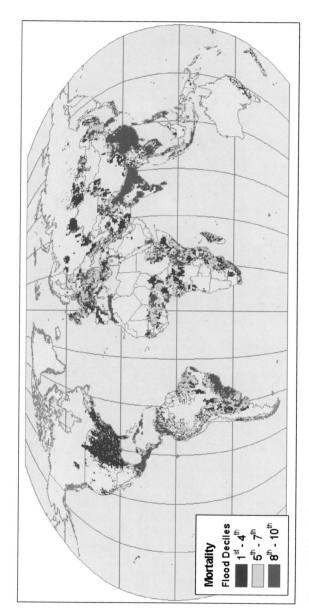

Figure 7.2(c) Global distribution of flood mortality risk.
Source: Dilley et al., 2005.

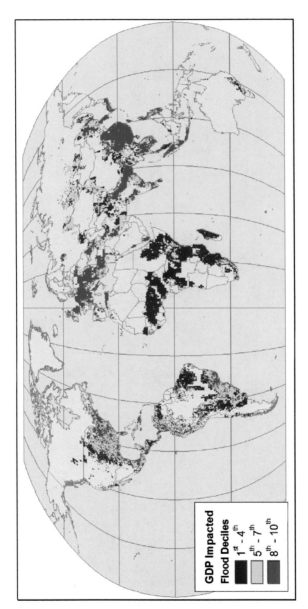

Figure 7.3(c) Global distribution of flood economic loss risk.
Source: Dilley et al., 2005.

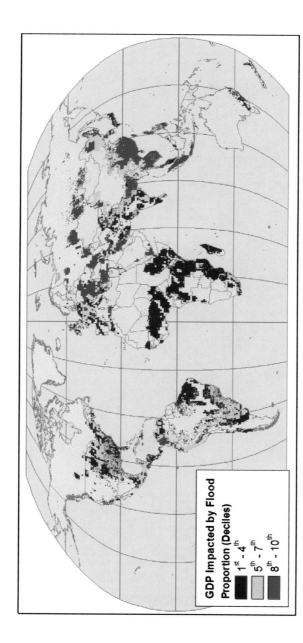

Figure 7.4(c) Global distribution of flood economic loss risk as a proportion of GDP.
Source: Dilley et al., 2005.

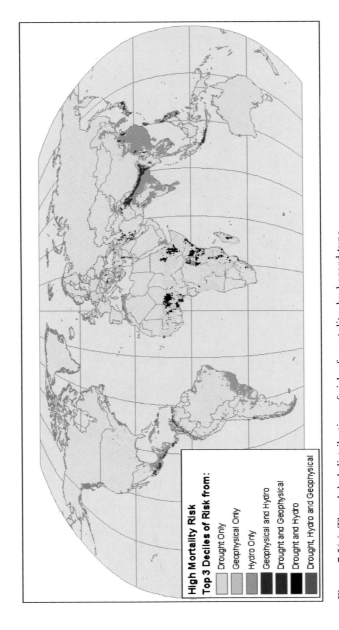

Figure 7.5(c) The global distribution of risk of mortality, by hazard type.
Source: Dilley et al., 2005.

High Mortality Risk
Top 3 Deciles of Risk from:

Drought Only
Geophysical Only
Hydro Only
Geophysical and Hydro
Drought and Geophysical
Drought and Hydro
Drought, Hydro and Geophysical

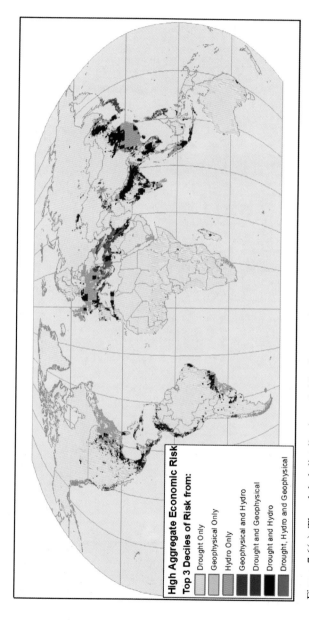

Figure 7.6(c) The global distribution of risk of economic loss, by hazard type.
Source: Dilley et al., 2005.

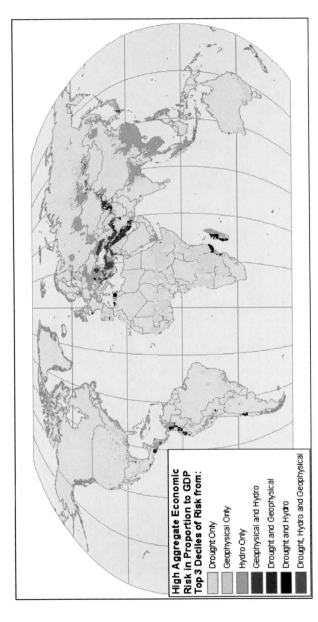

High Aggregate Economic Risk in Proportion to GDP Top 3 Deciles of Risk from:

- Drought Only
- Geophysical Only
- Hydro Only
- Geophysical and Hydro
- Drought and Geophysical
- Drought and Hydro
- Drought, Hydro and Geophysical

Figure 7.7(c) The global distribution of risk of economic loss as a proportion of GDP, by hazard.
Source: Dilley et al., 2005.

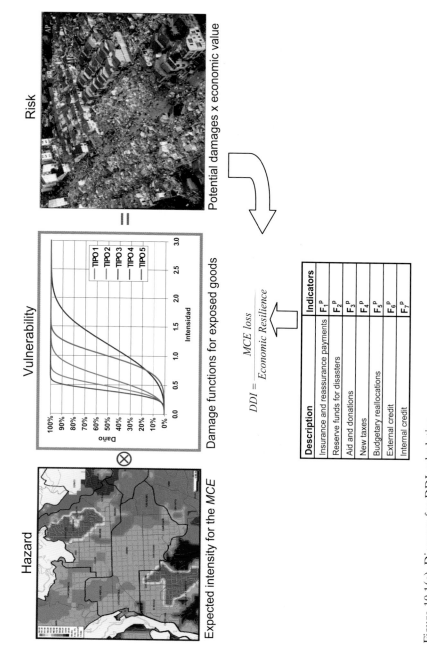

Hazard

Expected intensity for the *MCE*

⊗

Vulnerability

Damage functions for exposed goods

=

Risk

Potential damages x economic value

$$DDI = \frac{MCE\ loss}{Economic\ Resilience}$$

Description	Indicators
Insurance and reassurance payments	F_1^p
Reserve funds for disasters	F_2^p
Aid and donations	F_3^p
New taxes	F_4^p
Budgetary reallocations	F_5^p
External credit	F_6^p
Internal credit	F_7^p

Figure 10.1(c) Diagram for DDI calculation.
Source: Author.

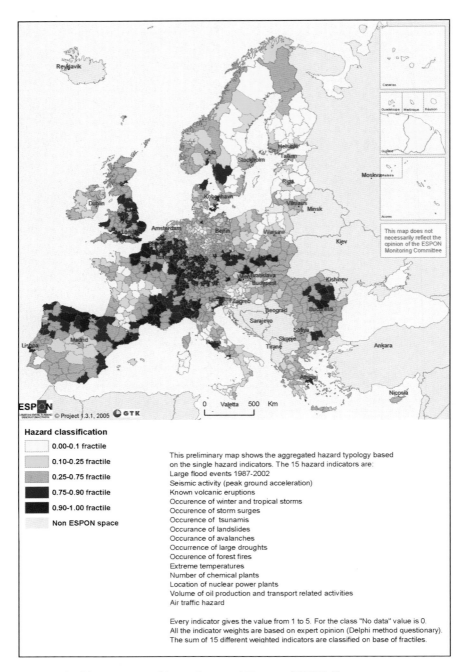

Hazard classification

- 0.00-0.1 fractile
- 0.10-0.25 fractile
- 0.25-0.75 fractile
- 0.75-0.90 fractile
- 0.90-1.00 fractile
- Non ESPON space

This preliminary map shows the aggregated hazard typology based on the single hazard indicators. The 15 hazard indicators are:
Large flood events 1987-2002
Seismic activity (peak ground acceleration)
Known volcanic eruptions
Occurence of winter and tropical storms
Occurence of storm surges
Occurence of tsunamis
Occurance of landslides
Occurance of avalanches
Occurrence of large droughts
Occurence of forest fires
Extreme temperatures
Number of chemical plants
Location of nuclear power plants
Volume of oil production and transport related activities
Air traffic hazard

Every indicator gives the value from 1 to 5. For the class "No data" value is 0.
All the indicator weights are based on expert opinion (Delphi method questionary).
The sum of 15 different weighted indicators are classified on base of fractiles.

Figure 11.4(c) Aggregated hazard map of Europe (NUTS 3).
Source: Schmidt-Thomé, 2005: 75.

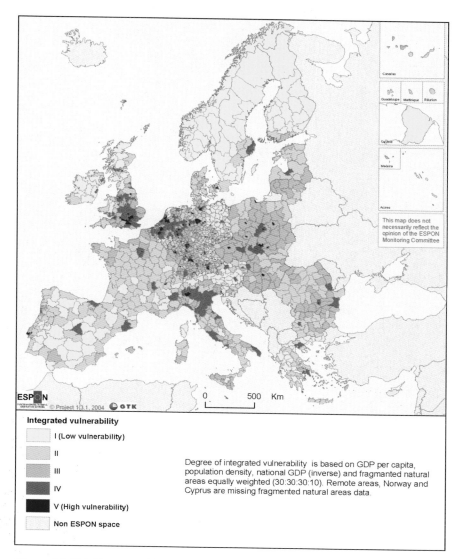

Integrated vulnerability

I (Low vulnerability)

II

III

IV

V (High vulnerability)

Non ESPON space

Degree of integrated vulnerability is based on GDP per capita, population density, national GDP (inverse) and fragmanted natural areas equally weighted (30:30:30:10). Remote areas, Norway and Cyprus are missing fragmented natural areas data.

Figure 11.5(c) Degree of interest vulnerability in Europe (NUTS 3).
Source: Schmidt-Thomé, 2005: 85.

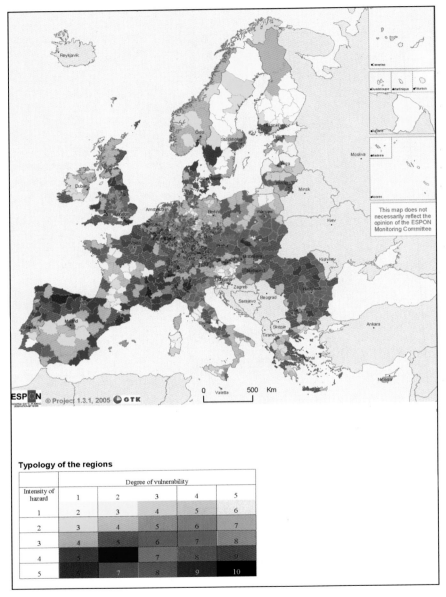

Typology of the regions

Intensity of hazard	Degree of vulnerability				
	1	2	3	4	5
1	2	3	4	5	6
2	3	4	5	6	7
3	4	5	6	7	8
4	5	6	7	8	9
5	6	7	8	9	10

Figure 11.6(c) Aggregated hazard risk in Europe (NUTS 3).
Source: Schmidt-Thomé, 2005: 88.

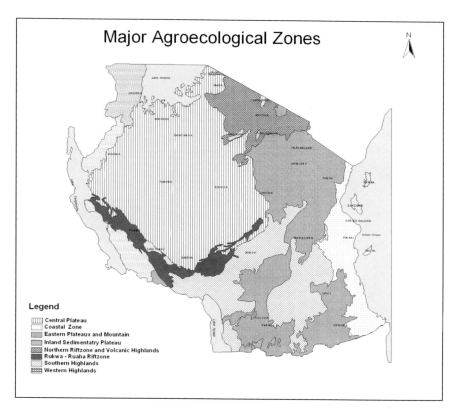

Major Agroecological Zones

N

Legend
- Central Plateau
- Coastal Zone
- Eastern Plateaux and Mountain
- Inland Sedimentatry Plateau
- Northern Riftzone and Volcanic Highlands
- Rukwa - Ruaha Riftzone
- Southern Highlands
- Western Highlands

Figure 12.1(c) Agro-ecological zones of Tanzania.
Source: de Pauw, 1984.

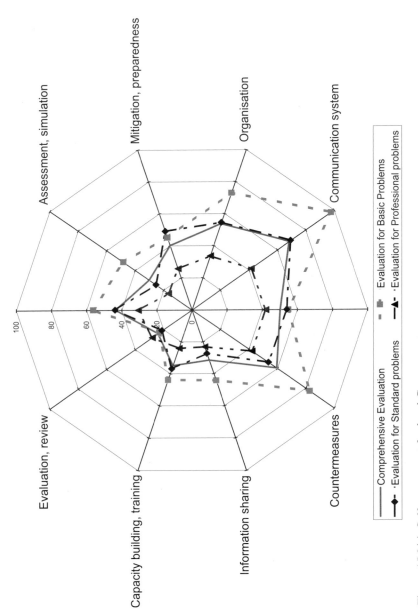

Figure 15.8(c) Self-assessment for local Government.
Source: Fire and Disaster Management Agency, Government of Japan.

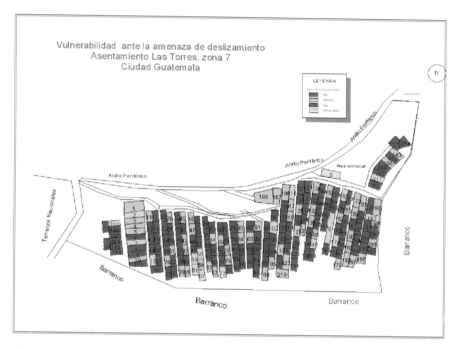

Figure 16.4(c) Vulnerability with respect to landslides in an urban settlement of Guatemala city.
Source: Pérez, 2002.

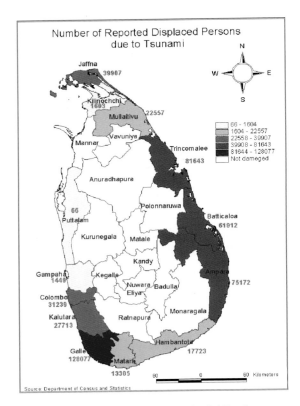

Figure 18.2(c) Overview of the tsunami impact in Sri Lanka.
Source: Own map, data Dep. of Census and Statistics.

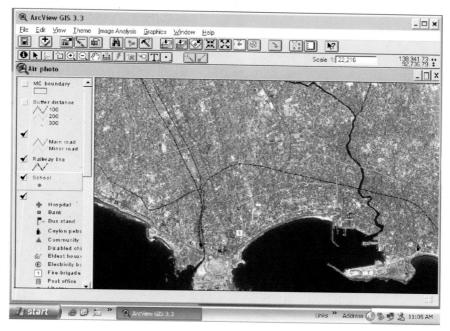

Figure 18.4(c) Spatial exposure of different critical infrastructures.
Source: Authors, based on satellite photo IKONOS.

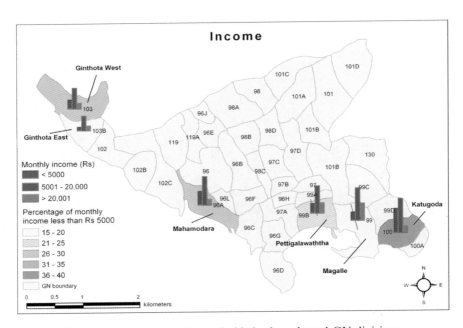

Figure 18.8(c) Income levels of households in the selected GN divisions.
Source: Authors.

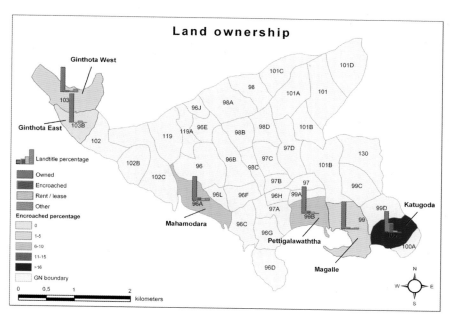

Figure 18.11(c) Landownership and squatting in the six GN divisions.
Source: Authors.

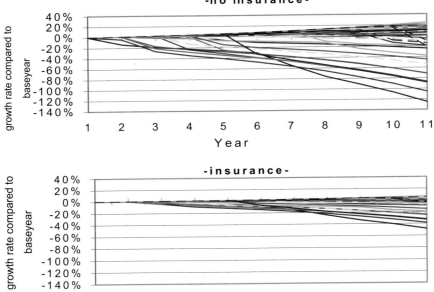

Figure 20.5(c) Simulated growth versus stability for El Salvador over a 10-year time horizon.
Source: Authors.

Figure 20.6(c) Wind hazards in Honduras.
Source: Swiss Re in Freeman et al., 2002a.

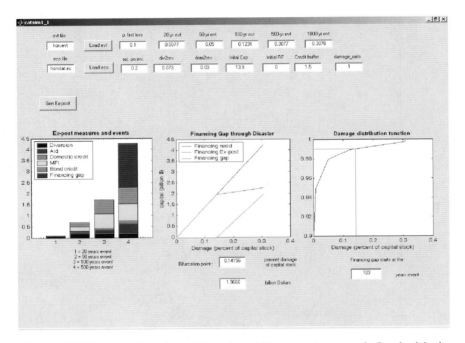

Figure 20.8(c) Assessing financial vulnerability to storm and flood risk in Honduras.
Source: Authors.

4

User needs: Why we need indicators

Angela Queste and Peter Lauwe

Abstract

This chapter gives an overview of the usefulness of indicators that measure vulnerability. It shows what the requirements are for practical applications, and how the geographical scale and target groups determine what kind of indicators can be used. As it is estimated that meteorological extreme events that cause floods like the Elbe Flood of 2002 will increase in frequency and intensity due to climate change, the main focus here will be on flood-related vulnerability, both to people and to critical infrastructures. Our examples for user needs have been taken from several international studies and from some applications in Germany.

Background

The last major water-related natural extreme event in Germany was the Elbe Flood, which occurred in August 2002 and was especially damaging in eastern German States like Saxony and Saxony-Anhalt (Figure 4.1(c)). The flood caused 21 deaths and affected about 330,000 people. It caused an estimated US$9,130,000 worth of damage (DKKV, 2003). Compared with other countries suffering similar events, a relatively small number of people lost their lives, but the financial cost was very high. Many privately-owned buildings were destroyed by the flood, and so too were critical infrastructures like the transport and water distribution systems, the telecommunication system and the health services. The flooding of the Elbe, and its consequences, showed dramatically that Germany is vulnerable to natural extreme events and demonstrated the importance of trying to identify the most vulnerable parts of society in order to be better prepared for emergency operations.

The Elbe Flood revealed major problems in disaster management, such as confusion in coordinating rescue workers and allocating the necessary emergency resources. Due to these lessons, a new strategy for civil protection was adopted in Germany in 2002, on the occasion of the conference of the Ministers of the Interior of the Federal Government and the States (IMK, 2002). This new strategy includes a plan for concerted action between the federal, regional and local levels of Government. One of the agreed measures was to conduct risk and vulnerability analyses on the national level. The objective is to use the analysis results to determine appropriate protection and intervention goals, and to identify the short, medium and long-term vulnerabilities of local and regional infrastructures. It is expected that this identification of major hazards and risks within the German States (*Länder*) will allow implementation of more effective protection measures, planning procedures and distribution of resources. In Germany, risk analyses are already undertaken by some cities and municipalities at the local level, but have to be developed at sub-national and national level.

As well as natural extreme events like floods, earthquakes and storms, the list of major hazards also encompasses military conflicts, national or international terrorism – including sabotage and major criminal acts – accidents and epidemics. The new strategy for civil protection foresees the development of a common hazard assessment of all German States for all the hazards cited above. In this context, a critical element is the assessment of vulnerability. Assessing vulnerability at national level can only be achieved by first identifying which indicators are appropriate for assessing the different aspects of vulnerability. But due to the fact that hazards occur at the local level, it is also necessary that the indicators be usable at both the sub-national and local level as well. The assessment of the vulnerability of critical infrastructures is also viewed as a key task that has to be implemented in the risk analysis.

Measuring vulnerability as an important element of risk assessment

The analysis of vulnerability must be integrated within the largest framework of risk assessment. The definition and application of vulnerability indicators is an important tool in this process, as Figure 4.2 shows.

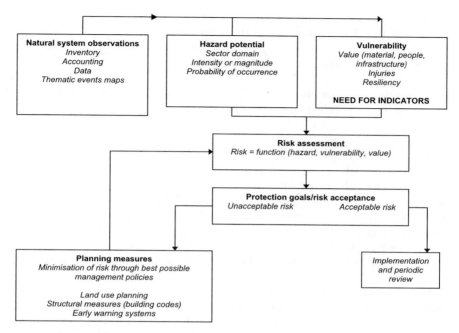

Figure 4.2 Framework for risk assessment.
Source: Adapted from UNESCO, 2003.

The gathering and systematic analysis of the vulnerability of specific objects and areas at sub-national level should enable disaster managers to be better prepared for emergency situations. Measuring the vulnerability of societies, and especially the vulnerability of critical infrastructure, should help to identify those weak points needing special attention during disaster situations. The information gathered about these vulnerabilities is an important basis for holding objective discussions as well as making decisions about which resources (financial and human) should be allocated in order to ensure preparedness and an effective response when an emergency strikes. This means developing tools and indicators to measure vulnerability that can also be used as an information base for decision makers.

Scale, functions and target groups

Before describing the scale, functions and target groups of indicators, we need to define such terms as vulnerability, indicator and user needs.

Other chapters in this book also offer definitions (see Birkmann, Chapter 2, for a detailed discussion about functions of indicators).

Vulnerability in disaster risk management

An overview of various vulnerability definitions is given by Birkmann (Chapter 1), Schneiderbauer and Ehrlich (Chapter 3), and Green (2004). (Additional definitions of key terms are also given by Thywissen, Chapter 8.)

Schneiderbauer and Ehrlich's (2004) definition of vulnerability as "the characteristics of a person or a group in terms of their capacity to anticipate, cope with, resist and recover from the impact of a natural or human-made disaster – noting that vulnerability is made up of many political-institutional, economic and socio-cultural factors" reflects the different aspects of human vulnerability very well, and is used mostly by the Federal Office of Civil Protection and Disaster Assistance in Germany (BBK – Bundesamt für Bevölkerungsschutz und Katastrophenhilfe).

Green (2004) offers a definition of vulnerability that also includes the vulnerability of systems of specific objects, such as critical infrastructures. He describes vulnerability as "the potential for attributes of a system to respond adversely to the occurrence of hazardous events".

For the practical concerns of the BBK and efforts to develop a national and sub-national disaster risk information system in future, indicators are viewed as tools to help identify and measure the vulnerability of specific objects like buildings, population subgroups or infrastructures to extreme events. Here, indicators can help to identify whether or not redundant systems exist and what resulting consequences are to be expected. In terms of analysis and emergency planning, the intended function of the indicators is to enable and advocate preventive actions in disadvantaged areas and to adapt the resource planning for emergency situations according to financial and personnel deficiencies. Furthermore, it is important to identify which user or user groups the indicators should be developed for (politicians, management of enterprises and so on). The selected indicators should fit the needs of the society and its public administration. There are already several vulnerability indicators that have been developed within specific contexts, but there is still a lack of vulnerability indicators that can be used for disaster risk management on different geographical scales and for different vulnerable population subgroups.

The scale matters

Indicators for (inter-)national approaches were developed to measure distinctions between countries. If the standard of living is broadly similar in all parts of a country, as in Germany, indicators like the illiteracy rate or the GDP are not useful, although some differences between the States do exist. Besides, for an indicator such as population density, its relationship to vulnerability is relatively complex. There seems to be no linear relationship, but rather one that depends on other structural effects of the society.

Vulnerability indicators at the local level

Normally, disasters occur at the local level, where flood plains, for example, tend to be narrow and short. Therefore, vulnerability indicators should be predominantly available for small geographical areas (Tapsell et al., 2002). On the local level, the working level for disaster first-responders (e.g. fire brigade, police) and weak points in the system are fairly obvious. Here, vulnerabilities can most effectively be reduced by including suitable measures in local planning and building codes, by public information campaigns and by raising the awareness of politicians and people and their ability to respond. If there is a systematic way to identify vulnerability and risk, then needs can be prioritised and resources allocated accordingly.

Vulnerability indicators at the regional level

In Germany, most disaster management is primarily established at the sub-national level. Each *Land* has its own law for disaster protection, the *Katastrophenschutzgesetz*. Effective disaster response requires knowledge about vulnerabilities at this level, in order to plan and distribute the resources needed to supplement what is locally available. The sub-national level is also the level at which regional planning should be carried out. Besides, a comparison of vulnerabilities within the local communities allows a more effective distribution of resources.

Vulnerability indicators at the national level

In Germany, as in other countries like the United States, the responsibility for disaster relief that cannot be dealt with on the State or local level is normally assumed by the national Government. In Germany, for example, the federal level can be asked by the *Länder* to provide resources for disaster management or to coordinate the distribution of resources. For that reason it is necessary also for the national agencies to have information about the risk and vulnerability status at local and sub-national level. Vulnerability indicators can help to identify the weakest points within the States, and thus support their preparedness for disaster situations. Besides, the protection of critical infrastructures, such as energy supplies, health care services or the railway system, is also dealt with at national level. Knowledge about the possible vulnerabilities of critical infrastructure and of vulnerable populations is essential for averting major dangers that can influence national safety, security or welfare.

The practitioner's view of important functions of vulnerability indicators

As already discussed by Birkmann in Chapter 2 and Schneiderbauer and Ehrlich in Chapter 3, there are specific requirements for indicators, such as hazard dependency and hazard independency. But successful indicators should also be simple and easy to collect, otherwise failures would be pre-programmed. The simpler data is to collect, the fewer mistakes that can be made by comparing data about different areas.

Good indicators should also be policy-relevant. Knowledge about vulnerabilities enables administrative decision makers to integrate vulnerability reduction policies and preventive measures into urban planning and urban development strategies.

Target groups and objects for vulnerability indicators

Vulnerability can be measured for specific objects or selected regions. It can be social, health-related, political, environmental, economic or technical in nature. All these vulnerability characteristics can be related to natural disasters, and especially to flood events. People who are socially deprived, elderly, disabled or in poor health are more vulnerable to flooding than others. Politically unstable and economically disadvantaged countries are also less resistant to floods than politically stable ones. Technical vulnerability can be a problem in highly engineered countries that are dependent on sophisticated infrastructures; but developing societies that depend on the functioning of a few simple, key infrastructures can also be vulnerable.

The vulnerability indicators would be used by politicians, the administration, relief organisations and operators of critical infrastructures on each geographical scale. Vulnerability maps that are based on the measured vulnerability values can be used to prevent disaster situations by prioritising activities and directing financial resources and personnel to the most vulnerable parts of the geographical region and the most vulnerable population subgroups.

Social vulnerability

Population subgroups that are vulnerable to the effects of flooding include the elderly, women, children, minorities, individuals with disabilities and those with low incomes (Hajat et al., 2003). Factors such as language, housing patterns, building construction, community isolation and the cultural insensitivity of the majority population may also affect the social vulnerability of these populations (Menne and Bertollini, forthcoming). Social vulnerability indicators for natural hazards were developed on behalf of the Australian Government by Dwyer et al. (2004). Within this approach, social vulnerability is reflected by the following indicators: age, income, gender, employment, residence type, household type, tenure type, health insurance, house insurance, car ownership, disability, English language skills and debts/savings. They also add to this catalogue two hazard indicators that reflect specific vulnerability: residential damage and injuries linked to a hazard context.

Other publications define which "community characteristics" reflect social vulnerability. The World Health Organization (WHO), for example, names characteristics like demographic aspects, culture, economy, infrastructure and the environment. Health indicators like the vaccination coverage rate or disease pattern after an emergency reflect demographic aspects (WHO, 1998). Tapsell et al. (2002) have identified indicators reflecting social vulnerability that consist of an index and other data: the elderly (aged 75+), lone parents, and those with pre-existing health problems and financial deprivation (following the Townsend index which includes parameters like the unemployment rate).

Vulnerability of critical infrastructure

Measurement of the vulnerability of critical infrastructures is one important issue related to disaster management. Critical infrastructures are organisations and systems with great importance for society that, if disrupted, would impact various supply chains and public safety and could lead to further dramatic consequences (BMI, 2005). Critical infrastructure sectors in Germany include energy supplies such as electric power, oil and gas delivery and storage, and nuclear power plants; the supply of essentials like water, food, health and emergency services; communication and information technology; transportation; hazardous materials (chemicals, toxic waste); banking and finance; Government services; the media; research institutes; and cultural assets. As is shown in Figure 4.3 all kinds of hazards can affect critical infrastructures. The resistance, resilience and susceptibility of the critical infrastructure components determine the degree of their vulnerability. The interdependencies between different infrastructure sectors are also important; for example, all are dependent on the energy sector. Strong coping capacities, like the availability of a redundant system, can reduce vulnerability. As society depends on the reliable functioning of critical infrastructures, disruptions and breakdowns can have negative effects, though society itself has its own coping capacities.

The National Oceanic and Atmospheric Administration (NOAA) of

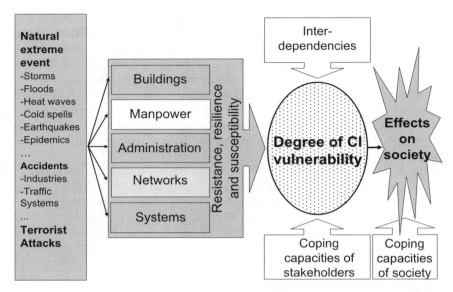

Figure 4.3 The vulnerability of critical infrastructures (CI holistic approach). Source: Authors.

the United States has defined which data are needed for assessing the vulnerability of critical infrastructures. The list includes structural integrity, construction type and quality, age, size, types of materials contained at or discharged from the facilities, and impacts the materials might have on environmental resources. The vulnerability of critical infrastructures can be measured on the above-mentioned different scales. At the local level, in companies like small waterworks, specific components or management factors can be analysed. At the regional level, vulnerability can be measured for components in sites like waterworks that are responsible for distributing water over large areas using long-distance pipelines. Examples of critical infrastructures on the national level would include the railway system and the energy system, for which indicators are needed to assess their vulnerability.

Example of vulnerability indicators at State level in Germany

One example, where indicators are already in place to measure the vulnerability of one German State towards extreme hazards is Mecklenburg-

Western Pomerania (LBK et al., 2002). Here, vulnerability is defined as the degree of damage that a population, buildings, industrial sites, economy, cultural assets and technical infrastructure might undergo through the occurrence of a hazardous event. This analysis includes vulnerability indicators that are based upon general population characteristics and structures. The most important indicators in this analysis are the number of inhabitants per community, the number of potentially affected farm animals and the number of potentially affected critical sites (sites for which a permit is needed). These indicators focus primarily on the density of people, farm animals, etc. as a characteristic to identify most vulnerable areas. Data were collected for the potential hazards posed by flooding from high tides in the Baltic Sea and flooding of the river Elbe.

For forest fire hazards, the indicator is *forest area*; for accidents with hazardous material along the coast, the *length of the coast* is the indicator used; for animal epidemics the indicator is the *number of potentially*

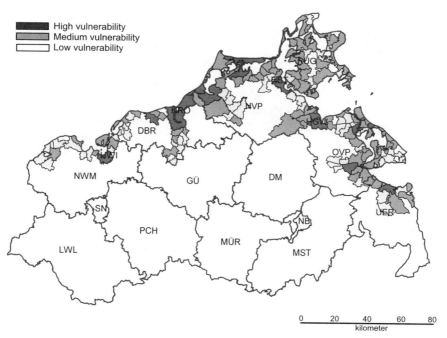

Figure 4.4 Vulnerability of persons in high tide flood-prone areas of the Baltic Sea.
Source: LBK et al., 2002.

affected farm animals (LBK et al., 2002). The vulnerability assessment uses a ranking scale between 0 (no damage) and 1 (total damage). How vulnerability related to high tide floods of the Baltic Sea was assessed is shown by the example of the indicator *number of flood-affected people*: Up to 100 people affected reflects a low vulnerability (the degree of damage is assessed as up to 0.3) and from 101 to 1,000 people affected indicates a medium vulnerability (up to 0.6). A high vulnerability is suggested for events where more than 10,000 people suffer from the flood event (damage degree between 0.7–0.9). The results of this vulnerability analysis are shown in Figure 4.4 and Figure 4.5. Figure 4.4 represents the number of people who are vulnerable to high tide floods of the Baltic Sea. Figure 4.5 shows the number of critical sites that could be damaged in such an event.

The example shows that it is possible to identify those areas that need more preventive measures than others. Additionally, they require more resources and sophisticated plans of action in order to establish an adequate preparedness for emergency situations.

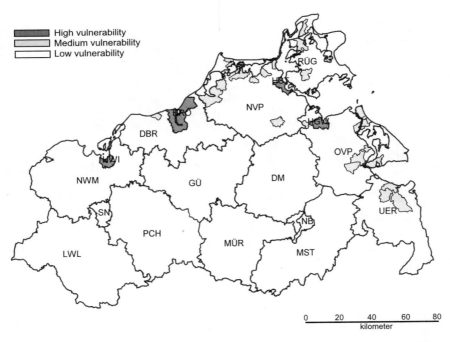

Figure 4.5 Vulnerability of critical sites in high tide flood-prone areas of the Baltic Sea.
Source: LBK et al., 2002.

Recommendations

Many approaches still remain in the scientific arena and more emphasis should be given as to how to integrate the tools and indicators into practical planning and decision-making processes, such as disaster emergency plans, strategic communication tools of risk and vulnerability to the public, awareness raising, systematic evaluation of the effectiveness of risk reduction and disaster management operations and also urban planning.

Cooperation between scientific institutions and governmental administrations (local, regional and national) would enhance the development of useful vulnerability indicators. Here, it is important to motivate all stakeholders to support the collection and analysis of data as well as to establish emergency prevention and action plans for emergency situations.

One of the main foci for defining vulnerability should be on the most vulnerable population groups within the society. However, the needs of the operators of critical infrastructures should also be taken into account in order to foster the implementation of preventive strategies and to allow for an assessment of necessary substitution measures in emergency situations.

REFERENCES

Bundesministerium des Innern (BMI) (2005) *Schutz Kritischer Infrastrukturen – Basisschutzkonzept. Empfehlungen für Unternehmen*, Berlin: BMI, available at http://www.bmi.bund.de/cln_012/nn_122688/Internet/Content/Common/ Anlagen/Broschueren/2005/Basiskonzept__kritische__Infrastrukturen, templateId=raw,property=publicationFile.pdf/Basiskonzept_kritische_ Infrastrukturen/.

Deutsches Komitee für Katastrophenvorsorge (DKKV) (2003) *Hochwasservorsorge in Deutschland. Lernen aus der Katastrophe 2002 im Elbegebiet.* Schriftenreihe des DKKV 29, Bonn: DKKV.

German Remote Sensing Data Centre of DLR (2005) available at http:// www.dlr.de/.

Dwyer, A., C. Zoppou, O. Nielsen, S. Day, and S. Roberts (2004) *Quantifying Social Vulnerability: A Methodology for Identifying Those at Risk to Natural Hazards*, Geoscience Australia Record 2004/14, available at http://www.ga. gov.au/image_cache/GA4267.pdf.

Green, C. (2004) "The Evaluation of Vulnerability to Flooding", *Disaster Prevention and Management* 13(4): 323–329.

Hajat, S., K.L. Ebi, R.S. Kovats, B. Menne, S. Edwards and A. Haines (2003) "The Human Health Consequences of Flooding in Europe and the Implications for Public Health: A Review of the Evidence", *Applied Environmental Science and Public Health* 1(1): 13–21.

Innenministerkonferenz (IMK) (2002) *Neue Strategie zum Schutz der Bevölker-
ung in Deutschland*, (Beschluss der Ständigen Konferenz der Innenminister
und -senatoren der Länder vom 06.12.2002 in Bremen, Top 36); Schriftenreihe:
WissenschaftsForum, Akademie für Krisenmanagement, Notfallplanung und
Zivilschutz, Band 4, Bonn: Bundesverwaltungsamt, available at http://www.
bva.bund.de/imperia/md/content/abteilungen/abteilungv/vb3/publikationen/
wissenschaftsforum/4.pdf.

Landesamt für Brand- und Katastrophenschutz Mecklenburg-Vorpommern und
TÜV Berlin (2002) Aktualisierung der Gefährdungsanalyse Mecklenburg-
Vorpommern – Teil II. Projektabschnitt 2001/2002: Bestimmung der Vulnera-
bilität des Landes gegenüber besonderen Gefährdungslagen und Katastrophen.
Schwerin 2002, available at http://www.lbk.mv-regierung.de/download/gfa2.pdf.

Menne, B. and R. Bertollini, eds, *Climate Change and Adaptation Strategies for
Human Health*, Section VI: Final report to the European Commission (forthcoming).

National Oceanic and Atmospheric Administration (NOAA) Coastal Services
Centre (no date) *Risk and Vulnerability Assessment Steps. Environmental Analysis. Risk and Vulnerability Assessment Tool (RVAT)*, available at http://
www.csc.noaa.gov/rvat/environ.html.

Schneiderbauer, S. and D. Ehrlich (2004) *Risk, Hazard and People's Vulnerability
to Natural Hazards: A Review of Definitions, Concepts and Data*, Brussels:
European Commission–Joint Research Centre (EC-JRC).

Tapsell, S.M., E.C. Penning-Rowsell, S.M. Tunstall and T.L. Wilson (2002) "Vulnerability to Flooding: Health and Social Dimensions", in *Phil. Trans. R. Soc.
Lond.* A. 360, pp. 1511–1525.

United Nations Educational, Scientific and Cultural Organization (UNESCO)
(2003) *The UN World Water Development Report (WWDR): Water for People,
Water for Life*. Paris/Oxford/New York: UNESCO, available at http://www.
unesco.org/water/wwap/wwdr/table_contents.shtml.

World Health Organization (WHO) (1998) *Health Sector Emergency Preparedness Guide: Making a Difference to Vulnerability*, Geneva: WHO.

Part II

Vulnerability and environment

5

Environmental components of vulnerability

Fabrice G. Renaud

Introduction

Vulnerability is a complex concept and the term is used very loosely depending on an individual's background and the context within which it is used (Thywissen, 2006; Chapter 24). One definition among others is that vulnerability is the intrinsic and dynamic feature of an element at risk that determines the expected damage or harm resulting from a given hazardous event and is often even affected by the harmful event itself. At the United Nations University, Institute for Environment and Human Security (UNU-EHS), the vulnerability of communities to natural and anthropogenic hazards is approached from a multidimensional perspective, which encompasses environmental, social and economic spheres and incorporates features such as susceptibility, exposure and coping capacities (Chapter 1).

When assessing the vulnerability of communities, the environmental sphere cannot be separated from the social and economic spheres because of the mutuality between human beings and the environment: human beings shape their environment and in turn the environment plays a major role in shaping the economic activities and social norms of human beings. The concept of mutuality allows for a better understanding of how humans create their vulnerability to given hazards, and if this conceptual approach is used, vulnerability assessment captures the multidimensionality of disasters (Oliver-Smith, 2004). Several existing vulnerability assessment frameworks incorporate environmental components and their interactions with social and economic systems within their conceptualisation (e.g. Turner et al., 2003 – see Chapter 6, Chapter 1) but the extent to which this is done and the nature of the interactions considered vary from framework to framework.

Below are preliminary concepts of how, from a practical perspective, the environmental dimension of vulnerability is considered and integrated with the other dimensions within the BBC vulnerability assessment framework. Examples from work carried out in the aftermath of the December 2004 earthquake and tsunami in Sri Lanka are used to illustrate these concepts.

The environment as a service provider

Vulnerability assessment within the BBC framework (Chapter 1) is carried out at the community level (or local scale). The environmental dimension of vulnerability is primarily seen through an anthropocentric lens whereby the environment is a provider of services to human beings, and it is the loss of capacity to satisfy human needs that is considered as a potential to increase the vulnerability of communities to external or internal stresses (see text box "Definition of archetypes of vulnerability", Chapter 6). The Millennium Ecosystem Assessment report (MA, 2005) highlights the fact that human impacts on various ecosystems are such that the capacity of the latter to provide services is seriously affected in many parts of the world. By its direct impact on vital resources (e.g. water, soil), environmental degradation increases the vulnerability of communities. The concept of mutuality mentioned above (Oliver-Smith, 2004) or the coupled human–environment systems described by Turner et al. (2003) require more than just a description of loss of services from various ecosystems. However, at this stage of development of the BBC framework and for practical reasons linked to field data availability – the time needed to carry out specific local assessments, and costs linked to the overall assessment – only three aspects of the linkages between humans and the environment they live in are considered:
• loss of ecosystem services as described above
• dependencies of communities on specific services provided by a limited number of ecosystem components
• vulnerability of the components of the ecosystems to a specific threat.
The environment can be divided into various components, such as air, land, soil, vegetation and water (groundwater, surface water, coastal waters, etc.), called resources hereafter. These resources play different roles depending on the type of hazard or threat considered and whether

we consider rural or urban communities. For example, rural communities rely much more on soil resources for their livelihoods (e.g. to grow crops or sustain pastures) than do urban settlers. Furthermore, a vegetation strip (e.g. a mangrove ecosystem) plays a different role depending on whether we are considering a tsunami or an earthquake hazard.

If vulnerability assessment is carried out before the impact of an event (which is its primary objective) then the role and state of each of the resources can be determined, which will indicate their capacity to provide the required services to communities. In some cases, vulnerability assessment is only carried out after an impact, particularly for hazards with long return periods. This was, for example, the case following the December 2004 tsunami that destroyed many coastal regions in South East and South Asia and in Eastern Africa. Although people knew before the day of the disaster that a tsunami could take place in the Indian Ocean, nobody paid much, if any, attention to this hazard before it tragically manifested itself. Now many organisations are involved in post-tsunami activities including vulnerability assessment, hoping to draw valuable lessons from this disaster. Within this context an assessment reveals the new vulnerability of the affected communities as, according to the definition given in the introduction, vulnerability itself can be affected by a hazardous event. From an environmental perspective, an impact assessment (which is different from a vulnerability assessment) can also reveal valuable information as to the role, exposure and vulnerability of important resources in relation to a given hazard.

Examples of environmental services

Examples of services provided by the environment are presented here using tsunami-affected Sri Lanka as a case study.

First, coastal and river bank vegetation plays many important roles, not only in terms of erosion control and provision of resources (e.g. mangrove ecosystems) but also in terms of buffering populations against hazards such as floods, storm surges and tsunami waves. If we consider the latter hazard, there are reported and anecdotal cases where communities were at least partially protected by mangroves during the December 2004 tsunami. However, the exact nature of the protection afforded by coastal vegetation is still being debated (e.g. Kathiresan and Rajendran, 2005, 2006; Kerr et al., 2006), as it is difficult to distinguish the many coastal and settlement features that can play a role; these include the effects of distance between affected communities and the coastline, local topography, local bathymetry and the wave-energy dissipation potential of the vegetation strips, among other parameters. Nevertheless, it can be argued

Figure 5.1 Destruction along the coast, Galle, Sri Lanka.
Source: Author.

that coastal vegetation protects populations, even if only because the existence of such vegetation means that populations settle slightly away from the coastline, thus reducing their exposure.

Unfortunately many affected communities in Sri Lanka were not buffered by vegetation during the tsunami as many infrastructures and houses were located close to the sea or right on the seashore (Figure 5.1). This pattern of settlement is, of course, not specific to Sri Lanka, as everywhere around the world people have settled along coasts and rivers, benefiting from all the services provided by respective water bodies (Affeltranger et al., 2005), but also being exposed to the hazards inherent to these water bodies. In Sri Lanka and in all non-landlocked countries, environmental degradation from destruction of coastal vegetation increases the exposure of communities to coastal hazards.

Second, freshwater aquifers play an important role in coastal peri-urban and rural areas of Sri Lanka as they provide water for all domestic uses, including drinking water, and in some cases also for irrigation. Before the tsunami, there were increasing pressures on this resource from greater abstraction for domestic and agricultural uses on the one hand, and the increasing pollution from domestic, agricultural and industrial sources on the other (e.g. Villhoth et al., 2005; UNICEF, 2005). The tsunami affected many wells along the coast: completely or partially destroying them, polluting the groundwater with high salt concentrations (and possibly other organic and inorganic pollutants) and depositing debris of all kinds in the wells. A situation that was already delicate before the event became critical for many communities. The vital role of groundwater was such that many attempts by Government and non-governmental organisations were made to restore the water quality of the wells. These activities allowed the reclamation of some of the wells, but in many cases, the cleaning operations (generally pumping out the well water once or several times) may have aggravated the situation by encouraging saline intrusion into the freshwater lenses (e.g. UNICEF, 2005).

More than a year after the tsunami, water quality had still not been restored in many coastal wells. A comprehensive monitoring programme put in place by the International Water Management Institute showed that salinity in monitored wells on the east coast was higher than pre-tsunami levels some seven months after the impact of the tsunami – coinciding mostly with the dry season (Villholth et al., 2005). One of our studies, conducted jointly with Ruhuna University, which covered communal wells in the Galle District (ongoing at the time of writing) showed that communities reported salinity, nauseating odours and coloured water as continuing problems more than a year after the event. Coastal communities around the country thus still relied on water purification systems or, more frequently, on water brought in by tanker lorries from outside, on a discontinuous and unreliable piped water system, or on wells tapping unaffected portions of the aquifers, depending on the location of the community and the impact of the tsunami. Impacts on drinking water supplies were much less severe in cities like Galle where the water distribution system that brings water from an unaffected reservoir to houses and businesses was restored relatively rapidly after the event.

At least two pieces of information from this groundwater-related post-impact assessment are relevant for vulnerability studies. The first is the dependency of some communities on a single environmental resource (here freshwater aquifers), and the second is the exposure of this resource to a specific hazard (here the tsunami and possibly storm surges). It is therefore crucial to rapidly find ways both to alleviate the vulnerability of the aquifer itself (e.g. by reducing the pressures on it) and to provide alternative sources of freshwater (e.g. by increasing the area covered by piped water systems, although clearly this would require large-scale investments, would take time, and should not be done at the expense of environmental degradation in non-coastal zone areas). Since the coastal areas of Sri Lanka will remain exposed to future tsunamis and storm surges, a survey should be carried out to locate good-quality freshwater aquifers close to the coast that could be used for emergency situations (as advised, for example, by UNESCO's Groundwater for Emergency Situations (GWES) programme; see Vrba and Verhagen, 2006).

A final example can be taken from a rural setting where the soil and in many cases the groundwater (for irrigation) play major roles in providing livelihoods to farming communities. Agricultural fields were damaged by the tsunami (Figure 5.2) whose effects included the destruction of crops and the salinisation of the resource base (soil and groundwater).

However, preliminary investigations by the International Rice Research Institute[1] established that in the case of well-drained soils, these effects would be short-lived, as rain water and/or supplementary irrigation would leach the salts from the soil. Through a rapid rural assessment study carried out in collaboration with Eastern University one year after the tsunami, it was established that in two rural communities in the coastal region of Batticaloa District (east of Sri Lanka), some fields may have been experiencing salinity problems. The problems of resuming agricultural activities were compounded by the loss of family members during the tsunami, and the reduction in labour availability due to resettlement and daily employment offers by Government and non-government organisations (GOs and NGOs), and loss of equipment and tools, to name only a few factors. Aid from GOs and NGOs helped some of the farming families, but more support was expected in the area. This example shows that although a vulnerability assessment can consider a specific part of an ecosystem, the links with the social and economic dimensions need also to be understood in order to obtain a clear picture of vulnera-

Figure 5.2 Tsunami-impacted paddy fields in Sri Lanka.
Source: Author.

bility and of which corrective tools are best suited to the specific circumstances of the communities.

The examples above relate to a post-impact analysis, but in a pre-impact assessment, the quality, quantity and availability of resources, and the dependency of communities on these resources, also need to be determined. Speaking more generally, because of increasing pressure by humans on natural resources such as land (soil) and surface- and groundwater, the environment's capacity to provide essential services (in terms of both quantity and quality) is being compromised worldwide. This in turn increases the vulnerability of communities, as they cannot rely on specific environmental resources to (1) sustain their way of life, and (2) allow them to minimise the impact or recover after a major hazardous event. Many examples from developed and developing countries and from rural and urban environments can be cited:

- Conversion of wetlands is a major driving force behind land degradation and can be a loss of a buffer zone for water flow regulation (UNEP, 2002). In some cases this can have direct implications for floodwater regulation.
- Overexploitation of water resources in water-scarce countries induces problems of water quality and future availability (UNEP, 2002), making communities reliant on scarcer resources and reducing the diversity of water supplies.
- Land degradation harms the physical, chemical and biological properties of soils (Marshall et al., 1996; Vlek, 2005), and therefore their productivity potential. This loss of productivity affects rural communities year-to-year, but also increases the impact of extreme events such as climatic droughts.

These trends in environmental degradation can create a vicious circle whereby over-exploited or mismanaged natural resources cannot provide communities with required services. This can be followed by a tendency to increase the pressure on the resources in an attempt to make a living or recover from a preceding drought, flood or other event, further degrading the resource (Figure 5.1). A degraded environmental resource then has more difficulty withstanding a future external impact and is very difficult to restore.

Environmental degradation and hazards

In addition to affecting the vulnerability of people, environmental degradation can contribute to the amplification or increase in frequency of certain types of hazards. The IPCC report (2001) indicated that climate change could generate more extreme weather patterns in many parts of the world in the future. Such changes would include higher rainfall intensities in some places and longer periods without rain in others. There would therefore be a direct impact on the local and regional hydrological cycles, with the prospect of hydrometeorological events such as floods and droughts of higher magnitude and frequency.

In addition to climate change, land use changes throughout the world have affected the characteristics and/or the likelihood of manifestation of some types of extreme events. The most cited example is deforestation in mountainous regions, which can increase erosion and decrease the infiltration capacity of soils, thus generating more runoff and more floods locally (however, great care needs to be taken when assuming direct links between deforestation and floods as the cause–effect relationship will vary from basin to basin). Deforestation can also be a major factor in landslides in hilly areas, as tragically shown by the 2004 mud floods in Haiti. However, deforestation is by no means the only example where land use changes affect the characteristics of hazardous events as, for example, conversion of pastures to urban land uses also has an impact on local hydrological cycles and flooding patterns (Trocherie et al., 2004).

The BBC vulnerability assessment framework recognises that actions can be taken (through actuation tools) in order to reduce the magnitude or frequency of hazards. This implies that policies such as improved land use or stricter emission controls can reduce the overall risks that communities are facing by acting directly on the hazard side of the model and not only on the vulnerability side as discussed in the previous sections.

Environmental degradation and society

By affecting people's livelihood, environmental degradation increases the vulnerability of some communities, and can also contribute to increasing the vulnerability of others through migrations. By affecting land productivity, land degradation worsens rural poverty, particularly when coping mechanisms are weak in rural areas, and poverty is often a driver for migrations from rural areas to urban centres (IFAD, 2001; Vlek, 2005; Figure 5.1). This migration may temporarily reduce pressure on the farmland (if remittances from outside allow rural families to make a better living than in the pre-migration situation) but may increase pressure on urban centres. Poor people who migrate generally settle in the poorest and often most exposed neighbourhoods in large cities. There, not only are they likely to become "urban poor", but they are exposed to hazards that they may not be familiar with and they may not acquire the new culture of risk necessary to cope with these hazards rapidly enough.

Alternatively, communities facing environmental degradation may be

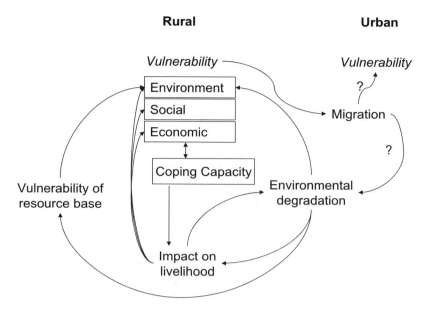

Figure 5.3 Potential effects of land degradation on rural and urban vulnerabilities (economic, social and environmental vulnerability; coping capacity).
Source: Author.

able to adapt to the new conditions, a strategy that has been used with varied success since the beginnings of humanity. Global environmental change is a reality but not a fatality. Many technical, institutional and political solutions exist to alleviate some of the pressures placed on environmental resources. Implementing these solutions is, however, often constrained by economic costs, institutional capacity or political willingness (as seen, for instance, in the difficulties linked to the ratification of the Kyoto Protocol).

Capturing the environmental dimension of vulnerability

The following concepts are integrated into the BBC vulnerability assessment framework in order to capture the environmental component of vulnerability:

- Determining what major services are provided by different environmental resources to the community. This can be achieved by various means: for example, by discussions with key informants (e.g. local Government agents, local NGOs, local religious leaders) or through participatory exercises with the local communities. This information can then be used by decision makers to determine the dependence of the community on one or several resources, and to recommend specific protection measures for those resources and/or find alternatives that could provide similar services.

- Determining the current pressure exerted on these resources (quality, quantity, access). This should lead to policies aimed at reducing the pressures.
- Determining the vulnerability of the resource itself. This in turn can lead to actions to either protect the resource further or improve its ability to recover to a satisfactory state (i.e. one that can provide the usual services) more rapidly.
- Determining the institutional capacity needed to deal rapidly with the impact of a disaster on affected environmental resources.
- Analysing environmental degradation processes that represent driving forces for migrations, such as soil erosion in some rural areas. Specific policies can then be put in place to tackle the problem at the source.

A wide range of tools can be used for this effect. These may vary from community to community but include participatory appraisals, questionnaire surveys, in-depth surveys and use of remote sensing. Several vulnerability assessment case studies are being carried out by UNU-EHS and collaborators throughout the world. Information, both qualitative and quantitative, is captured at the household level via questionnaires, with more in-depth surveys, and through local and national statistics. These test case studies will allow for the refinement of the methodology and will also reveal what is really practically achievable and what is not for a comprehensive vulnerability assessment at the local level.

Note

1. http://www.knowledgebank.irri.org/regionalSites/sriLanka/default.htm

REFERENCES

Affeltranger, B., J. Bogardi, and F. Renaud (2005) "Living with Water", in United Nations, eds, *Know Risk*, Geneva: UN, pp. 258–261.

Intergovernmental Panel on Climate Change (IPCC) (2001) *Climate Change 2001: Synthesis Report*. A Contribution of Working Groups I, II, and III to the Third Assessment Report of the IPCC, Watson, R.T., ed., Cambridge/New York: Cambridge University Press.

International for Agricultural Development (IFAD) (2001) *Rural Poverty Report 2001: The Challenge of Ending Rural Poverty*, Oxford: Oxford University Press.

Kathiresan, K. and N. Rajendran (2005) "Coastal Mangrove Forests Mitigated Tsunami", *Estuarine, Coastal and Shelf Science* 65: 601–606.

Kathiresan, K. and N. Rajendran (2006) "Reply to Comments of Kerr et al. 'Coastal Mangrove Forests Mitigated Tsunami'", *Estuarine, Coastal and Shelf Science* 67: 542.

Kerr, A.M., A.H. Baird, and S.J. Campbell (2006) "Comments on 'Coastal Mangrove Forests Mitigated Tsunami' by K. Kathiresan and N. Rajendran", *Estuarine, Coastal and Shelf Science* 67: 539–541.

Marshall, T.J., J.W. Holmes, and C.W. Rose (1996) *Soil Physics*, 3rd edition, Cambridge: Cambridge University Press.

Millennium Ecosystem Assessment (MA) (2005) *Ecosystems and Human Well-being: Synthesis*, Washington D.C.: Island Press, available at http://www.millenniumassessment.org//en/Products.aspx?.

Oliver-Smith, A. (2004) "Theorizing Vulnerability in a Globalized World: A Political Ecological Perspective", in G. Bankoff, G. Frerks and D. Hilhorst, eds, *Mapping Vulnerability: Disasters, Development and People*, London: Earthscan, pp. 10–24.

Thywissen, K. (2006) "Components of Risk: A Comparative Glossary", *Source 2*, Bonn: Publication Series of UNU-EHS.

Trocherie, F., N. Eckert, X. Morvan, and R. Spadone (2004) "Inondations Récentes: Quelques Éclairages", *Lettre Thématique Mensuelle de l'Institut Français de l'Environnement* 92: 1–4.

Turner II, B.L., R.E. Kaspersonb, P.A. Matsone, J.J. McCarthyf, R.W. Corellg, L. Christensene, N. Eckleyg, J.X. Kaspersonb, A. Luerse, M.L. Martellog, C. Polskya, A., Pulsiphera, and A. Schillerb (2003) "A Framework for Vulnerability Analysis in Sustainability Science", *PNAS* 100(14): 8074–8079.

UNICEF (2005) "Guidelines for the Rehabilitation of Tsunami Affected Wells", UNICEF-Sri Lanka, WASH Section (compiled by Roberto Saltori), Colombo.

United Nations Environment Programme (UNEP) (2002) *Global Environmental Outlook 3. Past, Present and Future Perspectives*, London: Earthscan.

Villholth, K.G., P.H. Amerasinghe, P. Jeyakumar, C.R. Panabokke, O. Woolley, M.D. Weerasinghe, N. Amalraj, S. Prathepaan, N. Bürgi, D.M.D. Sarath Lionelrathne, N. G. Indrajith, and S.R.K. Pathirana (2005) "Tsunami Impacts on Shallow Groundwater and Associated Water Supply on the East Coast of Sri Lanka", Report from the International Water Management Institute, Colombo, Sri Lanka: IWMI.

Vlek, P.L.G. (2005) *Nothing Begets Nothing: The Creeping Disaster of Land Degradation.* InterSecTions No. 1/2005, Bonn: United Nations University, Institute for Environment and Human Security.

Vrba, J. and B.Th. Verhagen, Eds (2006) "Groundwater for Emergency Situations: A Framework Document", *IHP VI – Series on Groundwater No. 12*, Paris: UNESCO.

6

Human vulnerability to environmental change: An approach for UNEP's Global Environment Outlook

Marcel Kok, Vishal Narain, Steven Wonink, Jill Jäger

Abstract

This chapter examines the application of the concept of human vulnerability to environmental change and addresses issues of human well-being in the context of sustainable development. It presents some of the work done in the context of the United Nations Environment Programme's (UNEP) Global Environmental Outlook (GEO). UNEP has conventionally employed the driving forces–pressure–state–impact–response (DPSIR) framework for integrated environmental assessment, but made a start towards analysing vulnerability in GEO-3, published in 2002. This chapter shows how frameworks for vulnerability assessment can be employed to examine human well-being from an environmental perspective, and to identify challenges and opportunities for policy makers to integrate environmental concerns into non-environmental policy domains. Initial ideas for vulnerability analysis in GEO-4 (due in 2007) are presented.

Introduction

Human vulnerability to environmental change is not new. More than 9,000 years ago, the Sumerians of Mesopotamia started irrigating their land to meet increased demands for food from a growing population. Despite this, their civilisation collapsed, partly because of waterlogging and salinisation. Soil erosion, loss of agro-ecosystem viability and silting up of rivers contributed to the collapse of the Mayan civilisation around 900 AD. More recent examples from the twentieth century are the Dust Bowl phenomenon that resulted from massive soil erosion in the United States during the 1930s and London's great smog of 1952 that killed some 4,000 people (UNEP, 2002a).

Recent scientific reports (Steffen et al., 2004; MA, 2005) have shown that we are now living in an era in which negative human influences on the earth system are happening on an unprecedented scale. The provision of ecosystem services, such as food production, clean air and water, or a stable climate, are under pressure. The rate of global environmental change that we are currently witnessing has not been experienced before in human history. This "no-analogue" situation of environmental change has an increasing impact on the well-being of people and communities. In the face of this ongoing environmental change, however, different people and communities face different consequences. Some people may gain while others may lose, but all are to a certain extent vulnerable to these changes in the environment. However, environmental change is only one of the factors influencing the vulnerability of people, and that is why, from a sustainable development perspective, other factors such as globalisation, equity and governance issues must also be taken into account in vulnerability analysis.

The Brundtland Commission stressed the interdependence of environment and development, and defined sustainable development as "development that meets the needs of the present without compromising the ability of future generations to meet their own needs" (WCED, 1987). Sustainable development is thus about the quality of life and about the possibilities of maintaining this quality here and now, elsewhere and in the future. Sustainable development requires the integrated analysis of the economic, social and environmental domains (see also Chapters 1 and 2). However, this often proves difficult to realise in research, and in national and international policy-making. By showing the vulnerabilities of specific people, groups or places as part of a specific environment or ecosystem exposed to environmental and non-environmental threats, an indication of "unsustainable" development patterns can be derived. This analysis can serve as a basis for the identification of challenges to and opportunities for enhancing human well-being and the environment, without losing sight of the needs of future generations.

The aim of this chapter is twofold. First the application of the concept of human vulnerability to environmental change will be examined to identify challenges and opportunities for enhancing human well-being and improving the environment. The second aim is to present how this analysis will be included in the next GEO report (GEO-4) of the UNEP.

This chapter first provides an overview of approaches to analysing human vulnerability to environmental change. Next, the evolving analysis of vulnerability in the GEO reports and a description of some of the work currently underway for GEO-4 are considered.[1] A number of relevant contributions to this analysis from UNEP's programme of work are included throughout this chapter. At the end of the chapter, conclusions are drawn and suggestions made for further research.

Human vulnerability to environmental change

Scholars have been increasingly applying an integrated approach in their analysis of environmental problems, recognising that these cannot be looked at in isolation. Understanding environmental trends requires the analysis of underlying pressures as well as of how society responds to them. A framework that is conventionally employed for the integrated analysis of environmental problems is the DPSIR framework, applied for instance, in the GEO of UNEP. The DPSIR framework seeks to connect causes (drivers and pressures) to environmental outcomes (state and impacts) and to activities (responses) that shape the environment.

The strength of the framework is that it makes possible the integration of driving forces, environmental trends and policy responses. As a guide for policy makers, it serves to identify both the proximate and ultimate causes of environmental change and to identify and evaluate policy responses. Typically, for instance, a phenomenon of falling water tables (state) can lead to rising costs of groundwater extraction (impacts). This (state) could be attributed to increasing numbers of tube wells (pressures) and would necessitate interventions, for example in the form of legislation, better pricing or collective action (responses).

An intrinsic weakness of the DPSIR approach, however, is that it assumes a somewhat linear approach to human–environment interactions that are in fact much more complex and cyclical. From an analytic and policy perspective, therefore, this approach may be unable to correctly identify causal factors for environmental change. The approach is also limited in its ability to examine why certain policy outcomes may not succeed in reversing trends in environmental change, for which one needs to revert to other interactive models of policy analysis. For example, to understand, in the above instance, why groundwater is a difficult resource to manage, one needs to understand why groundwater legislation is difficult to enforce, which requires a more process-oriented study of the intervention of groundwater legislation, of pricing and the patterns of social differentiation that may inhibit collective action.

An opportunity to further deconstruct the impacts of environmental change on human systems is provided by the vulnerability approach. Human vulnerability represents the interface between hazards and environmental change to human well-being and the capacity of people and communities to cope with those hazards. It is increasingly recognised that many of the social and economic problems in the world cannot be seen as separate from environmental problems (and vice versa), and that the human–environment system through which humans interact with their environment should be approached in an integrated manner. In this chapter we therefore look at both environmental and non-environmental pressures on human vulnerability and consider how these shape human well-being.

The concept of vulnerability is important in many different fields of research. In general terms, vulnerability refers to the potential of a system to be harmed by an external stress (i.e. threat). Many different approaches to assessing vulnerability have been developed, differing in the ways they define vulnerability, the scale of analysis or their thematic focus. In GEO-3 (UNEP, 2002a), for example, vulnerability was defined as "the interface between exposure to physical threats to human well-being and the capacity of people and communities to cope with those threats". The Intergovernmental Panel on Climate Change (IPCC) (2001) defines vulnerability in relation to climate change as "the degree to which a system is susceptible to, or unable to cope with, adverse effects of climate change, including climate variability". An overview of different definitions and approaches to vulnerability is provided in UNEP (2003), and by Thywissen in Chapter 8.

In studies of vulnerability over the last few decades at least two main strands of research can be distinguished. The first concentrated on the field of natural hazard research, looking at human vulnerability related to physical threats and disaster reduction (e.g Cutter, 1996 or World Bank, 2005). It has focused on vulnerability in relation to environmental threats, such as flooding, droughts or earthquakes. Vulnerability to these extreme events depends on their likelihood and the place where they occur. In the face of global environmental change it is not only the occurrence that matters but also changes in frequency and magnitude (e.g. changes in flood frequency and magnitude), which can be drastically altered by global trends. This field also examines the environmental threat posed by the slower, long-term process of climate change. Most research has resulted in analysing the dynamics in hazardous areas and impacts that occurred.

The second strand of research looked at socio-economic factors in relation to human vulnerability (e.g. Adger and Kelly, 1999; Watts and Bohle, 1993). It has shown that in the face of (non-)environmental threats, socio-economic factors are equally important. Exposure to the threats is to a large extent determined by socio-economic factors, as is the ability to cope with them. This has been shown in many cases where different communities and people have been exposed to the same threats, but the impacts have varied enormously. Poverty, conflict and lack of entitlements are some of the principle determinants.

In recent years, these two strands of research have been combined in a number of studies, in recognition that both aspects – namely, natural hazards and environmental changes and socio-economic factors – together determine human vulnerability to environmental change. This emerging, more comprehensive approach looks at multiple stresses from different domains and in this way comes closer to the concept of sustainable development, which requires integrating the economic, environmental and social dimensions within one framework. Integrated studies have, for example, looked at the vulnerability of communities in drylands in West Africa to climate change (Dietz et al., 2004) or the vulnerability of Indian agriculture to global change (TERI, 2003). An important element for human vulnerability studies is the spatial heterogeneity of people. Poor people tend to be clustered in specific places. Aggregated data mask much of this variation, which is extremely pertinent for analysing human vulnerability to environmental change. Biophysical maps and poverty maps could be combined to find out where the people are who are at risk from sea-level rise, extreme weather events or other environmental stresses (Henninger and Snel, 2002).[2]

Although there are differences of terminology, the different analytic frameworks distinguish between exposure, sensitivity and coping capacity/resilience, which are the main components of vulnerability. Exposure refers to the external stress (threat) to the system (community or individual), which can be caused by extreme events such as flooding, and now increasingly by changes in the magnitude and intensity of such events as a consequence of climate changes. It could also be caused by such socio-economic "events" as economic collapse or changes in the price of commodities. Sensitivity determines the extent to which each system is susceptible to exposure to external stress – for example, entitlement to land or resources, or proximity of an environmental threat, such as a flood-plain. Coping capacity/resilience determines the ability to deal with or recover from the impact of an external stress. It depends on factors such as education and insurance.

The three components that shape human vulnerability vary among communities and individuals, making human vulnerability to environmental change inherently different for each community or individual (Vogel and O'Brien, 2004). In addition, human vulnerability is:

- *Multi-dimensional.* Communities and people can be subject to different stresses at the same time. For instance, climate change and globalisation inflict multiple stresses on farmers who face changing weather patterns and a new economic reality (O'Brien and Leichenko, 2000).

- *Scale dependent.* Factors determining vulnerability operate over different time and space scales. They can be global and take place over a longer period (e.g. climate change or trade liberalisation) or at local or individual level (e.g. lack of entitlement to a natural resource or land) and take place during the short time scale of extreme events (e.g. earthquakes).
- *Dynamic.* Vulnerability is also a dynamic process. Stresses on the human–environment system are constantly subject to change in response to environmental change and socio-economic developments.

Few frameworks have incorporated all these different aspects. An example of a framework that tries to capture these aspects is the vulnerability framework recently developed by Turner et al. (2003). It assesses the human–environment system as a whole, describing its vulnerability as a combination of exposure, sensitivity and resilience. It also takes a multi-scale and multidimensional perspective making it an elaborate, though complex, framework to use (see Figure 6.1).

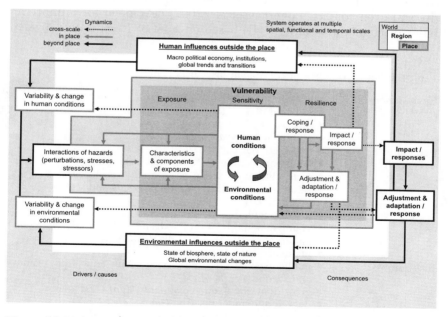

Figure 6.1 Vulnerability Framework.
Source: Turner et al., 2003.

Another approach to vulnerability comes from the perspective of resilience.[3] Although resilience is also used as a component of other vulnerability concepts, the resilience approach focuses particularly on this characteristic of a system. Resilience is defined here as the potential of a system to remain in a particular configuration and to maintain its feedbacks and functions; it involves the ability of the system to reorganise following disturbance-driven change (Walker et al., 2002). It determines the capacity to cope with the impact of a stressor, and depends on such factors as institutional capacity or financial resources. This approach is not focused on the desired future outcome, given that drivers are largely unpredictable, but on creating a system that is able to cope with this unpredictability in many different situations.

There is also a growing interest in human security as a part of vulnerability analysis. Human security is viewed as an umbrella concept that embraces overall economic development, social justice, environmental protection, democratisation, disarmament and respect for human rights. Research into this links the human dimensions of environmental change with a re-conceptualisation of security (UNEP/Woodrow Wilson International Center for Scholars, 2004). It builds on the assumption that environmental stress, often the result of global environmental change, coupled with increasingly vulnerable societies, may contribute to insecurity and even conflict. Lonergan et al. (2000) developed for instance an Index of Human Insecurity (IHI) based on a set of 16 indicators divided

Table 6.1 Selected indicators of human insecurity comprising the standard set

Environment	• net energy imports (% *of commercial energy use*)
	• soil degradation (*tonnes'yr.*)
	• safe water (% *of population with access*)
	• arable land (*hectares per person*)
Economy	• real GDP per capita (*US$*)
	• GNP per capita growth (*annual %*)
	• adult illiteracy rate (% *of population 15+*)
	• value of imports and exports of goods and services (% *of GDP*)
Society	• urban population growth (*annual %*)
	• young male population (% *aged 0–14 of total population*)
	• maternal mortality ratio (*per 100,000 live births*)
	• life expectancy (*yrs.*)
Institutions	• public expenditures on defence versus education, primary and secondary (% *of GDP*)
	• gross domestic fixed investment (% *of GDP*)
	• degree of democratisation (*on a scale of 1–7*)
	• human freedoms index (*on a scale of 0–40*)

Source: Lonergan et al., 2000.

over the four domains of sustainable development. See Table 6.1 for an overview of indicators included in the IHI.

The concept of human security is closely tied to the notion of poverty, in terms of a lack of basic needs. Poverty is an important component of human (in)security, since poor people are usually hit hardest by environmental change and are among the most vulnerable groups in society. Since poor people are more dependent on their natural resource base, they also have the greatest need for the ecosystem services this base provides.

The UNEP/GEO approach

In this part of the chapter we introduce the Global Environmental Outlook (GEO), the major global report of UNEP, as well as its regional and national State of the Environment reports. These are published to fulfil UNEP's mandate to keep under review the state of the environment and to strengthen the scientific basis of international environmental governance. The reports are ultimately intended to provide policy makers with an early-warning and monitoring capacity.

The GEO process got under way in 1995 with a mandate from the Governing Council of the United Nations Environment Programme (UNEP). GEO evolved into a consultative and participatory assessment process, involving a worldwide network of collaborating centres. The intention was to approach sustainable development from the environmental perspective. The first GEO report was published in 1997, the second in the year 2000 and the third in 2002.

Since 2002, UNEP has been mandated to produce a Global Environment Outlook every five years (with GEO-4 to be published in 2007), accompanied by annual GEO Yearbooks (starting in 2003) to report on trends and changes on an annual basis. The GEOs report on environmental trends, driving forces and policy responses across UNEP's seven regions: Asia and the Pacific, West Asia, Europe, Latin America and the Caribbean, North America, Africa and the Polar Regions. The global reports are accompanied by regional state of the environment reports for greater in-depth treatment of environmental issues and change at a regional level. The GEO reports employ the DPSIR framework for Integrated Environment Assessment, which links environmental pressures with trends and policy responses.

GEO-3 made a start towards analysing vulnerability, noting that this is shaped by a mix of social, ecological and economic forces: "Human vulnerability to environmental conditions has social, economic and ecological dimensions" (UNEP 2002a: 303). GEO-3 recognised that vulnerability has both spatial and temporal dimensions. The extent of vulnerability varies spatially. For instance, developing countries are more vulnerable to the impacts of climate change than developed ones. Likewise, locations such as high altitudes, flood plains, river banks, small islands and coastal areas may be more exposed to environmental risks than others. The temporal dimension of vulnerability is illustrated by the fact that in many countries coping capacity that was strong in the past has not kept pace with environmental change. GEO-3 identified some of the causes: the reduction or elimination of traditional methods, the emergence of new hazards for which no coping mechanism exists, lack of resources or absence of technology and skills.

At the same time, vulnerability varies across groups: men and women, poor and rich, rural and urban, and so on. Refugees, migrants, displaced groups, the very young and old, women and children are often the most vulnerable groups, subject to multiple stresses (UNEP 2002a). GEO-3 identified three critical areas as being closely related to vulnerability: human health, food security and economic losses. During the preparations for GEO-3, an attempt was made to capture the different aspects of vulnerability in one composite indicator, with the aim of comparing countries and regions. However, the work was not published in GEO-3 (see also Text Box 6.1). Given the complexities in assessing vulnerability, there are inherent problems in working with composite indicators of vulnerability.

GEO-3 also noted that "no standard framework exists for identifying all these factors" (UNEP, 2002a: 303). However, an important message was the need for "a significant policy response and action on several fronts" (UNEP, 2002a: 309). It identified two types of policy responses:

Box 6.1 A composite indicator for vulnerability

GEO-3 attempted to come up with a composite indicator for human vulnerability to environmental change. In this process a number of potential indicators were selected based on a number of categories (see Table 6.2). Although all indicators relevant to vulnerability are included, they have not been grouped on the basis of the vulnerability concept (exposure, sensitivity and coping capacity/resilience). In our opinion it remains to be seen if it is possible to combine all these indicators in a meaningful and useful way. In GEO-4, the analysis of vulnerability will be more location and context specific, focusing on different archetypes of vulnerability, compared with the more general approach that was explored for GEO-3 (UNEP, 2003).

Table 6.2 Environmental causes and indicators related to categories of human vulnerability

Human vulnerability	Environmental causes	Indicators
Health	– Urban air pollution – Water pollution/ sanitation – Toxic chemicals/food contaminants	– Number of people affected by environmental diseases (pollutants, chemicals), microbial infection, diarrhoea, chronic lung diseases – Number of people having access to safe drinking water and sanitation – Loss of DALY (Disability Adjusted Life Year)
Economic losses/ gains	– Environmental diseases – Soil erosion – Deforestation – Siltation	– Amount spent on treating environmental diseases – Amount spent on environmental clean up – Food productivity loss due to soil erosion, deforestation, etc. – Loss due to siltation of dams
Poverty	Depletion of natural resource base to meet the basic needs of food, fibre, firewood, income and employment	– Different income categories affected by natural resource degradation – Different income categories affected by air pollution and sea level rise – Different income categories affected by water contamination and lack of sanitation
Food security	– Loss of natural vegetation and biological diversity – Soil erosion – Surface and groundwater depletion – Rainfall distribution	– Percentage of natural vegetation cover – Percentage of people directly dependent upon land resources – Extent and distribution of degraded land – Freshwater availability – Rainfall variability
Loss of natural heritage and experience	Depletion of natural flora and fauna	– Areas designated as Protected Areas; natural recreation areas – Rate of deforestation – Rate of habitat loss

Table 6.2 (cont.)

Human vulnerability	Environmental causes	Indicators
Loss of IPR	Depletion of endemic species	– Number and distribution of endemic species – Number of Patent Rights
Conflicts	– Scarcity of water – Depletion of natural resource base	– Number of people living in water-scarce areas – Number of people dependent upon vegetation resources
Extreme events/ climate change impacts	– Flood, drought, fire, cyclone and other disasters – Global warming/Sea level rise	– Number of people living in disaster-prone areas – Number of people living within the 100 km of coast – Amount of greenhouse gases emission

Source: UNEP, 2003.

reducing the hazards through prevention and preparedness initiatives, and improving the coping capacity of vulnerable groups to enable them to deal with them. A case was also made for assessing and measuring vulnerability and developing systems of early warning.

In addition to the Global Environment Outlook, the issue of human vulnerability to environmental change also featured in many of the regional GEOs. The amount of attention given to this topic varies for each report, but when comparing the most recent reports with earlier publication it is clear that human vulnerability is receiving increased attention. In the reports on small island development states, human vulnerability to environmental change is an important topic (UNEP, 2005a/b/c). It is considered in the face of a growing threat of natural disasters attributed to climate change. The issue of human vulnerability was also taken up in the first *African Environment Outlook* (UNEP, 2002b), which specifically looked at human vulnerability to environmental change. Poverty and the direct dependence of people in Africa on their natural resource base were major themes of the report. The detailed case studies that provided the basis for the analyses can be found in the third *African Environment Outlook* (UNEP, 2004). Another regional report elaborating on the issue of human vulnerability was *North America's Environment Outlook* (UNEP 2002c). Here health and human settlement were dominant themes.

Vulnerability analysis in the next GEO report

In 2004 preparations started for GEO-4, to be published in 2007. The overall theme of the report will be "Environment for Development". The vulnerability approach will be used in a chapter on "Challenges and Opportunities". This chapter takes a sustainable development perspective and uses the vulnerability approach to show the combined implications of environmental and non-environmental changes for human well-being. It seeks to identify policy options within the environmental policy domain, but also in non-environmental policy domains such as trade and poverty alleviation (referred to as cross-cutting issues). In this way, it seeks to inform policy for improving coping capacity and making development more sustainable.

A set of six cross-cutting issues (non-environmental policy domains) that are important from a sustainable development perspective is used as the entry point for the policy analysis. These are: 1) health; 2) poverty, equity and livelihoods; 3) institutions and governance; 4) science and technology; 5) trade; and 6) conflict and cooperation. These will be explored by analysing a set of archetypes and patterns of vulnerability of human–environment systems. The chapter adopts the human–environment system as the unit of analysis, bringing together the notions of human vulnerability and environmental change.

The chapter on "Challenges and Opportunities" addresses the following key questions:

- Within the context of overall goals and strategies for sustainable development, how do the environmental state, variability, hazards and trends described in the chapters on the state of the environment affect human well-being?
- What factors shape the vulnerability of human–environment systems to multiple and interacting stresses?
- What are the response options?
  In addition, it addresses two sub-questions:
- What progress has been made on major initiatives in recent years to address human well-being and reduce vulnerability?
- What opportunities are provided by the six entry points for the analysis of policy (mentioned above) for reducing vulnerability of human–environment systems and improving human well-being?

Much of the vulnerability research is rather qualitative and of a case-study nature; upscaling is thus a challenge. The idea for GEO-4 is to achieve a satisfactory level of generalisation and upscaling through examination of archetypes and patterns of vulnerability. This idea is inspired by the "syndrome approach", which looks at non-sustainable patterns of

Box 6.2 Definition of archetypes of vulnerability

Archetypes of vulnerability are used to illustrate the various ways in which human well-being can be affected by environmental and non-environmental change. They offer a framework to assess the context specificity of environmental change, multiple stresses and human vulnerability. Archetypes of vulnerability are defined as specific, representative patterns of the interactions between environmental change and human well-being. Looking at the diversity of human–environment systems (as the major units of analysis for assessing vulnerability) throughout the world, it is evident that some situations share certain vulnerability-creating conditions. In this sense, the archetypes presented here are simplifications of real cases, which should help to demonstrate the basic processes whereby vulnerability is produced within a context of multiple stressors.

There is no unique or objective way to formulate a set of archetypes. Instead, the set will be developed as a whole, representing the most important and insightful processes and contexts, including the most vulnerable population groups throughout the world, such as indigenous people and the urban or rural poor, or economic sectors heavily dependent on environmental services. Also, the archetypes will have to reflect vulnerabilities across the full range of geographic and economic contexts that require attention in the GEO context: developing countries, industrialised countries and countries in transition. Finally, the set of archetypes should allow detailed and elaborate analysis of the way in which issues such as poverty, human health, institutions and governance, science and technology, trade and globalisation, and conflict and cooperation influence or interact with human–environment systems. The fact that these issues play out differently in different contexts is one of the major motivations for choosing the archetype approach.

An example of a pattern of vulnerability is the "drylands archetype" in developing countries. This relates to subsistence agriculture on marginal lands in developing countries, including dry forests. The marginality induced, for example, in steep areas under production, or areas with water stress or unsuitable soil conditions, causes a low agricultural potential that implies a high risk of overuse of the natural resources and subsequently declining yields. If the population pressure increases or agricultural prices fall, the subsistence farmers are forced to increase their production either by intensifying the agricultural production or by extending the area under production – involving yet more marginal areas. These forces cause further environmental degradation, including soil erosion, increased water stress or desertification,

Box 6.2 (cont.)

which again provokes declining yields. In this context, the loss of traditional knowledge and the lack of technology inputs are important catalysts of environmental degradation. The declining yields ultimately threaten the food supply and income of the farmers who have only limited access to alternative sources of income and, as a consequence of poor access to international markets, are not in a position to produce for export. The consequences for human well-being, including extreme poverty and a deterioration of health due to malnutrition, are severe. Under these circumstances, any increasing variability in precipitation – one result of climate change – is a particularly challenge to subsistence farmers. Other major stressors that put human well-being under pressure are population growth, the spread of diseases like HIV/AIDS, which reduce the capacity to cope with changes, and increasing soil degradation in the larger areas. In many instances migration remains the only possible way to escape the decline in human well-being. The archetypical situation where this situation occurs is the Sahel (see Dietz et al, 2003), but it can also be recognised in arid zones of South America, Africa and Asia.

Other archetypes could be developed for water stress and its technological solutions, heavily urbanising coastal areas, the resource-rich countries, (post-) conflict-induced vulnerability, rapid economic growth countries, and management of common pool resources.

human–environment interaction and analyses the dynamics behind them (Lüdeke et al., 2004; Wonink, Kok, Hilderink, 2005). The approach is illustrated in Text Box 6.2.

For the archetypes of vulnerability, the link with human well-being will be made by using a set of human well-being indicators, from which a selection can be made depending on the context. We thus take a pragmatic approach, building for example on the outcomes of the Millennium Ecosystem Assessment (MA). The following measures of well-being might be considered: material needs (such as access to resources, income), human health (such as nutrition, environment related diseases), security (such as personal security and disaster preparedness) and freedom of choice.[4] These aspects of human well-being are critical to increase and enhance the capacity to adapt and manage environmental change. In the MA an attempt has been made to link ecosystem services to human well-being (MA, 2005). Changes in ecosystem services influence different components of well-being. See Figure 6.2 for a detailed description of the

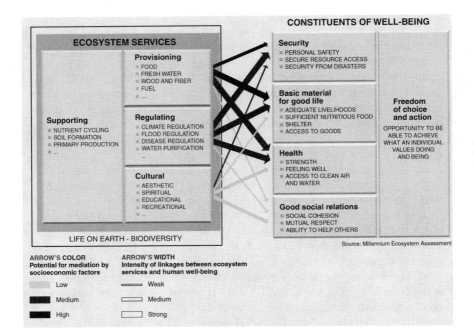

Figure 6.2 Linkages between ecosystem services and human well-being, the MA framework.

Source: MA, 2005: vi.

Note: This figure depicts the strength of linkages between categories of ecosystem services and components of human well-being that are commonly encountered, and includes indications of the extent to which it is possible for socio-economic factors to mediate the linkage. (For example, if it is possible to purchase a substitute for a degraded ecosystem service, then there is a high potential for mediation.) The strength of the linkages and the potential for mediation differ in different ecosystems and regions. In addition to the influence of ecosystem services on human well-being depicted here, other factors – including other environmental factors as well as economic, social, technological and cultural factors – influence human well-being, and ecosystems are in turn affected by changes in human well-being.

conceptual framework developed within the MA. The relation between environmental degradation and human well-being is, however, difficult to analyse, since it is mostly not a linear relation.

The MA framework shows that an important dimension of well-being is health; here the linkages with environmental change are strong. The WHO has made an assessment of risk factors (exposures) which contribute to diseases, expressed in so-called DALYs: disability-adjusted life years, a measure for the burden of disease (see Figure 6.2). While health directly influences the vulnerability of exposed people to environmental change, socio-economic status (income, education), the age structure of the population and the organisation of the health system (Hilderink and Lucas, 2006) are important factors in determining the sensitivity and coping capacity of humans from a health perspective.

The vulnerability analysis will provide the basis to identify and analyse challenges and opportunities to enhance human well-being and improve the environment through other policy domains. In this way the chapter will contribute to realising the overall aim of GEO-4 of showing the importance of environment for other policy domains. These policy domains will include the six cross-cutting issues mentioned before, and the challenge will be to identify options that could be taken in these domains that are beneficial from both a development/human well-being and environmental perspective. In environmental policy this process of internalising or integrating environment in other policy domains is often referred to as "mainstreaming".

Concluding remarks

In this chapter we have reviewed some of the research that has recently been undertaken on the assessment of human vulnerability to environmental change in combination with non-environmental changes. The chapter has examined different approaches to the analysis of vulnerability and how the vulnerability analysis framework will be used in GEO-4 to examine vulnerability to environmental change.

A potential weakness of the approach outlined in this chapter (which is being adopted in GEO-4) relates to the inclusion of non-environmental stresses and finding the appropriate responses inside and outside the environmental domain. Six issues that will be focused on have been mentioned: institutions and governance, health, science and technology, poverty, equity and livelihoods, conflicts and cooperation, and trade. It could be argued that these issues are not exhaustive. However, the intention is not to provide an exhaustive list but instead to provide an indication of non-environmental issues that have a bearing on human well-being and to build a framework for analysis of vulnerability. This framework is built on the premise that the vulnerability of human–environment systems is very context and place specific. The same could be said about the archetypes of vulnerability. Here, too, the intention is not to be exhaustive, but to provide an analysis of patterns of vulnerability.

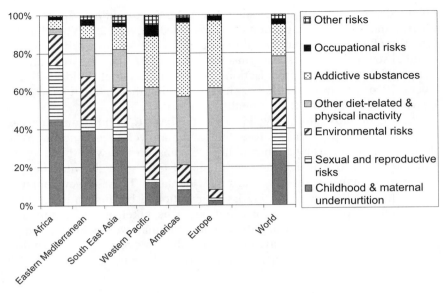

Figure 6.3 Risk factors for attributable DALYs for selected regions.
Source: MNP, 2005.

Future research should focus on the layering of responses to environmental change, seeing responses as being nested in a hierarchy. There is a case for process-based research that captures the varying domains and scales of social interaction in which actors respond to environmental change. A methodological implication of this is that research should focus on how adaptation and mitigation take place in different social and geographical contexts, and how these responses are shaped by the range of response options available and interactions between different scales.

However, one should be cautious about drawing generalisations from such studies. Such research will necessarily have to be very local and context specific and should not be used to develop blueprint policy solutions. Instead, research should focus on what kind of interventions are needed in very local and specific contexts (corresponding to the notion of archetypes) and how such interventions can be best mainstreamed into development planning and policy-making.

A further message of the chapter is that the susceptibility of the environment needs to be seen as an integral part of human vulnerability. However, the form that this vulnerability takes varies from one setting to another and is shaped by a combination of various natural, social and technological factors. These factors also shape responses to environmental change and there is a case for further research at a micro level to capture these interactions. In particular, there is value in understanding how different groups are vulnerable to environmental change and how they differ in their ability to adapt to it. Further interventions at improving the adaptive capacity of populations should start from an appreciation and understanding of existing practices and realities.

A related, though somewhat indirect, message of this chapter is the need for more qualitative, process-based approaches to the analysis of policy. As mentioned earlier, frameworks such as the DPSIR, while useful in integrating driving pressures, state-impact and policy responses, assume a somewhat linear approach to the analyses of phenomena that are seemingly more complex and circular. There is a case for a more qualitative, process-based approach that provides greater scope for the analysis of processes of policy implementation, to better understand the processes through which policies aimed at reducing vulnerability translate into practice.

Acknowledgements

The authors wish to express their thanks to the participants in two preparatory meetings for the chapter "Challenges and Opportunities" in GEO-4, held in Nicoya, Costa Rica, in January 2005 and Scheveningen, the Netherlands, in March 2005. We especially want to acknowledge the contributions of Dhari Al-Ajami, Richard Filcak, Des Gasper, Henk Hilderink, Sylvia Karlsson, Jennifer Mohamed-Katerere, Marybeth Long Martello, Vikrom Mathur, Ana Rosa Moreno, Gerhard Petschel-Held, Indra de Soysa and Frank Thomalla.

Notes

1. See http://www.unep.org/geo/.
2. See http://www.povertymap.net/.
3. See for instance the Resilience Alliance at http://www.resalliance.org/ev_en.php/.
4. Based on a presentation by Des Gasper at the GEO meeting held at Scheveningen, the Netherlands, in March 2005.

REFERENCES

Adger, W.N. and P.M. Kelly (1999) "Social Vulnerability to Climate Change and the Architecture of Entitlement", *Mitigation and Adaptation Strategies for Global Change* 4: 253–266.

Cutter, S.L. (1996) "Vulnerability to Environmental Hazards", *Progress in Human Geography* 20(4): 529–539.

Dietz, A.J., R. Ruben and A. Verhagen, eds (2004) *The Impact of Climate Change on Drylands, with a Focus on West Africa*, Dordrecht: Kluwer Academic.

Henninger, N. and M. Snel (2002) *Where Are the Poor: Experiences with the Development and Use of Poverty Maps*, Washington D.C.: World Resource Institute.

Hilderink, H.B.M. and P. Lucas (2006) *Global Environmental Change and Human Health: A Framework for Vulnerability Assessments*. Conference Proceedings, 12th Annual International Sustainable Development Research Conference, 6–8 April 2006, Hong Kong: Inderscience Publishers.

Intergovernmental Panel on Climate Change (IPCC) (2001) *Climate Change 2001: Impacts, Adaptation, and Vulnerability*, Contribution of Working Group II to the 3rd Assessment Report of the IPCC, Cambridge: Cambridge University Press.

Lonergan, S., K. Gustavson and B. Carter (2000) "The Index of Human Insecurity", *AVISO* 6: 1–11.

Lüdeke, M.K.B., G. Petschel-Held and H. J. Schellnhuber (2004) Syndromes of Global Change: The First Panoramic View, *GAIA: Ecological Perspectives for Science and Society* 13(1): 42–49.

Millennium Ecosystem Assessment (MA) (2005) *Ecosystems and Human Wellbeing: Synthesis*, Washington D.C.: Island Press, available at http://www.millenniumassessment.org//en/Products.aspx?

MNP (2005) *Outstanding environmental issues from human development*. A review of the role of the environment in achieving the Millenium Development Goals, Netherlands Environmental Assessment Agency, Bilthoven, The Netherlands.

O'Brien, K. and R. Leichenko (2000) "Double Exposure: Assessing the Impacts of Climate Change within the Context of Economic Globalization", *Global Environmental Change* 10: 221–232.

Steffen, W., B.L. Turner II, R.J. Wasson, A. Sanderson, P. Tyson, J. Jäger, P. Matson, B. Moore III, F. Oldfield, K. Richardson, and H.J. Schellnhuber (2004) *Global Change and the Earth System: A Planet Under Pressure*, Global Change – The IGBP Series, Berlin: Springer.

TERI (The Energy Research Institute)(2003) *Coping with Global Change: Vulnerability and Adaptation in Indian Agriculture*, New Delhi: TERI.

Turner, B.L., R.E. Kasperson, P.A. Matson, J.J. McCarthy, R.W. Corell, L. Christensen, N. Eckley, J.X. Kasperson, A. Luers, M.L. Martello, C. Polsky, A. Pulsipher and A. Schiller (2003) "A Framework for Vulnerability Analysis in Sustainability Science", *PNAS*, 100(14): 8074–8079.

United Nations Environment Programme (UNEP) (1997) *Global Environment Outlook*, New York/Oxford: Oxford University Press.

UNEP (1999) *Global Environment Outlook 2000*, London/New York: Earthscan.

UNEP (2002a) *Global Environmental Outlook 3: Past, Present and Future Perspectives*, London: Earthscan.

UNEP (2002b) *African Environmental Outlook: Past, Present and Future Perspectives*, Nairobi: DEWA/UNEP.

UNEP (2002c) *North America's Environment: A Thirty-Year State of the Environment and Policy Retrospective*, Nairobi: DEWA/UNEP.

UNEP (2003) *Assessing Human Vulnerability to Environmental Change; Concepts, Issues, methods and Case Studies*, Nairobi: DEWA/UNEP.

UNEP (2004) *African Environment Outlook. Case studies: Human Vulnerability to Environmental Change*, Nairobi: DEWA/UNEP.

UNEP (2005) *GEO Year Book 2004/2005: An Overview of our Changing Environment*, Nairobi: DEWA/UNEP.

UNEP (2005a) *Caribbean Environment Outlook*, Nairobi: DEWA/UNEP.

UNEP (2005b) *Pacific Environment Outlook*, Nairobi: DEWA/UNEP.

UNEP (2005c) *Atlantic and Indian Ocean Environment Outlook*, Nairobi: DEWA/UNEP.

UNEP and Woodrow Wilson Center for Scholars (2004) *Understanding Environment, Conflict, and Cooperation*, Nairobi: DEWA/UNEP.

Vogel, C. and K. O'Brien (2004) "Vulnerability and Global Environmental Change: Rhetoric and Reality", *AVISO* 13: 1–7.

Walker, B., S. Carpenter, J. Anderies, N. Abel, G. Cumming, M. Janssen, L. Lebel, J. Norberg, G.D. Peterson, and R. Pritchard (2002) "Resilience Management in Social-Ecological Systems: a Working Hypothesis for a Participatory Approach", *Conservation Ecology* 6(1): 14, available at http://www.ecologyandsociety.org/vol6/iss1/art14/.

Watts, M.J. and H.G. Bohle (1993) "The Space of Vulnerability: The Causal Structure of Hunger and Famine", *Progress in Human Geography* 17(1): 43–67.

Wonink, S.J., M.T.J. Kok and H.B.M. Hilderink (2005) *Vulnerability and Human-Well-Being*, Workshop Report, No. 50001903, MNP/RIVM, Bilthoven.

World Bank (2005) *Natural Disaster Hotspots: A Global Risk Analysis*, Washington D.C.: The World Bank, Hazard Management Unit.

World Commission on Environment and Development (WCED) (1987) *Our Common Future: Report of the World Commission on Environment and Development*, Oxford: Oxford University Press.

World Health Organization (WHO) (2002) *The World Health Report 2002: Reducing Risks, Promoting Healthy Life*, Geneva: World Health Organization.

Part III

Global, national and sub-national index approaches

7

Review of global risk index projects: conclusions for sub-national and local approaches

Mark Pelling

Abstract

The Disaster Risk Index, Hotspots and the Americas Indexing Programme are international initiatives for measuring disaster risk and risk management performance. This chapter reviews these initiatives and their implications for sub-national and local approaches.

Those involved in sub-national and local initiatives can learn from the methodological innovations and the gaps in knowledge and data identified by various international initiatives. These programmes have also raised a number of important issues for consideration: the limitations of using mortality as an indicator of human loss; the value of measuring economic loss in absolute terms and also as a proportion of economic capacity; difficulties in identifying the impacts of drought, which are often associated with complex emergencies; and the importance of producing outputs that are meaningful for development actors, if measurements are to contribute to development planning. In return, of course, the contributions made by local and sub-national efforts to measure vulnerability and coping capacity will help improve the collection and quality of input variables for international tools.

Finally, we propose that our work should investigate the potential for the aggregation and scaling-up of local vulnerability and capacity assessments. This will benefit local and regional risk planning and might also prove a resource for complementing or feeding into international indexes of vulnerability and risk.

Introduction

Between the spring of 2004 and the summer of 2005 three international indexes of disaster risk and its management were published: the Disaster Risk Index (DRI), Hotspots and the Americas Indexing Programme (see also Peduzzi, Chapter 8; Dilley, Chapter 9; Cardona, Chapter 10). The development of these initiatives offers a learning opportunity for planned sub-national and local approaches. But this is potentially a reciprocal relationship. Future work at the sub-national and local levels can enhance the information base for international indexes.

This chapter offers a review of the three international indexing initiatives, presenting a summary of the conceptual orientation, methods and results of each initiative in turn. We then discuss the challenges and opportunities for developing sub-national and local measurements of disaster risk on the basis of the approaches developed in the three international indexing programmes.

Background, structure and methodology of the three approaches

The indexing initiatives reviewed have two distinct methodological orientations. The DRI and Hotspots are deductive. Their measurements of vulnerability and risk are hazard specific and tied to disaster impact data. This adds realism to the analysis but means measurements cannot be undertaken where input data is lacking. The Americas Indexing Programme is inductive. Its measurement of vulnerability and capacity is built on and constrained by available socio-economic and performance variables. This means measurements can be undertaken even where disaster loss data is hard to come by – for example in places exposed to low-frequency high-impact hazards – but means that results are shaped by the choice and quality of input variables and not grounded in recorded loss data.

The following review focuses on those aspects of methodology and results that are most relevant to the measurement of vulnerability and capacity. The DRI, Hotspots and Americas programme are presented in turn with examples presented from flood hazard related assessment unless otherwise stated.

The Disaster Risk Index

The Disaster Risk Index (DRI) of the United Nations Development Programme (UNDP), in partnership with UNEP-GRID, aims to demonstrate the ways in which development influences disaster risk and vulnerability. While expert judgement can be used to identify such linkages, the DRI represents the first effort to produce a statistical methodology. The DRI has global coverage and a national scale of resolution. Some 22 tributarian States are also included.[1] The DRI is applied in full to earthquakes, tropical cyclones and flooding. Preliminary analysis was also undertaken for volcanoes, landslides and drought. The starting point for the DRI is to obtain or produce hazard maps for earthquakes, cyclones and flooding (and also drought), which are then overlain by population maps in a GIS system to identify national human exposure to each hazard type.

The DRI produces two measures of human vulnerability. The first, relative vulnerability, is calculated by dividing the number of people killed by the number of people exposed to a particular hazard type. Higher relative mortality equates to higher relative vulnerability. The simplicity of the model means that no country is excluded for showing outlier characteristics.

Relative vulnerability is highest in the top left-hand corner of Figure 7.1(c). The high relative vulnerability displayed by Venezuela is a result of the large number of deaths associated with catastrophic flooding in 1999; in this case landslides were an immediate cause of many of the deaths.

The second measure of vulnerability aims to identify those socio-economic variables that best explain recorded mortality for individual hazard types. A stepwise multiple regression is used with disaster mortality from the Emergency Disasters Data Base (EM-DAT) as the dependent variable. Independent variables include physical exposure and a list of 24 socio-economic variables selected by an expert group to represent: economic status, type of economic activities, environmental quality, demography, health and sanitation, education and human development. Those independent variables that best explain the variation in the dependent variable are chosen to describe the global characteristics of vulnerability for each hazard type. The time period of mortality data availability (21 years for flooding and cyclones) is extended for earthquakes (36 years) to compensate for the low frequency of this hazard type, thus allowing a longer time period for the registering of mortality within EM-DAT. Volcanic hazard requires a longer time span, for which reliable loss data is not available, leading to the dropping of volcanic hazard from the DRI index. The DRI analysis identified the following variables for flood risk in addition to physical exposure:
- low GDP per capita
- low density of population.

In other words, according to the DRI, the risk of dying in a flood is greatest in countries with high physical exposure to flooding, small national economies and low densities of population.[2] This may reflect the greater difficulty of preparing for floods in low-density rural societies where large-scale public works such as river and sea defences, which require collective labour or large financial investments are not easily delivered, and the difficulty in providing adequate emergency assistance and recovery support for low-density and widespread rural populations, such as those hit by flooding in Mozambique 2000.[3]

A DRI multi-hazard index combines values for hazard-specific socioeconomic variables. Hazard-specific models based on identified global vulnerability variables are run at the national level. For each hazard this allows the calculation of expected mortality for each country and territory based on the values of the globally selected vulnerability variables. The multiple-hazard risk index for each country is made by adding modelled deaths from individual hazard types. Some 39 countries have been excluded from the model. Countries marginally affected by a hazard, countries known to be exposed but with no loss data, and countries where the distribution of risk could not be explained by the model (for example, for drought in Sudan, where food insecurity and famine are more an outcome of armed conflict than of meteorological drought as defined in the model) are excluded. A final stage in the modelling process is to run a Boolean process to allocate one of five statistically defined categories of multi-hazard risk to each country. This is preferable to giving each country a raw numerical multi-hazard risk value. In order to examine the fit between modelled mortality and mortality recorded in EM-DAT, data from both sources are categorised into five country-risk classes and a cluster analysis performed to assess the closeness of fit.

Hotspots

The Hotspots project was implemented by Columbia University and the World Bank, under the umbrella of the ProVention Consortium. It aims to identify those places where risks of disaster-related mortality and economic losses are highest, on the basis of the exposure of people and GDP to major hazards, and on historical loss rates. Hotspots operates at the global level with a sub-national scale of resolution.[4] For Hotspots, which uses GIS grid cells as a unit of analysis, one challenge is where to draw the line and whether to include lightly populated or economically unproductive areas in the analysis. A decision was made to exclude grid cells with less than five people per km^2 and with no significant agricultural production. This reduces the number of grid cells in the global analysis from 8.7 million to 4.1 million, significantly reducing processing time and preventing these low-risk cells from biasing results.[5] Earthquakes, volcanoes, landslides, floods, drought and cyclones are included in the analysis. Hazard severity is indicated by event frequency or probability. Exposure for each grid cell faced with hazard is calculated on the basis of the population and economic assets of that cell. It is assumed that all people and economic assets within the individual grid cell are equally exposed to hazard.

Two sets of vulnerability coefficients have been calculated; one based on historical disaster mortality rates per hazard event, the other on historical rates of economic losses. Both vulnerability measures follow the same logic: 28 mortality and economic loss coefficients are calculated for each hazard. For both mortality and economic losses there is one loss rate for each of seven regions,[6] and four country wealth classes (high, upper-middle, lower-middle and low), defined according to standard classifications of the World Bank. For each hazard, historical mortality or economic losses per event for all countries in each region/wealth class are aggregated to obtain a loss rate for the hazard for the region/wealth class.

These rates, or weights, are aggregated for each of the 28 regions/wealth classes rather than calculated for each country individually because there is an insufficient number of hazard/loss events and, therefore, loss data, to calculate them for most individual countries. In an earthquake-prone country, for example, unless an earthquake occurred during the period covered by EM-DAT, the loss rate would be zero. Furthermore, only approximately 30 per cent of the events recorded in EM-DAT include data on economic losses. Calculating the loss rates across groups of similar countries creates a larger pool of events across which to calculate them. Nonetheless, the historical loss data used to calculate the rates is thin for some hazard region/wealth class combinations. A vigorous effort to improve the global database on disaster losses is currently underway to address this deficiency in future analyses.

Once calculated, and in order to obtain risk, these loss rates, or vulnerability coefficients, are used to weight hazard exposure of population or GDP for each grid cell. The weight from the corresponding region/wealth class in which the grid cell is located is used for each grid cell.

The Hotspots results are presented as relative risk values. The risk values for each of the 4.1 million grid cells are sorted into 10 equally sized deciles for each hazard, and for all hazards combined. The top 30 per cent of the values are considered relatively high risk, the middle 30 per cent are considered as relatively medium risk and the lowest 40 per cent as relatively low risk.

Hotspots produced relative risk maps for mortality, economic loss and economic loss as a proportion of GDP. In Figures 7.2(c), 7.3(c), and 7.4(c), relative flood risk is shown as high (red), medium (yellow) or low (blue). Broadly speaking, South and South East Asia register high risks of both mortality and economic loss from flooding. In addition, Central and South America and sub-Saharan Africa show high mortality risk from flooding. Europe, North America and the Caucuses show high risk from flooding measured through absolute economic loss.

A multi-hazard Hotspots index aggregates single-hazard Hotspot values. A challenge for Hotspots is the lack of commensurability between measures of hazardousness for different hazard types. For example, frequency is used to measure severity for droughts and probability values for landslides. Aggregating these measures of severity would simply inflate the relative hazard values of those hazard types measured on a larger scale (e.g. on a frequency of 0 to infinity compared to a probability of 0 to 1). To allow aggregation, a uniform adjustment is made to all values within a given region/wealth class so that the total mortality or economic loss for the class equals the mortality or economic loss recorded in EM-DAT for that hazard type.

The Hotspots multi-hazard risk results for the highest-risk areas are presented below. Risk of mortality is presented in Figure 7.5(c), risk of economic loss in Figure Figure 7.6(c) and risk of economic loss as a proportion of GDP in Figure 7.7(c).

The multi-hazard mortality-risk assessment (Figure 7.5(c)) was influenced strongly by high-risk individual hazard hotspots, for example those associated with drought mortality in sub-Saharan Africa, and flood and cyclone-associated mortality in Central America, the Caribbean, the Bay of Bengal, China and the Philippines. The Himalayas, sub-Saharan Africa and Central America show risk from two hazard sources. A comparison of multi-hazard mortality risk with that for total economic loss (Figure 7.6(c)) produces a familiar picture of risk shifting from low-income sub-Saharan Africa to the high-income States of Europe and North America. When risks of economic losses are calculated as a proportion of GDP (Figure 7.7(c)) compared to absolute GDP loss, multi-hazard risk remains high for the Middle East, is increased for eastern Africa, including Madagascar, and reduced for the Mediterranean States, North America, Europe and the Himalayas.

The Americas programme

The Americas Indexing Programme of the Instituto de Estudios Ambientales, Universidad Nacional de Colombia – Sede Manizales, in partnership with the InterAmerican Development Bank, aims to aid national decision makers in assessing disaster risk and undertaking risk management. The system of indicators presents a benchmarking of each country in different periods from 1980 to 2000 and the basis for consistent cross-national comparisons. Four independent indexes have been developed; each represents disaster risk or disaster risk management in different ways and is targeted at specific audiences. Each index has a number of variables that are associated with it and empirically measured:

- The *disaster deficit index* (DDI) measures a country's financial exposure to disaster loss, and the financial resources available for recovery.
- The *local disaster index* (LDI) represents the proneness of a country to locally significant disaster events, and their cumulative impact. Spatial variability and sub-national dispersion of disaster risk is also indicated.
- The *prevalent vulnerability index* (PVI) represents prevailing conditions of national level human vulnerability.
- The *risk management index* (RMI) measures a county's performance in disaster risk management.

The suite of indexes was applied to 12 countries in Latin America and the Caribbean (Argentina, Chile, Colombia, Costa Rica, Dominican Republic, El Salvador, Ecuador, Guatemala, Jamaica, Mexico, Peru, and Trinidad and Tobago). The sub-indexes have national scales of resolution.

The DDI is a function of the expected losses suffered by the State and the capacity of the State to generate reconstruction funds from private, Government and international sources when hit by a maximum considered disaster event (MCE). MCEs with return periods of 50, 100 and 500 years related to rapid-onset hazards are considered. Vulnerability is formally included as part of the derivation of the DDI. It is used to represent the proportion of an asset that is calculated as likely to be lost in an event of a given intensity (the MCE). A DDI value greater than 1.0 indicates a lack of financial capacity to cover the costs of the disaster's impact. In a parallel presentation of this index, MCE losses are also expressed as a proportion of annual national current account budgets.

The DDI index has two elements. Figure 7.8 shows a ranked presentation of national financial capacity to cope with an MCE. Figure 7.9 presents calculated absolute economic losses. Both are for an MCE with a 50-year return period (an 18 per cent probability of occurring in any ten years). Peru and the Dominican Republic are shown not to be able to cope with such an event, with El Salvador a very marginal case. Absolute economic losses are greatest for Mexico.

With an MCE of a 100-year return period (5 per cent probability of occurring in any 10 years), seven countries were unable to cope. At a 500-year return period (2 per cent probability of occurring in any 10 years) only Costa Rica could cope.

A complementary assessment, called the "DDI prime", was developed to indicate MCE losses as a proportion of current annual investment. In

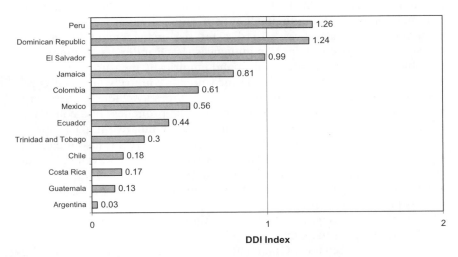

Figure 7.8 National financial exposure to catastrophic disaster.
Source: Cardona, 2005.

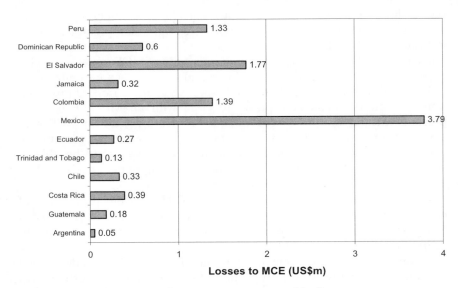

Figure 7.9 Absolute economic exposure to catastrophic disaster.
Source: Cardona, 2005.

El Salvador, for example, future disaster losses are the equivalent of 32 per cent of the annual capital budget; in Chile the figure is 12.5 per cent, with only four countries below 5 per cent.

The LDI includes four hazard types (landslides and debris flows, seismo-tectonic disturbances, floods and storms, and other events[7]), based on the categorisation of hazard used in the data source for this index: the DesInventar database, managed by La Red.[8] Values of local disaster magnitude and geographical distribution are calculated from three sub-indexes: mortality, people affected and physical loss (housing and crops) applied to sub-national regions or municipalities. Local data is combined to build the national LDI. A high LDI indicates high regularity in the magnitude and geographical distribution of disaster events recognised in the local reports and media across the country.

Figure 7.10 presents recorded mortality, people affected and economic loss associated with disaster events recorded in local and national media and reports, from 1996 to 2000. Colombia and Ecuador show a high incidence of deaths, with Guatemala and the Dominican Republic showing high numbers of people affected. Within the LDI an additional measure of the geographical concentration of disaster losses was calculated. This shows that losses were most evenly distributed within El Salvador. On the other hand, Ecuador, Chile, Colombia and Peru had the most geographically uneven distribution of losses.

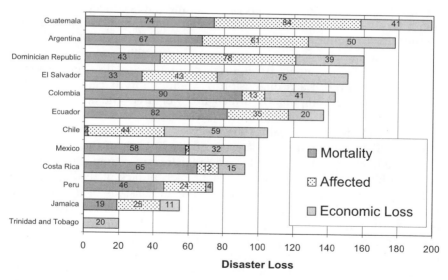

Figure 7.10 Loss from locally and nationally recognised disasters, 1996–2000.
Source: Cardona et al., 2004 (in UNDP, 2004).

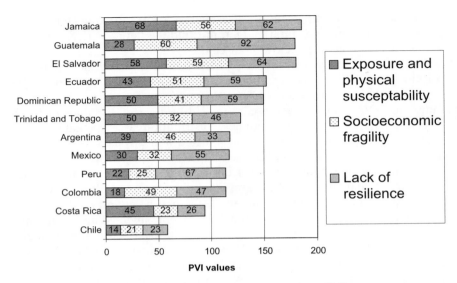

Figure 7.11 Socioeconomic vulnerability in the Americas, 2000.
Source: Cardona et al., 2004 (in UNDP, 2004).

The PVI is a composite index of inherent vulnerability at national level. It is derived from the aggregation of measures collected at the national level for three dimensions of human vulnerability: exposure and physical susceptibility, socio-economic fragility and lack of resilience. The PVI measures inherent (or intrinsic) vulnerability – no specific hazard type or scale of impact is required, neither is any disaster response capacity considered. Each dimension of vulnerability is calculated from eight quantitative components, which are weighted and aggregated to provide a final index value.

Figure 7.11 shows PVI values for the year 2000. Jamaica is shown to have the highest vulnerability, scoring highly in each of the three measures. Guatemala and El Salvador also register high composite vulnerability, with Guatemala showing very high levels of lack of resilience.

The RMI is also a composite index. Four dimensions of disaster risk management are included in its calculation: risk identification, risk reduction, disaster management and governance, and financial protection. Each dimension has six qualitative components, to be valued at the national level by expert judgement. The components are weighted and aggregated to arrive at the final index value. A sensitivity analysis is used to test for the influence on the results of the chosen weightings.

Figure 7.12 shows RMI values for the year 2000. Chile and Costa Rica

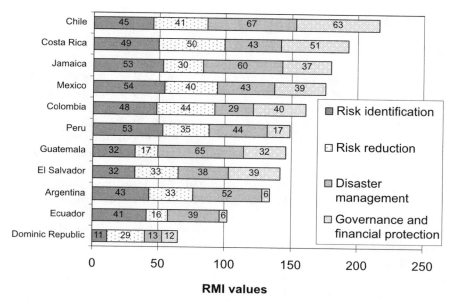

Figure 7.12 Disaster risk management performance in the Americas, 2000.
Source: Cardona et al., 2004 (in UNDP, 2004).

performed relatively well on all indicators. Chile returns particularly high
scores for disaster management and governance and financial protection.
Other countries returned a less even performance: Argentina and Ecuador, in particular, have weak scores for governance and financial protection, and the Dominican Republic for risk identification.

Lessons and open questions

The following discussion considers lessons to be learned and open questions from the international indexing initiatives for developing sub-
national and local measurements of vulnerability and coping capacity. In
DRI and Hotspots, vulnerability was calculated in relation to specific
hazard types before aggregating to multi-hazard analysis. Consequently,
challenges in representing hazard led to difficulties in measuring vulnerability. For the Americas programme vulnerability was measured as
an intrinsic status. Below, we look in turn for lessons and opportunities
in measuring hazards and vulnerability, and in aggregating for multiple-
hazard analysis.

Measuring hazard

Sub-national data is fed into each of the indexes. Hotspots and DRI mapped local data on hazards and people exposed to hazards into GIS systems and then aggregated to GIS cell and national levels respectively.[9] This data could be available to be verified by or used as input data for sub-national assessments.

Of the hazards for which analysis was attempted, flood and drought hazards proved the most difficult to map. For floods this was due to the lack of a global database. DRI was forced to overestimate; it used EM-DAT to identify floods and considered all those people living in floodplains as exposed. Hotspots was forced to underestimate; it used satellite imagery to identify flood events. However, the speed of local flash floods makes it likely that many of these may have been missed.

It is assumed that mapped hazard events will spatially overlap with sites of recorded losses. As the resolution of assessments increases, this assumption becomes harder to support. This is especially the case with slow-onset and long-duration events such as drought, where hydrological and socio-economic systems can spread mortality and economic loss attributed to a drought to distant areas. Greater input from local knowledge and the possibility of mapping indirect and secondary socio-economic impacts presents opportunities for local vulnerability and capacity assessments to refine international measurements in this regard.

Care must be taken not to use DRI and Hotspots to predict future risk distributions. Both models used past hazard exposure (as well as past data on disaster impact, population and socio-economic variables) to calculate vulnerability and risk values. The assumption is that places where hazard, vulnerability and disaster impacts were recorded in the past are those most likely to experience them in the future. This assumption becomes less tenable at finer resolutions where development pressures, such as rapid urbanisation and local environmental changes linked to global climate change, have the potential to radically alter local distributions of population, wealth, hazard and vulnerability over a short time period relative to hazard frequency. It is also possible that losses during past disasters will lead to a local learning process and the building of resilience, rather than the continuation of vulnerability, so that past impacts might locally be associated with future security, rather than vulnerability. Regular assessments accompanied by contextual analysis of pressures shaping hazard, vulnerability and disaster risk management can help overcome this challenge for measuring local risk.

An exception to the retrospective calculation of risk comes from work on landslides by the Norwegian Geotechnical Institute (NGI). Here, hazardousness was identified though the analysis of the geophysical and hydrometeorological characteristics of each grid cell in Hotspots, not by counting past events. In the inductive approach of the Americas programme's DDI, the use of an MCE instead of a past event meant that financial capacity was measured against hypothetical future risk, not past losses, allowing infrequent hazards such as earthquakes to be included in the analysis.

Drought is associated with more loss of life than any other hazard.[10] It has also proved to be the most difficult hazard to index. More than in other hazards, additional human and environmental processes can intervene between a hydrometeorological event and the recording of losses. Conceptually, this gap is not fundamentally different from that experienced in other hazard-specific disasters (which led Hotspots to include drought in their analysis). However, when the pilot DRI results showed that exposure to drought was not among the socio-economic variables explaining recorded drought losses as listed by EM-DAT, it was clear that the influence of other factors – particularly armed conflict, chronic illness and poor governance – was of a magnitude greater than that found with disasters triggered by other types of hazard. For this reason drought was left out of the final DRI analysis. Sub-national measurements will also have to contend with the politicised nature of deaths attributed to drought, although being closer to data collection and having the possibility of verifying data and index outputs with local actors can add validity to higher resolution studies.

Volcanic hazard was excluded from the DRI because the extremely low frequency of volcanic eruptions meant that many countries where the hazard was found had no records of loss in EM-DAT. Consequently it was not reasonable to undertake the regression analysis to identify socio-economic variables. Local assessments often include a review of past losses, but should not be tied to it, focusing instead on present conditions and trajectories of socio-economic and environmental change so that low frequency and potential future hazard types can be considered in multi-hazard vulnerability analysis.

Hazards can interact with each other. Hazard nesting occurs when one hazard triggers another. An example might be a landslide triggered by a flood, which was in turn caused by a cyclone. Hazard nesting was a challenge for international indexes where databases did not uniformly record the immediate and proximate hazard causes of loss. Figure 7.1(c) shows a DRI identifying Venezuela as a country highly vulnerable to flood risk. This result was greatly influenced by high mortality from landslides following a single episode of heavy rain. In this case the landslide factor was recognised by the DRI and incorporated in the analysis of results. Once again, the challenge is to develop local vulnerability measurements that can include much more contextual information, and use local knowledge to verify results. This also shows the necessity of multi-hazard based analysis. This is likely to be especially important in the multi-hazard environment of urban settlements (Pelling, 2005c).

Measuring vulnerability

DRI and Hotspots both use mortality in calculations of vulnerability. Mortality is arguably the most reliable comparative indicator of human loss at the global scale. Data on people affected, injured or made homeless are far less reliable. Reliance on mortality gives statistical rigour but limits policy impact. This can be seen most clearly in drought events, where complex interactions between drought, political violence, chronic disease and economic poverty can make it very difficult to ascribe causes of mortality. It is more reasonable to account for the livelihood impacts of a drought. But this can only be measured on the ground and at the local scale. A good deal of work in southern Africa, in particular that coordinated by the Famine Early Warning Systems Network (FEWS NET, see http://www.fews.net), has developed methodologies for measuring drought vulnerability up to the national scale.

In the Americas programme, the LDI included the number of people affected and economic loss alongside mortality. This extension is helpful, particularly because the three elements remain disaggregated. But the reliability of information on people affected is problematic: clear definitions of exactly what constitutes an affected person, and adherence to this definition, are required if meaningful national comparisons are to be made, but these are notoriously difficult to achieve. The same will be true for any studies hoping to aggregate local vulnerability or capacity assessments that include a measure of past losses in their assessments.

Hotspots also used economic loss as an indicator of disaster impact. Three constraints face the use of economic measures of loss:

- There is very rarely any account of long-term economic impacts (sometimes called secondary losses), including changes in national balance of payments, international debt or fluctuating levels of employment or price inflation in the years following a disaster.
- The focus on economic impacts excludes assessments of local economic loss, and thus the destruction or erosion of household livelihoods is not accounted for.
- A focus on GDP means losses to the informal sector – which can reach 50 per cent or more of the financial capacity of States in extreme cases, and often exceeds 50 per cent of the economic exchange in subnational units – will remain very difficult to account for. This is partially addressed in Hotspots through measuring impact as a proportion of GDP.

It is likely that local measures of economic loss will be able to respond to the latter two gaps but will find it difficult to track secondary disaster impacts through the macro-economy; this is a task for national-scale economic analysis undertaken some time after an event (Pelling et al., 2002).

The DDI and LDI indexes of the Americas programme also incorporate elements of economic loss. In the DDI, spending on the social sector (particularly housing) following disasters was included in measures of financial exposure and goes some way towards recognising the national consequences of damage to the informal sector.[11] The LDI measure of economic loss included an assessment of housing damage, which explicitly included estimates of loss in the informal housing sector.

Hotspots aimed to calculate vulnerability for individual grid cells where data was only available at the national level. Hotspots resolved this by allocating each grid cell to one of 28 wealth regions. This approach might be useful for sub-national calculations where local data is not uniformly available. A key problem with this type of approach is that exceptional areas within each group will be lost within the averaged vulnerability value of their group. In sub-national assessments the degree of suppression of extremes could be uncovered by rapid ground-truthing exercises and presented as a health warning on results. The overall approach is useful where local vulnerability data is lacking but where key indirect indicators are nonetheless available; when brought together with hazard data this method can help identify areas of high risk for more detailed local study – exactly the aim of Hotspots.

The inductive approach taken by the Americas programme is quite different from the deductive approach of the DRI and Hotspots. From the PVI and RMI five considerations are identified that can inform future inductive work at the sub-national and local levels:

- When choosing input variables there are dangers of overlap, leading to double counting of a particular attribute, and to omission when no suitable input variables can be found. The PVI includes many socio-economic variables but has been less successful in finding variables that capture the governance and political aspects of vulnerability. This leads to the measurement of a specific understanding of vulnerability that is shaped as much by variable availability as by vulnerability theory.
- The mechanism for choosing input variables must be transparent to prevent the political manipulation of findings. In the PVI, the indicator for exposure and susceptibility is built from three population, four macro-economic and one poverty input variables. One can imagine that a shift in policy priority could lead to the selection of alternative input variables producing different results. This can be an advantage, lending the method to policy flexibility. But care is needed to en-

sure that changes are based on technical rationality, not political expediency.

- The larger the number of components within each sub-index, the more difficult it becomes to attribute index characteristics to individual component indicators. This in turn makes it more difficult to provide clear policy advice.
- New variables may be needed and existing variables discarded as the context of vulnerability generation and data availability changes through time. The choice of input variables should be constantly under review.
- Vulnerability and coping capacity should not be expected to have a linear relationship. Good risk management performance does not lead necessarily to low recorded vulnerability. For example, Jamaica has a high PVI and a high RMI. It takes time for risk reduction policy to translate into reduced vulnerability, because disaster losses are influenced as much by variability in hazard frequency and severity as by vulnerability.

Building multi-hazard measurements

The DRI and Hotspots generated hazard-specific measures of vulnerability and risk and then, through aggregation, produced multi-hazard assessments. For these two approaches the biggest challenge was how to combine hazards measured on different metrics; for example, Hotspots devised a statistical method for combining the hazardousness of drought (measured by frequency) and landslides (measured by probability).

In the Americas programme, PVI and RMI measured vulnerability and capacity as intrinsic values, not specific to any hazard type. This avoided any problem of combining hazards but meant that measurements could not take the individual characteristics of particular hazard types into account. Both the hazard-specific and intrinsic measurements of vulnerability and capacity have advantages and disadvantages. Intrinsic measures are perhaps most useful for assessing capacity, which in this form can be presented as a generic value and resource for any future danger, including new hazards as yet unknown in a particular location.

Key contributions

What potential is there for developing sub-national risk indicators from the three approaches?

- In the DRI, a sub-national measure of relative vulnerability would not be difficult to generate. Exposed population is already mapped with fine resolution by DRI and Hotspots and could be combined with local loss data. Local loss data is harder to come by, but is available for those countries covered by La Red's DesInventar database, in addition, MunichRe records the local place of loss in its global NatCat database. UNDP has already begun to work with national representatives to build national DRI indicators.

The DRI approach to identifying socio-economic vulnerability indicators could be applied sub-nationally to any collection of socio-economic variables that are considered relevant and accessible. A common pool of variables is only needed when comparison across cases or time is required. This is helpful when constructing composite risk maps, but not necessary when looking to characterise the principal development pressures shaping risk in a specific location at a particular time.

- The Hotspots analysis already produces sub-national scale maps of risk. However, for many places the meaningfulness of a sub-national resolution is limited by large numbers of contiguous cells having identical values: large areas of China and India, for example, have common risk values. This suggests that data scarcity means local variability is not being fully represented in the current analysis; while data scarcity is less of a concern for a global analysis, it becomes more important as the resolution increases.

  Hotspots' separation of loss into mortality, economic loss and economic loss as a proportion of GDP is valuable and could be deployed at the sub-national level where sub-national measures of GDP exist or can be calculated. The Hotspots approach for calculating local area values from national data could also be deployed in sub-national level studies.

- Each of the Americas programme's indexes speaks to a particular policy community. This is especially valuable at the national level where disaster risk needs to be presented in a user-friendly way, using metrics already known and applied in the everyday planning processes of economic, social and infrastructural development. The methodologies for Americas programme indicators can be applied to lower scales where data is available and decentralised disaster risk planning authorities and resources have made sub-national analysis worthwhile. The conceptualisation of vulnerability and capacity as intrinsic properties can be useful at sub-national and local scale, possibly in conjunction with hazard-specific measures.

Conclusions and outlook

The DRI, Hotspots and the Americas Indexing Programme have shown us alternative approaches to the calculation of vulnerability and risk.

They also highlight the gaps in our data and understanding. Many of the challenges facing international disaster risk indexing are also relevant to sub-national and local measurement of vulnerability and risk.

From the deductive approach of the DRI and Hotspots we learn that:

- Local hazard maps exist, but the speed of flash flooding and the high secondary impacts of drought have yet to be incorporated.
- Care should be taken if retrospective data is used to assist forward-looking policy decision-making.
- Political and business interests can distort loss data, particularly in slow-onset and complex disasters such as drought, where losses are hard to attribute to any one pressure.
- It is difficult to incorporate low-frequency and future hazards when basing assessments of vulnerability on past experience.
- Hazard nesting means individual hazard phenomena can result in multiple hazard types: a cyclone can result in wind, flood and landslide damage.

Higher resolution assessments can add value by:

- exposing the trail of causality from physical event to recorded impact
- measuring livelihood loss and loss in the informal sector, alongside mortality and direct macro-economic impacts
- providing a detailed characterisation of vulnerability and capacity in high-risk locations.

From the inductive approach of the Americas indicator programme we can learn that:

- It is important to guard against double counting and omissions during selection of input variables.
- The choice of input variables must be transparent to prevent the manipulation of results.
- There is a tension between the comprehensive representation of vulnerability and capacity that comes from using a large number of input variables, and the clarity that comes from using a small number of variables.
- A constant review of input variables is needed as the causes of vulnerability and capacity and the availability of input variables change over time.
- The relationship between vulnerability and capacity is not linear.

High-resolution assessments can add value by:

- providing a verification check on input variables
- providing a detailed characterisation of vulnerability and capacity in high-risk locations
- showing how intrinsic measurements of vulnerability and capacity play out in particular places facing specific combinations of hazard.

Two key challenges for sub-national and local measurements of vulnerability and capacity come from the reviews undertaken in this chapter.

First, how can high-resolution assessments best feed into development and disaster risk reduction decision-making? There is much scope for local assessments to be part of early warning systems, but this connection influences the methodologies used and must be planned for from the outset. Second, can individual assessments be aggregated to upscale results and feed into or complement lower resolution assessments? Initial evidence suggests this is a possibility. ActionAid has already conducted aggregations of local vulnerability assessments in Sierra Leone and Zimbabwe.[12] More work on methods of aggregation can help this process, as can improved communication between the people involved in planning and undertaking local and international scale analyses of risk.

Acknowledgements

This chapter has drawn from work undertaken by the author in a consultancy for UNDP, Bureau for Crisis Prevention and Recovery, and in particular from the project report entitled *Visions of Risk: A Review of International Indicators of Disaster Risk and its Management*. Many thanks to UNDP for granting permission for material to be reproduced here. The views expressed in this chapter and any errors remain those of the author and not UNDP.

Notes

1. For example, Montserrat and Bermuda were treated independently rather than as part of the United Kingdom. Tributarian States will also be included when the term "country" is used in discussions of the DRI method and results.
2. Vulnerability and hazard exposure variables were identified through a correlation with mortality data from EM-DAT.
3. Christie, F. and Hanlon, J. (2001) *Mozambique and the Great Flood of 2000*, James Currey, Oxford.
4. The units of analysis are some 4.1 million 2.5 by 2.5 minute grid cells. With areas ranging from 21 km^2 at the equator to 11 km^2 at the poles, these cells cover most of the inhabited land area of the globe.
5. With a minimal number of people or assets exposed to hazard, calculated disaster risk would always appear low for these grid cells.
6. Africa, East Asia and the Pacific, Europe and Central Asia, Latin America and the Caribbean, Middle East and North Africa, North America, South Asia.
7. Other events include biological and technological phenomena.
8. http://www.desinventar.org/desinventar.html.
9. See Pelling, 2005a and 2005b for reviews of internationally available data with sub-national resolution that could be used to feed into sub-national measurements of risk.
10. For a discussion of the gaps in data and in the understanding and modelling drought risk see an ISDR report entitled "An Integrated Approach to Reducing Societal Vulnerability to Drought" on http://www.unisdr.org/.

11. Indexes aim to provide technical, non-political information for decision makers. In so doing, the DDI correctly includes social spending as an area of potential exposure from the perspective of national finances. It is hoped this might encourage financial mechanisms for risk and loss management and investment to reduce the fragility of informal housing. But such information is hostage to political prioritisation. A less progressive response to the DDI could be to cut back on Government support for the social sector. Such a decision would in effect transfer risk from the national exchequer to low-income groups. This concern shows the advantage of multi-dimensional benchmarking as a means of assessing national disaster risk reduction performance. In the Americas programme, any strategic shifting of disaster risk to low-income groups should register on the PVI, with a lack of comprehensive disaster management planning that such a strategy implies being flagged by the RMI.

12. Personal communication from Ethlet Chiwaka from Action Aid's International Emergencies Team in 2005.

REFERENCES

Cardona, O.D. et al. (2004) *Results of Application of the System of Indicators on Twelve Countries of the Americas.* IDB/IDEA Program of Indicators for Disaster Risk Management, Manizales: National University of Colombia, available at http://idea.unalmzl.edu.co/.

Cardona, O.D. et al. (2005) *Indicators of Disaster Risk and Risk Management. Program for Latin America and the Caribbean,* Summary Report for World Conference on Disaster Reduction, IDB/IDEA Program of Indicators for Disaster Risk Management, National University of Colombia / Inter-American Development Bank, available at http://www.iadb.org/int/DRP/Ing/Red6/Docs/IDEAR06-05eng.pdf.

Dilley, M., R.S. Chen, U. Deichmann, A.L. Lerner-Lam and M. Arnold (2005) *Natural Disaster Hotspots: A Global Risk Analysis,* Washington DC: World Bank, Hazard Management Unit.

Pelling, M. (2005a) *Visions of Risk: A Review of International Indicators of Disaster Risk and its Management,* Geneva: UNDP–Bureau for Crisis Prevention and Recovery (BRCP).

Pelling, M. (2005b) "Disaster Data: Building a Foundation for Disaster Risk Reduction", in IFRC, *World Disasters Report, 2005,* Geneva: International Federation of the Red Cross and Red Crescent Societies.

Pelling, M. (2005c) "Measuring Vulnerability to Urban Natural Disaster Risk", *Open House International,* Special Issue on Managing Urban Disasters, 31(1) 125–132.

Pelling, M., A. Özerdem and S. Barakat (2002) "The Macro-economic Impact of Disasters", *Progress in Development Studies* 2(4): 283–305.

United Nations Development Programme (UNDP) (2004) *Reducing Disaster Risk: A Challenge for Development. A Global Report,* New York: UNDP–Bureau for Crisis Prevention and Recovery (BRCP), available at http://www.undp.org/bcpr/disred/rdr.htm.

8

The Disaster Risk Index: Overview of a quantitative approach

Pascal Peduzzi

Introduction

So far, the international community's response to disasters has been mostly reactive, with only a limited budget invested in prevention. One reason might lie in the fact that disasters get much more attention and media coverage than preventive measures. Prevention programmes will never offer the striking images that disasters do. When tragic events do get the attention of decision makers, they are soon distracted by the next headlines. Even if there were a willingness to invest in prevention, the question would be: *where*? Obviously such decisions cannot be based on media coverage. The floods that hit India, Nepal and Bangladesh in August 2004 and killed 2,000 people were given a mere 9,000 words in British newspapers, whereas the same day, Hurricane Charley killed 16 people in Florida and 19,000 words were written (Adams, 2004). Clearly, there is a need for a more objective way of comparing countries at risk.

Background

In order to promote prevention and other risk reduction measures, and to avoid decisions based on risk perception, the United Nations Development Programme (UNDP) conceived the idea of creating an index based on a quantitative approach that would allow for comparisons between

171

countries. The challenge was how to compare countries hit by different hazard types, such as drought versus floods? The response was to build an index based on mortality. One person killed by a cyclone is comparable to one person killed by a flood. The other reason for this choice was that data on mortality is the most complete and the most reliable (the Emergency Disasters Data Base (EM-DAT) from the Centre for Research on the Epidemiology of Disasters (CRED) was used for this purpose). Other parameters such as economic losses, numbers of injured or lost livelihoods all suffer from either lack of data or lack of comparability potential, or both. The mandate provided to UNEP/DEWA/GRID-Europe by UNDP was to design an index that would be statistically robust, simple to understand, allow evolution in the incorporation of new hazard types, and be replicable. This chapter provides a summary of the research published in the report *Reducing Disaster Risk: A Challenge for Development* (UNDP, 2004).

Methodology and structure

The DRI took more than three years of study; the methodology and research was performed by a team of four people (Dr H. Dao, C. Herold, Dr F. Mouton and P. Peduzzi) with support and review from numerous experts for each hazard type, as well as other temporary staff for support when needed.

The formula used for estimating risk was based on a UN definition (UNDRO, 1979), which states that risk results from three components: the *hazard occurrence probability*, the *elements at risk* (in this case the population) and their *vulnerability* (Coburn et al., 1991). By multiplying the frequency of hazards by the population affected, the *physical exposure* was obtained. This figure represents the average number of people affected each year by a specific hazard. For example, in the Philippines around 77 per cent of the population lives in regions affected by tropical cyclones, and the average yearly number of cyclones equals 5.57; hence the physical exposure is 428 per cent (5.57 × 77 per cent).

The first task was to find all the requested geophysical data and then to model the different hazards in order to obtain the frequency for earthquakes, drought, floods and cyclones for each location on the globe. This search was made easier by the fact that UNEP/GRID-Europe already had a collection of data from their Project for Risk Evaluation, Vulnerability, Information and Early Warning (PREVIEW) which modelled and geo-referenced tropical cyclones, volcanoes and earthquakes (Peduzzi et al., 2005). However, two datasets of the most significant hazards were missing: namely floods and drought. Therefore a method was established to identify watersheds affected by floods, by intersecting text information from EM-Dat and geophysical datasets (Peduzzi et al., 2002, 2005). Close collaboration and exchange of data between the DRI project and the Hotspots project enabled the gap in knowledge on drought frequency to be closed. This dataset was elaborated by the International Research Institute for Climate Prediction, Columbia University, on the basis of a methodology from Brad Lyon.

Once the data was obtained, the CIESIN and UNEP models for population distribution were merged and the result was multiplied by frequency to compute the physical exposure (see Figure 8.1(c)). This already normalised the differences between populations highly affected by a selected hazard and those populations affected less frequently. Physical exposure alone, however, is not to be misunderstood as being equal to risk, because two countries with similar levels of physical exposure to a given hazard can experience divergent degrees of risk, as the application of the DRI showed. Similarly, physical exposure cannot be seen as an indicator of vulnerability but as a prerequisite for disaster risk to exist. Without people being exposed to hazardous events, there is no risk to human life.

Vulnerability, finally, is perceived as the concept that explains why people with the same level of physical exposure can be more or less at risk. Coping capacity and adaptive competence are the variables that modify the vulnerability. In order to compare the vulnerability levels of different countries, the DRI calculates the so-called relative vulnerability of a country to a given hazard. This figure is obtained by dividing the number of people killed by the number of people exposed. The more people killed in proportion to the people exposed, the more vulnerable a country is to the given hazard.

Selected results

The first result was remarkable: although the least developed countries together represented 11 per cent of the physical exposure to hazard, they accounted for 53 per cent of the casualties. In contrast, the most developed countries represent 15 per cent of the physical exposure to hazards, yet they only account for 1.8 per cent of the victims (see Figure 8.2). This disparity cannot be explained by geographic location. Haiti and the Dominican Republic are located on the same island and hence are affected by the same tropical cyclones, yet Haiti suffers on average 4.3 times more casualties than its more prosperous neighbour. This clearly illustrates that a population's vulnerability is linked to the level of development.

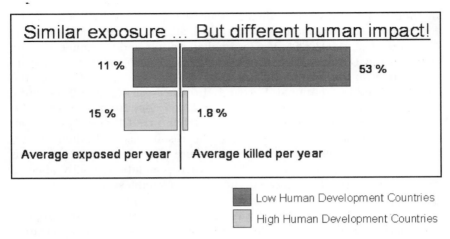

Figure 8.2 Comparing exposure and mortality rates in the most/least developed countries.
Source: Peduzzi, 2005.

By computing the ratio of those killed to those exposed, the average vulnerability of a population can be computed for each country for each hazard type. However, these figures do not explain why different countries are unequally vulnerable or what role the level of development plays in this context. For this reason, indicators on economic levels, type of economy, development level (e.g. Human Development Index (HDI)), wealth (GDP purchasing power parity), education, environment, health, corruption and demography (population growth, density, urban growth, etc.) were introduced into a database. A multiple logarithmic regression analysis was performed over the 26 socio-economic indicators in order to see which combination of parameters associated with physical exposure to a selected hazard best explains the number of casualties.

These analyses were carried out on earthquakes, tropical cyclones, floods and drought, and although the results for drought were not satisfying, the regression on the other three hazards produced interesting results.

Earthquakes

A total of 158,661 deaths were associated with earthquakes around the world between 1980 and 2000. The highest casualties were sustained in Iran where a total of 47,267 people were killed. Worldwide, about 130 million people on average were recorded as being exposed to earthquake risk on a yearly basis. Japan has by far the largest absolute numbers of people exposed to earthquakes, with 300 million people at risk. Indonesia and the Philippines follow with about 16 million each. Taiwan comes fourth in the top-15 list, followed by the Americas (USA, Chile and Mexico), China, Turkey, India, Guatemala, Colombia, Iran, Peru and Afghanistan. In terms of the number of exposed people in proportion to the population, however, most of these States (except Taiwan) are ranked lower. In this category, the small countries of Vanuatu and Guam are clearly the most vulnerable with, in the case of Vanuatu, every inhabitant being exposed to an earthquake 1.5 times a year on average.

As far as relative vulnerability to earthquakes is concerned, the results have to be seen in terms of the relatively short observation period of 20 years. Armenia was by far the most vulnerable country in this period, due to a single catastrophic event that occurred in 1997. Meanwhile Guatemala appears far less vulnerable because the big earthquake of 1976 happened to take place outside the observation period. However, the analysis shows that some countries, such as the Islamic Republic of Iran, Afghanistan and India, which experience frequent earthquakes, suffer proportionally far higher losses of life than others like Chile or the USA. Overall, a strong correlation between the number of victims and the physical exposure was observed. The regression analysis of vulnerability indicators showed that, statistically, physical exposure and the rate of urban growth acted together in being associated with the risk of death from earthquakes. This finding cannot be generalised for every country since building regulations, which vary from country to country, play a key role in generating physical vulnerability to earthquakes in urban areas.

Tropical cyclones

For the purpose of the DRI analysis, the term "tropical cyclone" was defined as encompassing tropical storms, hurricanes (alternatively termed typhoons, tropical cyclones or severe tropical storms) and super typhoons. Up to 119 million people on average are exposed to tropical cyclones every year and some populations experience an average of more than 4 events per year. About 250,000 people died worldwide from tropical cyclones between 1980 and 2000. Bangladesh alone accounts for more than 60 per cent of these victims. In terms of absolute numbers of physical exposure, China has the leading position with an average of almost 600 million people being exposed every year. But India, the Philippines, Japan and Bangladesh also have very high absolute numbers – between 350 and 120 million – with all of them having highly populated coastal areas, and particularly deltas. Altogether, 84 countries distributed over the tropics presented various levels of physical exposure to cyclones. Island States and archipelagos like the British Virgin Islands, Vanuatu, Mauritius and the Philippines show the highest exposure in proportion to the population.

In terms of *relative vulnerability* to cyclone's characteristics similar to the vulnerability to earthquakes can be observed. Due to the relatively short observation period, some stochastic events falling within the period caused peak values for specific countries. Hence, Honduras and Nicaragua, although not among the countries with the highest physical exposure, recorded the highest relative vulnerability as a result of Hurricane Mitch, which occurred in 1998; this hurricane highlighted a major problem with the category "tropical cyclones" because many of the casualties were killed not by the winds per se, but by floods, flash floods, landslides and debris flows triggered by the cyclone.

In general, a close correlation between the degree of physical exposure and number of victims can be observed, as was the case with earthquakes. The strong influence of human development status on disaster risk is also apparent again. Haiti, the island State most at risk, has low levels of human development while Mauritius and Cuba, for example, show lesser risk and higher development standards. Moreover, the regression analysis revealed a strong correlation between a combination of physical exposure, the proportion of arable land and the HDI. Arable land can be understood as a proxy for rural population: the more arable land, the higher the proportion of the rural population. Rural populations seem, according to the model, to be more vulnerable to this hazard than urban ones. Quite logically, the lower the country's HDI, the higher the risk from tropical cyclones. Poor rural housing, lack of rescue services and lack of access to disaster preparedness and early warning systems are mentioned in the study as some of the reasons.

Floods

The total number of deaths associated with floods worldwide was 170,010 between 1990 and 2000. On average, about 196 million people in more than 90 countries are exposed to catastrophic flooding every year. Populous South Asian countries figure strongly at the top of the list of absolute exposure, with India and China indicating around 150 million exposed people each. After India and China are Bangladesh, Indonesia, Pakistan, Myanmar, Iran, and Afghanistan, with at least 40 million people exposed in each, followed by Brazil, Nepal, Peru, the United States, Japan, Colombia and Viet Nam. It is obvious that all these countries are either characterised by a mountainous topography or have highly populated coastal, or particularly delta, areas. Regarding the physical exposure in proportion to the population, it becomes evident that smaller or less populated countries like Bhutan, Afghanistan, Ecuador and Nepal are to be found in the top ranks.

In terms of relative vulnerability, the decisive importance of stochastic events occurring within the observation period can be noted. Venezuela appears to be the country with highest relative vulnerability to flooding, mainly because of one single exceptional event that occurred in December 1999. Moreover, a large proportion of the deaths were associated with debris flows in dense urban communities not located in floodplains. As in the case of earthquakes and cyclones, a strong correlation between the number of victims and the physical exposure was observed. Furthermore, the regression analysis showed an inverse correlation of recorded deaths and GDP per capita, meaning that the poorest were the most vulnerable, a finding consistent with what could be expected. More surprising was the finding that the variable "local population density" was also inversely correlated. Hence, areas with low densities of population were more vulnerable, probably because of difficulties of access for rescue teams.

Drought and volcanoes

Originally, the study included both drought and volcanoes: global datasets were generated and physical exposure computed. This was intended to allow the computation of a vulnerability proxy, once statistical regression had been applied. However, the model for volcanoes appeared not to be relevant. This was explained by the hypothesis that, unlike other hazards, the smaller the frequency of volcanoes, the higher the risk. Moreover, the same volcano, depending on wind direction, could have very different impacts. This chaotic behaviour appears not to be robust enough for inclusion in this first release of the DRI.

Drought is a different matter. People do not die from drought, but from food insecurity, showing that drought vulnerability combines elements of an anthropogenic and a natural hazard. The analysis revealed that conflicts and political tensions played an important role. To better explain the vulnerability of affected countries, information on conflicts should be taken into account, which would require complex methodologies and case-by-case analysis. Although a physical model was achieved, it was not of comparable quality to the others, and it was decided not to include it in the first phase of the DRI. Drought was included as a case study and further refinements are planned in order to include it in the next computation of the Index.

Open questions and limitations

The results were surprising and very strong correlations were found. However, the approach does not take account of the fact that disasters affect people's lives and livelihoods in manifold ways besides the loss of life. People may also suffer injury, illness or stress (physically as well as psychologically), houses may be destroyed, and cattle and crops may be lost. Despite this, it was decided to use numbers of people killed as the main indicator because it was felt this would show the best comparison of countries. Indeed, the number of "injured" is not useful for comparative purposes; it appears that the greater the number of health infrastructures, the higher the number of people reported as "injured"! The variable "killed", in contrast, is not subject to discussion. Still, this is clearly a limitation, since what really makes an impact on development level is loss of livelihood, and to measure this variable would require the collection of data livelihood losses on a global scale.

The DRI accounts for large- and medium-scale disasters, defined as those events involving more than ten deaths, 100 affected and/or a call for international assistance. These kinds of disasters are the only ones for which public data is available at a global scale (CRED data was used) so far. Hence, the DRI only includes such disasters and, in doing so, excludes all small-scale events that may occur frequently and be highly localised (small landslides or debris flows for example) or creeping.

Earthquakes, tropical cyclones, floods and droughts cause 94 per cent of the total hazard mortality. These four hazards were selected for the Index on the basis of time and resources available, considering that all the datasets had to be generated. However, other hazards not accounted for in the DRI might be of the same or even much greater significance in individual countries. For example, tsunamis account for 68 per cent of casualties in Papua New Guinea, and landslides account for 33 per cent of victims in Peru. The DRI is designed to accept additional hazard types, so that in the future new features can be included.

The relatively short time span (1980 to 2000) of the survey netted some surprising results, as the examples show, which probably would have looked quite different if the observation period had reflected the geological and climatologic time dimensions that are representative of the hazards examined. To decrease this bias, the frequency of earthquakes was computed for 40 years, but this timeframe is still negligible in terms of geological time. For floods and tropical cyclones, the observation period is less of an issue, since they recur over a smaller time period. For earthquakes, there exists a global dataset measuring "peak ground acceleration"; however, it is not possible to derive frequencies from this dataset.

By mandate, the DRI had to be generated on the basis of existing socio-economic indicators, availability and the overall completeness of country data. Some other indicators, for example of "efficiency of management" and "prevention measures", would have been very useful.

In terms of flood hazard calculation, a particular limitation of the approach can be observed. Entire watersheds were mapped as flood-prone areas despite the fact that usually only a small area of the watershed is flooded. The resolution of watersheds varies depending on their location, a feature that might be corrected in the future by using a digital elevation model with better resolution (such as Shuttle Radar Topography Mission (SRTM)), as well as by a better modelling method. All these questions offer opportunities for further improvements of the DRI.

Outlook

This analysis provides a useful and neutral tool for the evaluation of countries facing risk from natural hazards. It is hoped that this tool will help countries with both high vulnerability and high exposure to adopt more risk reduction measures.

Future developments of the DRI are under discussion. Improving the risk assessment for drought is a priority, since without this hazard, Africa would be out of the picture entirely. Landslides are the next hazard that will be added, using research conducted in collaboration with the Norwegian Geotechnical Institute (NGI) for the World Bank Natural Disaster Hotspots project (World Bank, 2005). Now that new datasets are available, assessment of flood hazards will also be improved. Once developed, a computation of multiple risk can be achieved by adding the casualties as modelled. The methodology already exists. This study proved the connection between the level of development and vulnerability. It demonstrated that natural disasters do not exist as such. Only natural hazards occur, and thus the challenge now facing countries is how to better incorporate the risk in land management.

Notes

- A web-based interactive tool for comparing countries is provided at http://gridca.grid.unep.ch/undp/.
- The location of frequency and physical exposure can be visualised at http://grid.unep.ch/preview/.
- The detailed methodology for the project can be found at http://www.grid.unep.ch/product/publication/earlywarning_articles_reports.php/.

REFERENCES

Adams, R. (2004) "Where Death Really Counts", *The Guardian*, 20 August.

Anand, S. and A. Sen (2000) "The Income Component of the Human Development Index", *Journal of Human Development* 1(1): 83–106.

Anderson, M. and P. Woodrow (1989) *Rising from the Ashes: Development Strategies in Times of Disaster*, Boulder: Westview.

Blaikie, P., T. Cannon, I. Davis and B. Wisner (2004) *At Risk: Natural Hazards, People's Vulnerability, and Disasters*, 2nd edn, London: Routledge.

Bolt, B.A., W.L. Horn, G.A. Macdonald and R.F. Scott (1975) *Geological Hazards*, Berlin/New York: Springer-Verlag.

Burton, I., R.W. Kates and G.F. White (1993) *The Environment as Hazard*, 2nd edn, New York/London: Guilford Press.

Carter, N. (1991) *Disaster Management: A Disaster Manager's Handbook*, Manila: Asian Development Bank.

Chen, R.S., U. Deichmann, Dilley, M. et al. (2005) *Natural Disaster Hotspots: A Global Risk Analysis*, New York/Washington: Columbia University/World Bank.

Coburn, A.W., R.J.S. Spence and A. Pomonis (1991) *Vulnerability and Risk Assessment*, UNDP Disaster Management Training Programme. Cambridge: Cambridge Architectural Research Limited.

Dao, H. and P. Peduzzi (2004) "Global Evaluation of Human Risk and Vulnerability to Natural Hazards", in *Sh@ring*, Proceedings of the 18th International Conference on "Informatics for Environmental Protection", EnviroInfo 2004, Geneva, 21–23 October 2004, Vol. 1, pp. 435–446.

Giardini, D. (1999) "Annali di Geofisica: The Global Seismic Hazard Assessment Program (GSHAP) 1992–1999", *Instituto Nazionale di Geofisica* 42(6).

Peduzzi, P. (2000) "*Insight on Common/Key Indicators for Global Vulnerability Mapping*", Summary of presentation made at the Expert Meeting on Vulnerability and Risk Analysis and Indexing, 11–12 September 2000, Geneva: UNEP, available at http://www.grid.unep.ch/product/publication/earlywarning_articles_reports.php/.

Peduzzi, P., H. Dao and C. Herold (2002) *Global Risk And Vulnerability Index Trends per Year (GRAVITY), Phase II: Development, Analysis and Results*, Geneva: UNDP/BCPR, available at http://www.grid.unep.ch/product/publication/download/ew_gravity2.pdf.

Peduzzi, P., H. Dao and C. Herold (2005) "Mapping Disastrous Natural Hazards Using Global Datasets", *Natural Hazards* 35(2): 265–289.

Peduzzi, P., H. Dao, C. Herold and F. Mouton (2003a) *Global Risk And Vulnerability Index Trends per Year (GRAVITY), Phase IIIa: Drought analysis*, Geneva: UNDP/BCPR.

Peduzzi, P., H. Dao, C. Herold and F. Mouton (2003b) *Global Risk And Vulnerability Index Trends per Year (GRAVITY), Annex to WVR and Multi Risk Integration: Phase IIIb*, UNEP.

Peduzzi, P., H. Dao, C. Herold and D. Rochette (2001) *Feasibility Study Report on Global Risk and Vulnerability Index Trends per Year (GRAVITY)*, Geneva: UNDP/BCPR.

Smith, K. (1996) *Environmental Hazards: Assessing Risk and Reducing Disaster*, London/New York: Routledge.

Tobin, G.A. and B.E. Montz (1997) *Natural Hazards, Explanation and Integration*, London/New York: Guildford Press.

United Nations Development Programme (UNDP) (2004) *Reducing Disaster Risk: A Challenge for Development. A Global Report*, New York: UNDP–Bureau for Crisis Prevention and Recovery (BRCP), available at http://www.undp.org/bcpr/disred/rdr.htm.

UNDP/UNEP/GRID (Global Resource Information Database) (2005) *DRI Disasters Reduction Information Analysis Tool*, available at http://www.gridca.grid.unep.ch/undp/.

United Nations Disasters Relief Co-ordinator (UNDRO) (1979) *Natural Disasters and Vulnerability Analysis*, Report of Expert Group Meeting, 9–12 July, Geneva: UNDRO.

United Nations Environment Programme (UNEP) (2002) *Global Environmental Outlook 3: Past, Present and Future Perspectives*, London: Earthscan.

UNEP/DEWA (Division of Early Warning and Assessment)/GRID (2005) *Project of Risk Evaluation, Vulnerability, Information and Early Warning (PREVIEW)*, available at http://grid.unep.ch/preview/.

World Bank (2005) *Natural Disaster Hotspots: A Global Risk Analysis*, Dilley, M. et al., eds, Washington D.C., The World Bank, Hazard Management Unit.

9

Disaster risk hotspots: A project summary

Maxx Dilley

Introduction and rationale[1]

The Global Natural Disaster Risk Hotspots project was undertaken to assess disaster risks globally. The assessment focused on two disaster-related outcomes: mortality and economic losses. Relative risks of these two outcomes were assessed for major natural hazards on a 5 × 5 km global grid.

The project also generated a set of more localised and/or hazard-specific case studies. The case studies demonstrate that the same theory of disaster causality that underpins the global analysis also applies at local scales, and that more localised analyses can inform national and local disaster risk management planning.

The Hotspots analysis was intended to provide evidence about disaster risk patterns to improve disaster preparedness and prevent losses. High-risk areas are those in which disasters are expected to occur most frequently and losses are expected to be highest. Making risks foreseeable provides motivation for risk reduction (Glantz, 2002). Identification of risk levels and risk factors creates possibilities for shifting emphasis from reliance on ex post relief and reconstruction after disasters towards ex ante prevention and preparedness in order to reduce losses and recovery time. The Hotspots results provide an evidence base for prioritising risk management efforts and bringing attention to areas where risk management is most needed.

The Hotspots project was initiated by the ProVention Consortium with funding from the United Kingdom's Department for International Development. Additional support for the case studies was provided by the Norwegian Ministry of Foreign Affairs and the US Agency for International Development. The project was implemented by more than a dozen institutions, led by Columbia University and the World Bank, and involved perhaps a hundred scientists. The Hotspots project benefited enormously from interactions with the project on *Reducing Disaster Risk: A Challenge for Development* (UNDP, 2004), a collaborative effort involving the United Nations Development Program (UNDP), the United Nations Environment Program (UNEP) and others.

Structure and methodology

Starting from the understanding that disaster losses are caused by interactions between hazard events and the characteristics of exposed elements that make them susceptible to being damaged (vulnerabilities), the Hotspots project estimated risk levels by combining hazard exposure with historical vulnerability for two indicators of elements at risk – gridded population and gross domestic product (GDP) per unit area – for six major natural hazards: earthquakes, volcanoes, landslides, floods, droughts and cyclones. Hazard destructive potential is a function of the magnitude, duration, location and timing of the event (Burton et al., 1993). To be damaged, however, elements exposed to a given type of hazard must also be vulnerable to that hazard; that is, the elements must have intrinsic characteristics, or vulnerabilities, that allow them to be damaged or destroyed (UNDRO, 1979). Elements of value that may have such vulnerabilities include people, infrastructure, and economically or environmentally important land uses.

Relative levels of the risks of disaster-related mortality and economic losses were calculated for population and GDP based on $2.5' \times 2.5'$ latitude–longitude grid cells, providing an estimate of relative risk levels at sub-national scales. Since the objective of the analysis was to identify hotspots where natural hazard impacts are expected to be large, it was clear that a large proportion of the earth's land surface, which is sparsely populated and not intensively used, did not need to be included. Therefore, grid cells with population densities of less than five people per km^2 and without a significant agricultural land use were masked out. The remaining grid cells (coloured orange, blue or green in Figure 9.1(c)) represented only slightly more than half of the total landmass (about 55 per cent) but most of the world's population (about 6 billion people).

Any global analysis is clearly limited by issues of scale as well as by the availability and quality of data. For a number of hazards, records for the entire globe are only available for the last 15 to 25 years and the spatial information for geolocating these events, and the spatial resolution, are relatively crude. Data on historical disaster losses, and particularly on economic losses, are also limited.

Keeping these constraints in mind, three indices of disaster risk were developed:

- disaster-related mortality risks, assessed for global gridded population
- risks of total economic losses, assessed for global gridded GDP per unit area
- risks of economic losses expressed as a proportion of the GDP per unit area for each grid cell.

Three types of data were used to calculate risks of the above outcomes:

- data on the elements at risk (population and economic product, that is, GDP)
- data on the six hazards
- data on vulnerability.

The hazard exposure of the population and GDP in each grid cell was determined by multiplying the population or GDP in each cell by the historical hazard frequency or probability for each hazard. Risk levels were calculated by weighting the result with a vulnerability coefficient.

These vulnerability coefficients were estimated from hazard-specific historical mortality and economic loss data, obtained from the Emergency Disasters Data Base (EM-DAT, see www.cred.be). Mortality and economic loss rates for each hazard were calculated for each of seven geographic regions, further subdivided into four country/wealth groups. This gave 28 vulnerability weights for each hazard, one for each region and country/wealth class combination. The appropriate historical loss rate for each hazard was used to weight the hazard exposure in each cell to arrive at the risk of mortality or economic losses.

Due to the limited time period and quality of the input data one can say that overall it is appropriate to use the results to identify those areas at *relatively* high risk due to a particular natural hazard. Data quality and resolution dictate that the results are inadequate for assessing *absolute* levels of risk or for detailed *comparisons* of levels of risk across hazards. For a number of the available hazard datasets, such as those based on media reports, relatively small or modest events may be substantially undercounted, especially in developing countries where reporting is likely to be less complete.

To estimate relative risks, therefore, the total number of grid cells was divided into *deciles* (10 groups of approximately equal number of cells) based on the value of each calculated risk indicator (expected mortality or economic losses for each hazard). Cells with the value of zero for an indicator were excluded. When a risk indicator had large numbers of cells with the same values (cyclones, drought, floods and earthquakes), deciles were grouped together.

Results

Maps of the results can be found in Dilley et al. (2005) and the review of Pelling in Chapter 7. In these maps, the relative risk levels are shown with high-risk cells in red, medium in yellow, low in blue and undetectable in white. For each hazard, one map shows the relative risks of disaster-related mortality associated with the hazard, one the relative risks of disaster-related total (aggregate) economic losses, and one the relative risks of disaster-related economic losses in proportion to the GDP present in each grid cell. Patterns of risk change depending on which outcome is being assessed, with mortality risks generally higher in developing countries and risks of total economic losses higher in wealthier areas.

Some general results from the Hotspot assessment include:

- Cyclone-related risks are concentrated in coastal areas along the east sides of continents.
- Drought risk areas are much more spatially extensive, with mortality risks and risks of economic losses in proportion to GDP highest in semi-arid Africa. The risks of total economic losses are generally highest in the Americas, Europe and Asia.
- For floods the spatial extent of the risky areas is also very extensive. Flood-related mortality risks are high on all continents, and in Asia, Europe and the Americas they are high in terms of mortality, total economic losses and proportional economic losses.
- The extent of the areas at high risk from drought and flooding suggests that managing climate-related risks is a high priority in many areas.
- The risks associated with earthquakes are more localised, largely restricted to tectonic plate boundaries: the west coast of the United States, the east coast of Asia, and across central Asia. In the latter region, relative risks of all three outcomes are high.
- The risks posed by volcanoes are very localised, although wind-borne ash can affect larger areas beyond the immediate area of an eruption.
- Landslide-related risks are highest in mountainous areas.

In addition to human and direct economic losses, disasters impose costs as well. These include expenditures for disaster relief and recovery, and for rehabilitation and reconstruction of damaged and destroyed assets. In the case of major disasters, meeting these additional costs can require external financing or international humanitarian assistance.

This combination of human and economic losses plus the additional costs of relief, rehabilitation and reconstruction makes disasters an economic issue as well as a humanitarian one. Disaster relief costs drain development resources from productive investments to support consumption over short periods of time. Disaster-related losses offset economic growth and contribute to poverty.

Until vulnerability, and consequently risks, are reduced, countries with high proportions of population or GDP in hotspot areas are especially likely to incur repeated disaster-related losses and costs. In order to quantify this phenomenon, the World Bank provided data on emergency loans and reallocation of existing loans to meet disaster reconstruction needs from 1980 to 2003 for this study (http://www.worldbank.org/dmf). The total of emergency lending and loan reallocation from 1980 to 2003 was $14.4 billion. Of this, $12 billion went to the 20 countries listed in Table 9.1.

Table 9.1 Countries receiving emergency loans and reallocation of existing loans to meet disaster reconstruction needs, 1980–2003

Country	Earthquake	Floods	Storms	Drought
India	X		X	X
Turkey	X	X		
Bangladesh		X	X	
Mexico	X	X		
Argentina		X		
Brazil		X		
Poland		X		
Colombia	X	X		
Iran	X			
Honduras		X	X	
China	X	X		
Chile	X			
Zimbabwe				X
Dominican Republic			X	
El Salvador	X			
Algeria	X	X		
Ecuador	X	X		
Mozambique		X		X
Philippines	X			
Viet Nam		X		

Source: World Bank Hazard Management Unit (http://www.worldbank.org/dmf).

Conclusions and next steps

Disaster risk management therefore deserves serious consideration as an issue for sustainable development. Through the identification of risk factors, and through analysis of the correspondence between assessed risks and historical disaster patterns, the approach presented here makes these risks foreseeable, creating an incentive for action to reduce risks and losses through preemptive action rather than perpetuating a repetitive cycle of disaster, relief and recovery.

Following the Hotspots project and related efforts, plans are being made to seek the support of systematic analysis of disaster risks in high-risk areas to inform risk management planning. UNDP, the ProVention and a group of collaborating institutions have initiated a preparatory project for a Global Risk Identification Programme (GRIP). The GRIP will build on the results of previous analyses, with a priority on supporting the creation of evidence on disaster risk levels and factors at national to local scales, working with local authorities and experts. The GRIP will promote and support improvement in the availability and quality of data on disaster losses as well as the integration of higher-resolution, better-quality data on hazards, exposure and vulnerability in high-risk areas to support risk management decision-making.

Note

1. Portions of this paper were taken from Dilley et al. (2005). A synthesis of the Hotspots project report may be downloaded from http://www.ldeo.columbia.edu/chrr/research/hotspots/. The full report is *Natural Disaster Hotspots: A Global Risk Analysis*, 150 pages 8.5 × 11 April 2005, Price $20.00. ISBN: 0-8213-5930-4, http://publications.worldbank.org/ecommerce/catalog/product?item_id=4302005/.

REFERENCES

Burton, I., R.W. Kates and G.F. White (1993) *The Environment as Hazard*, 2nd edn, New York/London: Guilford Press.

Dilley, M, R.S. Chen, U. Deichmann, A.L. Lerner-Lam and M. Arnold (2005) *Natural Disaster Hotspots: A Global Risk Analysis*, Washington D.C.: The World Bank, Hazard Management Unit.

Glantz, M.H. (2002) *Flashpoints Informal Planning Meeting (IPM) 4–5 April 2002: Annotated agenda with discussion results, available at* http://www.esig.ucar.edu/flash/summary.html.

United Nations Development Programme (UNDP) (2004) *Reducing Disaster*

Risk: A Challenge for Development. A Global Report, New York: UNDP, Bureau for Crisis Prevention and Recovery (BRCP), available at http://www.undp.org/bcpr/disred/rdr.htm.

United Nations Disasters Relief Co-Ordinator (UNDRO) (1979) *Natural Disasters and Vulnerability Analysis*, Report of Expert Group Meeting, 9–12 July 1979, Geneva: UNDRO.

10

A system of indicators for disaster risk management in the Americas

Omar D. Cardona

Introduction

Disaster risk management requires risk "dimensioning", and risk measuring should take into account not only the expected physical damage, victims and economic equivalent loss, but also social, organisational and institutional factors. The difficulty in *achieving* effective disaster risk management has been, in part, the result of the lack of a comprehensive conceptual framework of disaster risk that could facilitate a multidisciplinary evaluation and intervention. Most existing indices and evaluation techniques do not adequately express risk and are not based on a holistic approach that invites intervention.

It is necessary to make risk "manifest" in different ways. The various planning agencies dealing with the economy, the environment, housing, infrastructure, agriculture or health, to mention but a few relevant areas, must be made aware of the risks that each sector faces. In addition, the concerns of different levels of Government should be addressed in a meaningful way. For example, risk is very different at the local level (a community or small town) than it is at the national level. If risk is not presented and explained in a way that attracts stakeholders' attention, it is not possible to make progress and reduce the impact of disasters.

Disaster risk is most detailed at a micro-social or territorial scale. As we aggregate and work at more macro scales, details are lost. However, decision-making and information needs at each level are quite different, as are the social actors and stakeholders. This means that appropriate evaluation tools are necessary to make it easy to understand the problem and guide the decision-making process. It is fundamentally important to understand how vulnerability is generated, how it increases and how it accumulates. Performance benchmarks are also needed to facilitate decision makers' access to relevant information as well as the identification and proposal of effective policies and actions.

The disaster risk management indicators programme in the Americas meets this need. The system of indicators proposed by the Instituto de Estudios Ambientales (IDEA) for the Inter-American Development Bank (IDB) permits a systematic and quantitative benchmarking of each country during different periods between 1980 and 2000, as well as comparisons across countries. It also provides a more analytically rigorous and data-driven approach to risk management decision-making. This system of indicators enables the depiction of disaster risk at the national level (but also at the sub-national and urban level, to illustrate its application in those scales), allowing the identification of key issues by economic and social category. It also makes possible the creation of national risk management performance benchmarks in order to establish performance targets for improving management effectiveness.

Creating a measurement system based on composite indicators is a major conceptual and technical challenge, which is made even more so when the aim is to produce indicators that are transparent, robust, representative, replicable, comparable and easy to understand. All methodologies have their limitations, which reflect the complexity of what is to be measured and what can be achieved. As a result, for example, the lack of data may make it necessary to accept approaches and criteria that are less exact or comprehensive than what would have been desired. These trade-offs are unavoidable when dealing with risk and may even be considered desirable. Based on the conceptual framework developed for the programme, a system of risk indicators is proposed that represents the current vulnerability and risk management situation in each country. The indicators proposed are transparent, relatively easy to update periodically, and easily understood by public policy makers.

The system of indicators, a product of the IDB-IDEA programme, provides a holistic approach to evaluation that is also flexible and compatible with other evaluation methods (Cardona, 2001 and 2004). As a result, it is likely to be increasingly used to measure risk and risk management conditions. The system's main advantage lies in its ability to disaggregate results and identify factors that should take priority in risk management actions, while measuring the effectiveness of those actions. The main objective is to facilitate the decision-making process. In other words, the concept underlying this methodology is one of controlling risk rather than obtaining a precise evaluation of it (physical truth). Four components or composite indicators have been designed to represent the main elements of vulnerability and show each country's progress in managing risk. They are described in the following sections. Programme reports, technical details and the application results for the countries in the Americas can be consulted at the following web page: http://idea.unalmzl.edu.co (Cardona et al., 2003a, 2003b, 2004a, 2004b, 2005; Carreño, Cardona and Barbat, 2005; IDEA, 2005).

The Disaster Deficit Index (DDI)

The DDI measures country risk from a macroeconomic and financial perspective in relation to possible catastrophic events. It requires the estimation of critical impacts during a given period of exposure, as well as the country's financial ability to cope with the situation. This index measures the economic loss that a particular country could suffer when a catastrophic event takes place, and the implications in terms of resources needed to address the situation. Construction of the DDI requires undertaking a forecast on the basis of historical and scientific evidence, as well as measuring the value of infrastructure and other goods and services that are likely to be affected. The DDI captures the relationship between the demand for contingent resources to cover the losses caused by the maximum considered event (MCE), and the public sector's economic resilience (that is, the availability of internal and external funds for restoring affected inventories).

$$DDI = \frac{MCE \text{ loss}}{Economic \text{ Resilience}} \tag{1}$$

Potential losses (index numerator) are calculated by using a model that takes into account different hazards (which are calculated in probabilistic form according to historical data on the intensity of past phenomena) and the actual physical vulnerability of the elements exposed to such phenomena (Ordaz and Santa-Cruz, 2003). Figure 10.1(c) shows a diagram illustrating how to obtain the DDI.

Economic resilience (the denominator of the index), on the other hand, represents internal and external resources that are available to the Government, in its role as a promoter of recovery and as owner of affected goods, when the evaluation is undertaken. Access to these resources has limitations and costs that must be taken into account as feasible values according to the macroeconomic and financial conditions of the country. In this evaluation the following aspects have been taken into account: the approximate *insurance and reinsurance payments* that the country would receive for goods and infrastructure insured by Government; the *reserve funds for disasters* that the country has available during the evaluation year; the funds that may be received as *aid and donations*, public or private, national or international; the possible value of *new taxes* that the country could collect in case of disasters; the *margin for budgetary reallocations* of the country, which usually corresponds to the margin of discretional expenses available to Government; the feasible value of *external credit* that the country could obtain from multilateral organisms and in the external capital market; and the *internal credit* the country may obtain from commercial and, at times, the Central Bank, when this is legal, signifying immediate liquidity.

A DDI greater than 1.0 reflects the country's inability to cope with extreme disasters even by going as much into debt as possible. The greater the DDI, the greater the gap between losses and the country's ability to face them. If constrictions for additional debt exist, this situation implies the impossibility to recover.

To help place the DDI in context, we have developed a complementary indicator, DDI′, to illustrate the portion of a country's annual capital expenditure (CE) that corresponds to the expected annual loss or the pure risk premium. That is, DDI′ shows the percentage of the annual investment budget that would be needed to pay for future disasters.

$$DDI' = \frac{\text{Expected annual loss}}{\text{Capital expenditures}} \tag{2}$$

The pure premium value is equivalent to the annual average investment or saving that a country would have to make in order to approximately cover losses associated with major future disasters.

These indicators provide a simple way of measuring a country's fiscal exposure and potential deficit (or contingency liabilities) in case of an extreme disaster. They allow national decision makers to measure the budgetary implications of such an event and highlight the importance of including this type of information in financial and budgetary processes (Freeman et al., 2002). These results substantiate the need to identify and propose effective policies and actions such as, for example, using insurance and reinsurance (transfer mechanisms) to protect Government resources or establishing reserves based on adequate loss estimation criteria. Other such actions include contracting contingency credits and, in particular, the need to invest in structural (retrofitting) and non-structural prevention and mitigation to reduce potential damage and losses as well as the potential economic impact of disasters.

The Local Disaster Index (LDI)

The LDI identifies the social and environmental risks resulting from more recurrent lower-level events (which are often chronic at the local and sub-national levels). These events have a disproportionate impact on more socially and economically vulnerable populations, and have highly damaging impacts on national development. This index represents the propensity of a country to experience small-scale disasters and shows their cumulative impact on local development. The index attempts to represent the spatial variability and dispersion of risk in a country resulting from small and recurrent events. This approach is concerned with the national significance of recurrent small-scale events that rarely enter international, or even national, disaster databases, but which pose a serious and cumulative development problem for local areas and, more than likely, also for the country as a whole. These events may be the result of socio-natural processes associated with environmental deterioration (Lavell, 2003a and 2003b) and are persistent or chronic in nature. They include landslides, avalanches, flooding, forest fires and drought as well as small earthquakes, hurricanes and volcanic eruptions.

The LDI is equal to the sum of three local disaster sub-indicators that are calculated based on data from the DesInventar database for number of deaths, number of people affected and losses in each municipality.[1]

$$LDI = LDI_{Deaths} + LDI_{Affected} + LDI_{Losses} \qquad (3)$$

The LDI captures simultaneously the incidence and uniformity of the distribution of local effects. That is, it accounts for the relative weight and persistence of the effects attributable to phenomena that give rise to municipal-scale disasters. The higher the relative value of the index, the more uniform the magnitude and distribution of the effects of various hazards among municipalities. A low LDI value means low spatial distribution of the effects among the municipalities where events have occurred. Figure 10.2 illustrates schematically how LDI is obtained for a country, based on the information of events in each municipality.

Similarly, we calculated an LDI that takes into account the concentration of losses (direct physical damage) at the municipal level and is aggregated for all events in all countries. This indicator shows the disparity of risk within a single country. An LDI value close to 1.0 means that most of the losses for the country are concentrated in a few municipalities.

The usefulness of these indices for economic analysts and sector officials in charge of establishing rural and urban policies lies in the fact that they make it possible to measure the persistence and cumulative impact of local disasters. Thus decision makers can prompt the con-

194

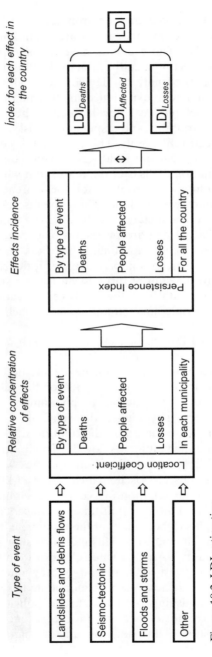

Figure 10.2 LDI estimation.
Source: Author.

sideration of risk in territorial planning at the local level, as well as the protection of hydrographic basins. The indices can also be used to justify resource transfers to the local level that are earmarked for risk management and the creation of social safety nets.

The Prevalent Vulnerability Index (PVI)

The PVI depicts predominant vulnerability conditions by measuring exposure in prone areas, socio-economic fragility and lack of social resilience. These items provide a measure of direct as well as indirect and intangible impacts of hazard events. The index is a composite indicator that provides a comparative measure of a country's pattern or situation. Inherent vulnerability conditions underscore the relationship between risk and development (UNDP, 2004).[2] Vulnerability, and therefore risk, are the result of inadequate economic growth and deficiencies, which may be corrected by means of adequate development processes. Although the indicators proposed are recognised as useful for measuring development (Holzmann and Jorgensen, 2000; Holzmann, 2001), their use here is intended to capture favourable conditions for direct physical impacts (exposure and susceptibility), as well as indirect and, at times, intangible impacts (socio-economic fragility and lack of resilience) of potential physical events (Masure, 2003; Davis, 2003). The PVI is an average of these three types of composite indicators:

$$PVI = (PVI_{Exposure} + PVI_{Fragility} + PVI_{Lack\ of\ Resilience})/3 \qquad (4)$$

The indicators used for describing exposure, prevalent socio-economic conditions and lack of resilience have been estimated in a consistent fashion (directly or in inverse fashion, respectively), recognising that their influence explains why adverse economic, social and environmental impacts take place following a dangerous event (Cardona and Barbat, 2000; Cardona, 2004). Each one is made up of a set of indicators that express situations, causes, susceptibilities, weaknesses or relative absences which affect the country, region or locality under study, and which would benefit from risk reduction actions. The indicators were based on figures, indices, existing rates or proportions derived from reliable databases available worldwide or in each country.

The best indicators of exposure and/or physical susceptibility (PVI_{ES}) are the susceptible population, assets, investment, production, livelihoods, historic monuments and human activities (Masure, 2003; Lavell, 2003b). Other indicators include population growth and density rates, as

Table 10.1 PVI_{ES} estimation

Description	Indicator	Weight	
Population growth, average annual rate (%)	ES1	w1	
Urban growth, avg. annual rate (%)	ES2	w2	
Population density, people/5 Km^2	ES3	w3	
Poverty-population below US$ 1 per day PPP	ES4	w4	
Capital stock, million US$ dollar/1000 km^2	ES5	w5	PVI_E
Imports and exports of goods and services, % GDP	ES6	w6	
Gross domestic fixed investment, % of GDP	ES7	w7	
Arable land and permanent crops, % land area	ES8	w8	

Source: Author.

well as agricultural and urban growth rates. Table 10.1 shows the PVI_{ES} composition.

These variables reflect the nation's susceptibility to dangerous events, whatever their nature or severity. Exposure and susceptibility are necessary conditions for the existence of risk. Although, in any strict sense, it would be necessary to establish if exposure is relevant for each potential type of event, we may nevertheless assert that certain variables reflect comparatively adverse situations where natural hazards can be deemed to be permanent external factors without needing to establish their exact nature.

Socio-economic fragility (PVI_{SF}) may be represented by indicators such as poverty, lack of personal safety, dependency, illiteracy, income inequality, unemployment, inflation, debt and environmental deterioration. These indicators reflect relative weaknesses that increase the direct effects of dangerous phenomena (Cannon, 2003; Davis, 2003; Wisner, 2003). Even though these effects are not necessarily cumulative (and in some cases may be superfluous or correlated), their influence is especially important at the social and economic levels (Benson, 2003). Table 10.2 shows the PVI_{SF} composition.

These indicators show that an intrinsic predisposition for adverse social impacts in the face of dangerous phenomena exists regardless of their nature or intensity (Lavell, 2003b; Wisner, 2003). The propensity to suffer negative impacts establishes a vulnerability condition of the population, although it would be necessary to establish the relevance of this propensity in the face of all types of hazard. Nevertheless, as with exposure, it is possible to suggest that certain values of specific variables reflect a relatively unfavourable situation in the eventuality of natural hazards, regardless of the exact characteristics of those hazards.

Lack of resilience (PVI_{LR}), seen as a vulnerability factor, may be rep-

Table 10.2 PVI$_{SF}$ estimation

Description	Indicator	Weight	
Human Poverty Index, HPI-1	**SF1**	w1	
Dependents as proportion of working age population	**SF2**	w2	
Social disparity, concentration of income measured using Gini index	**SF3**	w3	
Unemployment, as % of total labour force	**SF4**	w4	PVI$_S$
Inflation, food prices, annual %	**SF5**	w5	
Dependency of GDP growth of agriculture, annual %	**SF6**	w6	
Debt servicing, % of GDP	**SF7**	w7	
Human-induced soil degradation (GLASOD)	**SF8**	w8	

Source: Author.

resented by means of the complementary or inverse relationship of a number of variables that measure human development, human capital, economic redistribution, governance, financial protection, community awareness, the degree of preparedness to face crisis situations, and environmental protection.[3] These indicators are useful to identify and guide actions to improve personal safety (Cannon, 2003; Davis, 2003; Lavell, 2003a and 2003b; Wisner 2003). Table 10.3 shows the PVI$_{LR}$ composition.

These indicators capture the capacity to recover from or absorb the impact of dangerous phenomena, whatever their nature and severity (Briguglio, 2003). Not being able to cope adequately with disasters is a vulnerability condition, although in a strict sense it is necessary to estab-

Table 10.3 PVI$_{LR}$ estimation

Description	Indicator	Weight	
Human development index, HDI [Inv]	**LR1**	w1	
Gender-related development index, GDI [Inv]	**LR2**	w2	
Social expenditure; on pensions, health, and education, % of GDP [Inv]	**LR3**	w3	
Governance index (Kaufmann) [Inv]	**LR4**	w4	
Insurance of infrastructure and housing, % of GD [Inv]	**LR5**	w5	PVI$_{LR}$
Television sets per 1,000 people [Inv]	**LR6**	w6	
Hospital beds per 1,000 people [Inv]	**LR7**	w7	
Environmental sustainability index, ESI [Inv]	**LR8**	w8	

Source: Author.

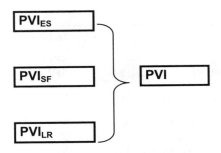

Figure 10.3 PVI evaluation
Source: Author.

lish this with reference to all potential types of hazard. Nevertheless, as with exposure and socio-economic fragility, we can posit that some economic and social variables reflect a comparatively unfavourable position if natural hazards exist (Benson, 2003). The factors of lack of resilience are not very dependent or conditioned by the action of the event.

In general, PVI reflects susceptibility due to the degree of physical exposure of goods and people, PVI_{ES}, which favour the direct impact in case of hazard events. In the same way, it reflects conditions of socio-economic fragility that favour the indirect and intangible impact, PVI_{SF}. Also, it reflects lack of capacity to absorb consequences, for efficient response and recovering, PVI_{LR}. Reduction of these kinds of factors, as the purpose of the human sustainable development process and explicit policies for risk reduction, is one of the aspects that should be emphasised. Figure 10.3 shows how PVI is obtained.

The PVI should form part of a system of indicators that allow the implementation of effective prevention, mitigation, preparedness and risk transfer measures to reduce risk. The information presented by an index such as the PVI should prove useful to ministries of housing and urban development, environment, agriculture, health and social welfare, economy and planning. Although the relationship between risk and development should be emphasised, it must be noted that activities to promote development do not, in and of themselves, automatically reduce vulnerability.

The Risk Management Index (RMI)

The RMI brings together a group of indicators that measure a country's risk management performance. These indicators reflect the organisational, development, capacity and institutional actions taken to reduce vulnerability and losses, to prepare for crisis and to recover efficiently from disasters. This index was designed to assess risk management *performance*. It provides a qualitative measure of management based on predefined *targets* or *benchmarks* that risk management efforts should aim to achieve. The design of the RMI involved establishing a scale of achievement levels (Davis, 2003; Masure, 2003) or determining the "distance" between current conditions and an objective threshold or conditions in a reference country (Munda, 2003).

The RMI was constructed by quantifying four public policies, each of which has six indicators. The policies include the identification of risk, risk reduction, disaster management, and governance and financial protection. Risk identification (RI) is a measure of individual perceptions, how those perceptions are understood by society as a whole, and the objective assessment of risk. Risk reduction (RR) involves prevention and mitigation measures. Disaster management (DM) involves measures of response and recovery. And, finally, governance and financial protection (FP) measures the degree of institutionalisation and risk transfer. The RMI is defined as the average of the four composite indicators:

$$RMI = (RMI_{RI} + RMI_{RR} + RMI_{DM} + RMI_{FP})/4 \qquad (5)$$

Each indicator was estimated on the basis of five performance levels (*low, incipient, significant, outstanding,* and *optimal*) that correspond to a range from 1 (low) to 5 (optimal).[4] This methodological approach permits the use of each reference level simultaneously as a "performance target" and allows for comparison and identification of results or achievements. Government efforts of formulating, implementing and evaluating policies should bear these performance targets in mind (Carreño et al., 2004; Carreño, 2006)

It is important to recognise and understand the collective risk to design prevention and mitigation measures. The design of prevention and mitigation measures depends on the individual and social risk awareness and the methodological approaches to assess risk. It is then necessary to measure risk and depict it by means of models, maps and indices capable of providing accurate information for society as a whole and, in particular, for decision makers. Methodologically, RMI_{RI} includes the evaluation of hazards, the characteristics of vulnerability in the face of these hazards and estimates of the potential impacts during a particular period of exposure. The measurement of risk seen as basis for risk mitigation (only) is relevant when the population recognises and understands that risk. Table 10.4 shows the RMI_{RI} composition.

Table 10.4 RMI_{RI} estimation

Description	Indicator	Weight	
Systematic disaster and loss inventory	**RI1**	w1	
Hazard monitoring and forecasting	**RI2**	w4	
Hazard evaluation and mapping	**RI3**	w5	
Vulnerability and risk assessment	**RI4**	w6	RMI_{RI}
Public information and community participation	**RI5**	w7	
Training and education on risk management	**RI6**	w8	

Source: Author.

The major aim of risk management is to reduce risk (RMI_{RR}). Reducing risk generally requires the implementation of structural and nonstructural prevention and mitigation measures. It implies a process of anticipating potential sources of risk, putting into practice procedures and other measures to either avoid hazard, when that is possible, or reduce the economic, social and environmental impacts through corrective and prospective interventions of existing and future vulnerability conditions. Table 10.5 shows the RMI_{RR} composition.

The goal of disaster management (RMI_{DM}) is to provide appropriate response and recovery efforts following a disaster. It is a function of the degree of preparedness of the responsible institutions as well as the community as a whole. The goal is to respond efficiently and appropriately when risk has become disaster. Effectiveness implies that the institutions

Table 10.5 RMI_{RR} estimation

Description	Indicator	Weight	
Risk consideration in land use and urban planning	**RR1**	w1	
Hydrographic basin intervention and environmental protection	**RR2**	w4	
Implementation of hazard-event control and protection techniques	**RR3**	w5	
Housing improvement and human settlement relocation from prone areas	**RR4**	w6	RMI_{RR}
Updating and enforcement of safety standards and construction codes	**RR5**	w7	
Reinforcement and retrofitting of public and private assets	**RR6**	w8	

Source: Author.

Table 10.6 RMI$_{DM}$ estimation

Description	Indicator	Weight	
Organisation and coordination of emergency operations	**DM1**	w1	
Emergency response planning and implementation of warning systems	**DM2**	w4	
Endowment of equipments, tools and infrastructure	**DM3**	w5	RMI$_{DM}$
Simulation, updating and test of interinstitutional response	**DM4**	w6	
Community preparedness and training	**DM5**	w7	
Rehabilitation and reconstruction planning	**DM6**	w8	

Source: Author.

(and other actors) involved have adequate organisational abilities, as well as the capacity and plans in place to address the consequences of disasters. Table 10.6 shows the RMI$_{DM}$ composition.

Adequate governance and financial protection (RMI$_{FP}$) are fundamental for sustainability, economic growth and development. They are also basic to risk management, which requires coordination among social actors as well as effective institutional actions and social participation. Governance also depends on an adequate allocation and use of financial resources to manage and implement appropriate retention and transfer strategies for dealing with disaster losses. Table 10.7 shows the RMI$_{FP}$ composition. Lastly, Figure 10.4 shows how to obtain RMI.

Table 10.7 RMI$_{FP}$ estimation

Description	Indicator	Weight	
Interinstitutional, multisectoral and decentralising organisation	**FP1**	w1	
Reserve funds for institutional strengthening	**FP2**	w4	
Budget allocation and mobilisation	**FP3**	w5	
Implementation of social safety nets and funds response	**FP4**	w6	RMI$_{FP}$
Insurance coverage and loss transfer strategies of public assets	**FP5**	w7	
Housing and private sector insurance and reinsurance coverage	**FP6**	w8	

Source: Author.

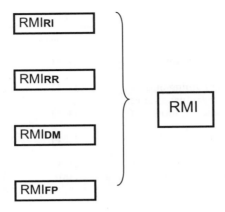

Figure 10.4 RMI evaluation
Source: Author.

Indicators at sub-national and urban level

Depending on the country, sub-national divisions (department, States or provinces) have different degrees of political, financial and administrative autonomy. Nevertheless, the system of indicators that was developed allows for individual or collective evaluation of sub-national areas and was developed using the same concepts and approaches outlined for the nation as a whole. All results for the indicators and for different periods are included in the reports of Barbat and Carreño (2004a, 2004b) and Carreño (2006). Risk analysis can further be disaggregated to metropolitan areas, which are usually made up of administrative units such as districts, municipalities, communes or localities, which will have different risk levels.

Dropping down the spatial and administrative scale, the need for evaluations within urban-metropolitan areas and large cities is also desirable. Taking into account the spatial scale at which urban risk analysis is undertaken, it is necessary to estimate or create scenarios for damage and loss that could occur for the different exposed elements that characterise the city (i.e. buildings, public works, roads). The estimation of an MCE for the city would allow us to evaluate in greater detail the potential direct damage and impacts so as to prioritise interventions and actions required to reduce risk in each area of the city.

The urban risk indicators are similar to those used at other levels but with the addition of two new indicators: the index of physical risk, R_P, and the impact factor, F. The former is based on hard data, while the latter is based on soft variables that depict social fragility and lack of resilience. In turn, these two indicators allow us to create a total risk index, R_T, for each unit of analysis. These indicators require greater detail than those used at the national or regional level and they focus on urban variables (Cardona and Barbat, 2000; Barbat, 2003a, 2003b; Barbat and Carreño, 2004a, 2004b; Carreño 2006). In other words, we have developed a methodology that combines the disaster deficit and the prevalent vulnerability indices used for the national and sub-national analyses. Table 10.8 shows how to obtain total risk indices for each analysis unit at urban level.

Conclusions and future analysis

The IDB-IDEA programme of indicators puts heavy emphasis on *developing a language of risk* that various types of decision makers can understand. The disaster deficit, local disaster and prevalent vulnerability indices (DDI, LDI and PVI) are risk proxies, which measure different factors that affect overall risk at the national and sub-national levels. By depicting existing risk conditions, the indicators highlight the need for intervention. This study indicates that the countries of the region face significant risks that have yet to be fully recognised or taken into account by individuals, decision makers and society as a whole. These indicators are a first step in correctly measuring risk so that it can be given the priority that it deserves in the development process. Once risk has been identified and measured, activities can then be implemented to reduce and control it. The first step in addressing risk is to recognise it as a significant socioeconomic and environmental problem. The RMI is also novel and far more wide reaching in its scope than other similar attempts have been in the past. In some ways this is the most sensitive and interesting indicator of all. It is certainly the one that can show the fastest rate of change, given improvements in political will or deterioration of governance. This index has the advantage of being composed of measures that more or less directly map sets of specific decisions/actions onto sets of desirable outcomes.

The indicators of risk and risk management described here have permitted an evaluation of 12 Latin American and Caribbean countries based on integrated criteria. The results show that it is possible to describe risk and risk management using coarse grain measures and classify countries according to a relative scale. An evaluation of individual countries allowed us to compare individual performance indicators for the period from 1980 to 2000.[5] The reports of the programme also estimated the indicators at the sub-national and urban level. This profile is a first

Table 10.8 Indicators of physical risk, social fragility and lack of resilience and their weights

$$R_T = R_P(1 + F)$$

Ind	Description	w		
F_{RF1}	Damaged area	w1		
F_{RF2}	Number of deceased	ws		
F_{RF3}	Number of injured	w3		
F_{RF4}	Ruptures in water mains	w4	R_P	Physical risk
F_{RF5}	Rupture in gas network	w5		
F_{RF6}	Fallen lengths on HT power lines	w6		
F_{RF7}	Telephone exchanges affected	w7		
F_{RF8}	Electricity substations affected	w8		

Ind	Description	w		
F_{FS1}	Slums-squatter neighbourhoods	w1		
F_{FS2}	Mortality rate	w2		
F_{FS3}	Delinquency rate	w3		
F_{FS4}	Social disparity index	w4		
F_{FS5}	Population density	w5		
F_{FR1}	Hospital beds	w6	F	Impact factor
F_{FR2}	Health human resources	w7		
F_{FR3}	Public space/shelter facilities	w8		
F_{FR4}	Rescue and firemen manpower	w9		
F_{FR5}	Development level	w10		
F_{FR6}	Preparedness/emergency planning	w11		

Source: Author.

step for creating a "common operating picture" of disaster risk reduction for the region. That is, it represents a common knowledge base that can be accessed, viewed and understood by all of the different policy makers responsible for disaster risk reduction in the region. Any group that is not included or that fails to comprehend the level and frequency of risk is likely to fail to engage actively in the risk reduction process. Consequently, the construction of an effective common knowledge base for the system of decision makers responsible for disaster risk reduction is fundamental for achieving change in practice.

Undoubtedly, the construction of the indicators is methodologically complex for run-of-the-mill professionals while the demands for information are relatively onerous in some cases, given access and identification problems. Certain variables or types of information are not readily available and require research, as opposed to rote collection where such information exists as a normal part of data systematisation at the national or international levels. Doubts exist as to the veracity and accuracy of some items of information, although overall the procedures used to "test" the information assure a very reasonable level of accuracy and veracity. In the same way, weighting procedures and decisions could be questioned at times but again, overall, the decisions taken seem to be well justified and lead to adequate levels of accuracy. The use of official employees of risk management institutions at the national level in order to undertake the qualitative analyses is open to revision given the clear bias, in some cases, in favour of positive qualifications. The alternative, using scientists, informed, independent persons and academics would resolve certain problems but might create others. Thus, a crosscheck double entry approach may be best, where both types of sectors are taken into consideration.

To date the system of indicators has been opened up to scrutiny and discussion by international advisors, academics, risk professionals and a limited number of national, technical and professional staff, but to few policy makers as such. In the short term it would thus be very wise to organise a series of national dialogues where the derived indicator results and implications are presented to a selected number of national level policy and decision makers. This would allow a testing of relevance and pertinence and offer conclusions as regards future work on the programme. It is very important to take into account the set of "next steps" that might be taken to improve the reliability and validity of the data collected and the analyses undertaken. In the future, sustainability for the programme and promoting its applicability at the decision-making level require, among other things:

- dissemination of the guidelines to easy analysis and indicator calculation
- transformation of indices into political indicators

- the diffusion and acceptance of the indicators and the method by national decision makers in analysed countries and in other countries
- an agreement as to procedures for future collection of information and analysis.

Lastly, perhaps the most important contribution of the programme was to initiate a systematic procedure of measuring and documenting disaster risk across the 12 nations engaged in this project. Once initiated, however, the programme itself becomes a process in which the participants learn by engaging in data collection, analysis and interpretation of findings. Some of the methods, adopted because no other measures existed, may now be re-examined and redesigned as cumulative data show new possibilities for refining the measures, or as data collection methods yield new possibilities for more complete and comprehensive documentation of risk and risk reduction practices.

Notes

1. DesInventar is a database implemented by La Red de Estudios Sociales en Prevención de Desastres de América Latina (LA RED).
2. Inherent vulnerability conditions are the predominant socio-economic conditions that favour or facilitate negative effects as a result of adverse physical phenomena (Briguglio, 2003).
3. The symbol [Inv] is used here to indicate an inverse variable ($\neg R = 1 - R$).
4. It is also possible to estimate the RMI by means of weighted sums of fixed values (such as 1 through 5, for example), instead of using fuzzy sets and linguistic descriptions. However, that simplification eliminates the non-linearity of risk management and yields less accurate results.
5. For obvious space limitations the results for each country cannot be included in this paper.

REFERENCES

Barbat, A. (2003a) *Vulnerability and Disaster Risk Indices from Engineering Perspective and Holistic Approach to Consider Hard and Soft Variables at Urban Level*, IDB/IDEA Program of Indicators for Risk Management, Manizales: National University of Colombia, available at http://idea.unalmzl.edu.co/.

Barbat, A. (2003b) *Detailed Application of the Holistic Approach for Seismic Risk Evaluation on an Urban Center using Relative Indices*, IDB/IDEA Program of Indicators for Risk Management, Manizales: National University of Colombia, available at http://idea.unalmzl.edu.co/.

Barbat, A.H. and M.L. Carreño (2004a) *Indicadores de Riesgo y Gestión a Nivel Subnacional: Aplicación Demostrativa en los Departamentos de Colombia*, IDB/IDEA Program of Indicators for Risk Management, Manizales: National University of Colombia, available at http://idea.unalmzl.edu.co/.

Barbat, A.H. and M.L. Carreño (2004b) *Análisis de Riesgo Urbano Utilizando Indicadores: Aplicación Demostrativa Para la Ciudad de Bogotá, Colombia*, IDB/IDEA Program of Indicators for Risk Management, Manizales: National University of Colombia, available at http://idea.unalmzl.edu.co/.

Benson, C. (2003) *Potential Approaches to the Development of Indicators for Measuring Risk from a Macroeconomic Perspective*, IDB/IDEA Program of Indicators for Risk Management, Manizales: National University of Colombia, available at http://idea.unalmzl.edu.co/.

Briguglio, L. (2003) *Methodological and Practical Considerations for Constructing Socio-economic Indicators to Evaluate Disaster Risk*, IDB/IDEA Program of Indicators for Risk Management, Manizales: National University of Colombia, available at http://idea.unalmzl.edu.co/.

Cannon, T. (2003) *Vulnerability Analysis, Livelihoods and Disasters Components and Variables of Vulnerability: Modelling and Analysis for Disaster Risk Management*, IDB/IDEA Program of Indicators for Risk Management, Manizales: National University of Colombia, available at http://idea.unalmzl.edu.co/.

Cardona, O.D. (2001) *Estimación Holística del Riesgo Sísmico Utilizando Sistemas Dinámicos Complejos*, Barcelona: Technical University of Catalonia, available at http://www.desenredando.org/public/varios/2001/ehrisusd/index.html.

Cardona, O.D. (2004) "The Need for Rethinking the Concepts of Vulnerability and Risk from a Holistic Perspective: A Necessary Review and Criticism for Effective Risk Management", in G. Bankoff, G. Frerks and D. Hilhorst, eds, *Mapping Vulnerability: Disasters, Development and People*, London: Earthscan.

Cardona, O.D. and A.H. Barbat (2000) *El Riesgo Sísmico y su Prevención*, Cuaderno Técnico 5, Madrid: Calidad Siderúrgica.

Cardona, O.D., J.E. Hurtado, G. Duque, A. Moreno, A.C. Chardon, L.S. Velásquez and S.D. Prieto (2003a) *The Notion of Disaster Risk: Conceptual Framework for Integrated Risk Management*, IADB/IDEA Program on Indicators for Disaster Risk Management, Manizales: National University of Colombia, available at http://idea.manizales.unal.edu.co/ProyectosEspeciales/adminIDEA/CentroDocumentacion/DocDigitales/documentos/01%20Conceptual%20Framework%20IADB-IDEA%20Phase%20I.pdf.

Cardona, O.D., J.E. Hurtado, G. Duque, A. Moreno, A.C. Chardon, L.S. Velásquez and S.D. Prieto (2003b) *Indicators for Risk Measurement: Methodological Fundamentals*, IDB/IDEA Program of Indicators for Disaster Risk Management, Manizales: National University of Colombia, available at http://idea.unalmzl.edu.co/.

Cardona, O.D., J.E. Hurtado, G. Duque, A. Moreno, A.C. Chardon, L.S. Velásquez and S.D. Prieto (2004a) *Disaster Risk and Risk Management Benchmarking: A Methodology Based on Indicators at National Level*. IDB/IDEA Program of Indicators for Disaster Risk Management, Manizales: National University of Colombia, available at http://idea.unalmzl.edu.co/.

Cardona, O.D., J.E. Hurtado, G. Duque, A. Moreno, A.C. Chardon, L.S. Velásquez and S.D. Prieto (2004b) *Results of Application of the System of Indicators on Twelve Countries of the Americas*, IDB/IDEA Program of Indicators for Disaster Risk Management, Manizales: National University of Colombia, available at http://idea.unalmzl.edu.co/.

Cardona, O.D., J.E. Hurtado, G. Duque, A. Moreno, A.C. Chardon, L.S. Velásquez and S.D. Prieto (2005) *Indicators of Disaster Risk and Risk Management: Program for Latin American and the Caribbean: Summary Report*, IDB/IDEA Program of Indicators for Disaster Risk Management, Manizales: National University of Colombia, available at http://www.iadb.org/int/DRP/Ing/Red6/Docs/IDEAR06-05eng.pdf.

Carreño, M.L. (2006) *Técnicas Innovadoras para la Evaluación del Riesgo Sísmico y su Gestión en Centros Urbanos: Acciones Ex Ante y Ex Post*, Doctoral Thesis, Barcelona: Technical University of Catalonia.

Carreño, M.L, O.D. Cardona and A.H. Barbat (2004) *Metodología para la Evaluación del Desempeño de la Gestión del Riesgo*, Monografías CIMNE, Barcelona: Technical University of Catalonia.

Carreño, M.L, O.D. Cardona and A.H. Barbat (2005) *Sistema de Indicadores para la Evaluación de Riesgos*, Monografía CIMNE IS-52, Barcelona: Technical University of Catalonia.

Davis, I. (2003) *The Effectiveness of Current Tools for the Identification, Measurement, Analysis and Synthesis of Vulnerability and Disaster Risk*, IDB/IDEA Program of Indicators for Disaster Risk Management, Manizales: National University of Colombia, available at http://idea.unalmzl.edu.co/.

Freeman, P.K., L.A. Martin, J. Linnerooth-Bayer, R. Mechler, G. Pflug and K. Warner (2002) *Disaster Risk Management: National Systems for the Comprehensive Management of Disaster Financial Strategies for Natural Disaster Reconstruction*, SDD/IRPD, Regional Policy Dialogue, Washington D.C.: Inter-American Development Bank.

Holzmann, R. (2001) "Risk and Vulnerability: The Forward Looking Role of Social Protection in a Globalizing World", Social Protection Working Paper No. 0109/June, Washington, D.C.: World Bank.

Holzmann, R. and S. Jorgensen (2000) "Manejo Social del Riesgo: Un Nuevo Marco Conceptual para la Protección Social y Más Allá", Social Protection Working Paper No.0006, Washington, D.C.: World Bank.

Instituto De Estudios Ambientales (IDEA) (2005) *System of Indicators for Disaster Risk Management: Program for Latin American and the Caribbean. Main Technical Report*, IDB/IDEA Program of Indicators for Disaster Risk Management, Manizales: National University of Colombia, available at http://idea.unalmzl.edu.co/.

Lavell, A. (2003a) *I. International Agency Concepts and Guidelines for Disaster Risk Management; II. The Transition from Risk Concepts to Risk Indicators*, IDB/IDEA Program of Indicators for Risk Management, Manizales: National University of Colombia, available at http://idea.unalmzl.edu.co/.

Lavell, A. (2003b) *Approaches to the Construction of Risk Indicators at Different Spatial or Territorial Scales and the Major Components of Indicator Systems: Conceptual Bases, Risk Construction Processes and Practical Implications*, IDB/IDEA Program of Indicators for Risk Management, Manizales: National University of Colombia, available at http://idea.unalmzl.edu.co/.

Masure, P. (2003) *Variables and Indicators of Vulnerability and Disaster Risk for Land-use and Urban or Territorial Planning*, IDB/IDEA Program of Indicators for Risk Management, Manizales: National University of Colombia, available at http://idea.unalmzl.edu.co/.

Munda, G. (2003) *Methodological Exploration for the Formulation of a Socio-Economic Indicators Model to Evaluate Disaster Risk Management at the National and Sub-National Levels: A Social Multi-Criterion Model*, IDB/IDEA Program of Indicators for Risk Management, Manizales: National University of Colombia, available at http://idea.unalmzl.edu.co/.

Ordaz, M. and S. Santa-Cruz (2003) *Computation of Physical Damage to Property Due to Natural Hazard Events*, IDB/IDEA Program of Indicators for Disaster Risk Management, Manizales: National University of Colombia, available at http://idea.unalmzl.edu.co/.

United Nations Development Programme (UNDP) (2004) *Reducing Disaster Risk: A Challenge for Development. A Global Report*, New York: UNDP–Bureau for Crisis Prevention and Recovery (BRCP), available at http://www.undp.org/bcpr/disred/rdr.htm.

Wisner, B. (2003) *Turning Knowledge into timely and Appropriate Action: Reflections on IADB/IDEA Program of Disaster Risk Indicators*, IDB/IDEA Program of Indicators for Risk Management, Manizales: National University of Colombia, available at http://idea.unalmzl.edu.co/.

11

Multi-risk assessment of Europe's regions

Stefan Greiving

Abstract

This chapter presents a methodology for assessing the risk potential of a certain area by means of aggregating all spatially relevant risks that are caused by natural and technological hazards. The approach was elaborated and applied Europe-wide in the context of the project "Spatial effects of natural and technological hazard in general and in relation to climate change", which is part of the European Spatial Planning Observation Network (ESPON, see www.espon.lu). An aggregated hazard map, an integrated vulnerability map and an aggregated risk map are the key results of this research project. Vulnerability is recognised as the key component of risk and consists of the two elements: degree of exposure to hazard, one the one hand, and coping capacity on the other.

Background

A risk is unavoidable whenever a decision is made whether it has spatial relevance or not. In this context, space can be defined as the area within which human beings and their artefacts may be threatened by spatially relevant hazards. The decision about whether to tolerate the risk or to try to alter it can be understood as an integral part of the existing socio-economic structures and institutions, with spatial planning representing one element in the total equation.

Spatial planning makes decisions for society about whether and how certain spaces will be used. Therefore, spatial planning influences vulnerability in cases of spatially relevant natural and technological hazards. The spatial character of a hazard can be defined by spatial effects that might occur if a hazard turns into a disaster, or by the possibility for an appropriate spatial planning response. This dual character also opens up questions about the relevance of different levels of spatial planning as well as the relationship to sectoral planning. Furthermore, the nature of spatial planning requires a determined, multi-risk approach, which considers all relevant hazards that threaten a certain area as well as the vulnerability of this area.

It is a fact that every hazard has a spatial dimension; it takes place somewhere. However, spatial relevance is not yet spatial planning relevance, but it might be of interest for a sectoral planning division or an emergency response unit.

One of the most serious problems in this context is represented by so-called external effects: a spatial and temporal inconsistency between opportunities (e.g. benefits from new settlements) and risks, which are necessarily part of any decision-making about future land use or whether to invest in construction at a certain location. The classic illustration of this planning problem is represented by the (intra-generational) conflict between actors who are located upstream and those who are downstream. A municipality located upstream may profit from the existence of an industrial area that is located in the flood plain of a river, and may protect this area by means of a dike. One direct consequence of this action would be an increased flood risk for areas located downstream, because of the reduced flood-plain capacity in combination with flood waves, which would occur faster and with a higher peak.

Beside this kind of intra-generational conflict, there are also intergenerational aspects to be considered, because decisions taken today affect the risks and quality of life experienced by those living tomorrow (Rawls, 1971, "theory of justice"). For this reason, some kind of regulative spatial planning is needed in order to protect the interests of future generations.

Spatial planning has to anticipate the consequences (or better, opportunities and risks) of actions from the very beginning of a planning process as part of the exercise to define goals. In addition to a continuous evaluation and reviewing of fixed planning goals, planners should also consider the effect of implemented measures on the environment.

Keeping in mind the core elements of sustainable development laid out in the Rio declaration in 1992, it is clear that societies cannot be sustainable if they face increasing risks from natural and technological hazards. The US National Science and Technology Council has pointed out that:

Sustainable development must be resilient with respect to the natural variability of the earth and the solar system. The natural variability includes such forces as floods and hurricanes and shows that much economic development is unacceptably brittle and fragile. (FEMA, 1997: 2)

Godschalk et al. (1999: 526) have argued that "a resilient community is one that lives in harmony with nature's varying cycles and processes." This includes events like earthquakes, storms and floods as natural events, which cause the most harm for a non-sustainable society.

In consequence, a fourth criterion should be added to sustainability's economic, social and environmental components (Greiving, 2002: 203). Sustainability can be understood as a responsibility to develop mechanisms to help societies enhance the capacity and resilience needed to adapt to the future consequences of present processes.

In view of the flexibility that is needed for a strategy aiming at resiliency, it would be quite problematic to develop a detailed set of instruments and measures as a kind of corset for planning practice. Such an approach would fail because of the very nature of planning: even from a theoretical point of view, given the unpredictability of both social development and natural processes, it is impossible to create measures which would be valid for each individual case of planning. Moreover the large number of relevant hazards, which may interact with the result of cumulative effects, has to be taken into account. Finally, the variety of planning systems, and of course the multitude of natural and socio-economic settings in general and especially the given differences in the national planning systems, make a formulation of coherent instruments or concrete mitigation measures almost impossible.

In consequence, the formulation of guidelines for harmonising a successful planning process and used methodologies seems to be more promising than the formulation of general measures intended to fit all hazards. Harmonised risk assessment methodologies can be understood in this context as crucial tools for achieving valid and comparable results within a threatened area.

A spatially oriented risk assessment has three main characteristics: first, it has to be multi-hazard oriented, which means that it must go beyond sectoral considerations of risks. Second, only those risks with spatial relevance are considered. This means that ubiquitous risks like epidemic diseases or traffic accidents are not the focus of the analysis. And third, only collective risks that threaten a community as a whole are relevant, not individual risks like driving in a car or smoking.

A spatial approach to risk is of high relevance for those authorities and stakeholders that act in a spatial context. This encompasses those persons or institutions that make spatially relevant decisions, typically involving large amounts of data and complex decision-making processes including normative weighting procedures. These actors may be interested in a spatial risk assessment approach because they are charged with ensuring spatial development (e.g. land-use planning, regional development funding) or with insuring spatial structures (e.g. offering insurance or re-insurance services).

In view of these requirements, it should be highlighted that several risk assessment approaches have been recently extended from a single to a multi-hazard perspective (e. g. software HAZUS MH, see FEMA http:// www.fema.gov/hazus/hz_index.shtm; UNDP Disaster Risk Index, see Peduzzi, Chapter 8, and Pelling, Chapter 7).

Integrative approaches for assessing hazards in their spatial context ("hazards of place") have been developed by geographers since the 1970s (Hewitt and Burton, 1971; Cutter and Solecki, 1989). Further methodological elaborations on this subject have rarely been attempted, as Cutter (1996) points out. Especially in Europe a multi-hazard approach has not been used in spatial planning for many years. Although there is a tradition of spatial planning research in the context of single hazards (coastal flooding, river flooding, earthquakes and nuclear power plants), a synthetic consideration of spatially relevant hazards has only recently been addressed by a few authors (Egli, 1996; Burby, 1998; Greiving, 2002; Fleischhauer, 2004). One reason for this recent change of perception is the realisation that risk potential is increasing and that it is not sufficient to restrict risk policies only to the response phase of the emergency management cycle. The mitigation of hazards is essential for promoting sustainable development, but appropriate spatial planning tools have still to be developed. Thus, a methodology for a spatial risk assessment has to take into account the following criteria in order to meet the required goals: (1) a multi-hazard perspective, (2) a spatial perspective, and (3) an integration of risk components (hazards and vulnerability). Figure 11.1 indicates how the different components are defined and interlinked.

The multi-risk approach described in this chapter is a harmonised assessment methodology. It aims to assess the risk potential of a certain area by means of aggregating all spatially relevant risks that are caused by natural and technological hazards.

Intention of the approach

The approach was developed at the Institute of Spatial Planning, University of Dortmund (IRPUD) and was first applied and adjusted for a supranational risk assessment at the regional level, assessing the integrated risk potentials of the approximately 1,500 NUTS-3 regions of the

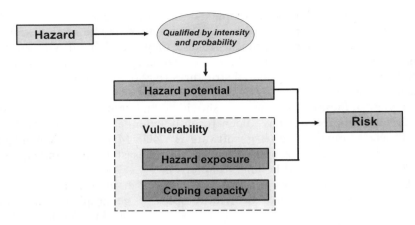

Figure 11.1 Components of risk.
Source: Author.

enlarged European Union (EU-27+2) (Schmidt-Thomé, 2005).[1] In principle, however, the methodology can be applied at any geographical level
and for any hazard and risk-related purpose (Batista et al., 2004).

This risk assessment approach tries to determine the total risk potential of a sub-national region. This means aggregating all relevant risks
(from earthquakes, floods, etc.) to assess the integrated risk potential.
The approach includes both natural and technological hazards, but excludes risks with no real spatial underpinning (e.g. epidemics). Hence it
is an integrated risk assessment of spatially relevant hazards.

Structure and methodology

The integrated risk assessment of multi-hazards comprises four elements:
Hazard maps: for each spatially relevant hazard a separate hazard map is
produced showing in which regions and with which intensity this hazard
occurs.
Integrated hazard map: the information on all individual hazards is integrated in one map showing the combined overall hazard potential for
each region.
Vulnerability map: information on the hazard exposure as well as coping
capacity in regard to potential hazards is combined to create a map showing the overall vulnerability of each region.
Integrated risk map: the information from the integrated hazard map and
the integrated vulnerability map are combined, thus producing a map

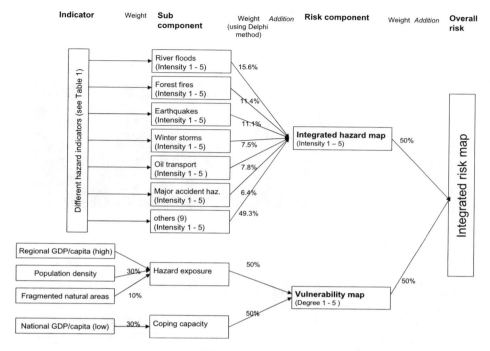

Figure 11.2 Calculation of the Integrated Risk Index (IRI).
Source: Author.

that shows the integrated risk to each region. Figure 11.2 explains how risk was calculated.

Hazard maps

Hazard maps show where and with what intensity individual hazards occur. These maps do not yet contain any information on regional vulnerability. Thus they are merely hazard maps and not risk maps. The intensity of a hazard is determined on the basis of data on, for example, a hazard's frequency and magnitude of occurrence. These differ due to the specific characteristics of each hazard, which makes it impossible to define or derive a single classification that is valid for all types of hazards. Therefore the intensity of each hazard is classified separately on an ordinal scale, using five relative hazard intensity classes (Figure 11.3). This relative scale provides a way around the apparently insurmountable differences in assessing risks between the various scientific disciplines involved, which is the main obstacle to integrated risk assessment. In addition, this

Degree of vulnerability / Overall hazard intensity	1	2	3	4	5
1	2	3	4	5	6
2	3	4	5	6	7
3	4	5	6	7	8
4	5	6	7	8	9
5	6	7	8	9	10

Figure 11.3 Integrated risk matrix.
Source: Schmidt-Thomé, 2005: 85.

relative scale allows the use of different hazard-related data regarding several spatially relevant hazards. Table 11.1 gives an example of indicators that have been used in the EU-27+2 study (Schmidt-Thomé, 2005). The table also indicates the relative importance of each hazard, generated from the experts' opinion, based on the Delphi method.

Integrated hazard maps

In the next step the individual hazard maps are aggregated into one integrated hazard map by adding together all single hazard intensities.[2] Mathematically this is possible and easy to do because the intensities of all hazards are classified into five ordinal classes. For seven common hazards the range of values therefore lies between 15 and 75 (15 hazards), which have to be converted to an overall hazard intensity of 1 to 5 (Figure 11.3). More problematic is the question of whether all hazards should be aggregated with equal or differing weights. Such weighting of hazards implies normative decisions, which of course have a crucial impact on the results of the integrated hazard values. Different weighting schemes can be justified, depending, for example, on recent disaster experiences and thus on a perception of heightened hazard. It is therefore proposed that the researchers involved and/or major stakeholders of the regions for which the risk assessment is conducted engage in a so-called Delphi process to assign different weights to the different hazards. The Delphi method, developed by Helmer (1966) has become widely accepted by a broad range of institutions (Turoff and Linstone, 1975; Cooke, 1991).

Table 11.1 Hazard indicators

Natural and technological hazards	Hazard indicators	Relative importance in %
Avalanches	Areas that have reported landslide potential (derived from several sources)	2,3
Droughts	Number of observed droughts 1904–1995	7,5
Earthquakes	Peak ground acceleration	11,1
	Earthquake casualties	
Extreme temperatures	Hot days	3,6
	Heat waves (7-day maximum temperature)	
	Cold days	
	Cold waves (7-day maximum temperature)	
Floods	Large river flood event recurrence (1985–2002)	15,6
Forest fires	Observed forest fires per 1,000 km^2 (1997–2003)	11,4
	Biogeographic regions	
Landslides	Expert opinion (questionnaire that was sent to all geological surveys of Europe)	6,0
Storm surges	Approximate probability of having winter storms	4,5
	Changes in annual wind speed	
Tsunamis	Areas that have experienced tsunamis that resulted mainly from gravitational landslides (terrestrial landslides)	1,4
	Areas in close vicinity to tectonically active zones	
	Areas in close proximity to tectonically active zones that have already experienced tsunamis following from earthquakes, volcanic eruptions and/or resulting (submarine) landslides	
Volcanic eruptions	Known volcanic eruptions within last 10,000 years	2,8
Air traffic hazards	Existence of airports in a 5 km radius	7,5
	Amount of passengers per year	
Major accident hazards	Number of chemical production plants per km^2 per NUTS 3 level	2,1
Hazards from nuclear power plants	Location of nuclear power plants	8,4
	The distance from nuclear power plants, based on fallout experience of the Chernobyl accident	
Oil production, processing, storage and transportation	Sum of refineries, oil harbours and pipelines in NUTS 3 region	7,8

Source: Schmidt-Thomé, 2005: 16.

Table 11.2 Possible indicators for measuring vulnerability in Europe

Indicator	dp/cc	econ/soc/ecol	Description	Data availability
Regional GDP/capita	dp	econ	High regional GDP/capita measures the value of endangered physical infrastructure and the extent of possible damage to the economy. Insurance company point of view.	+
Population density	dp	econ/soc	Measures the amount of people in danger.	+
Tourism (e.g. number of tourists/number of hotel beds)	dp/cc	econ/soc	Tourists or people outside their well-known environment are generally unaware of the risks and don't necessarily understand the seriousness of hazardous situations. Secondly, tourist dwellings are often located in high-risk areas and might not meet the requirements of structural risk mitigation.	–
Culturally significant sites	dp	econ	Such sites are unique and important for the cultural and historical identity of people.	–
Significant natural areas	dp	ecol	Areas with special natural values (e.g. national parks or other significant natural areas) can be considered vulnerable because they are unique and possibly home to rare species of flora of fauna.	–
Fragmented natural areas	dp	ecol	Natural areas that are small and fragmented are vulnerable, since they are likely to be totally destroyed if a hazard strikes.	+
National GDP/capita	cc	soc	Low national GDP/capita measures the capacity of people or regions to cope with a catastrophe. In the hazards project the national GDP/capita was used, since the presumption was that coping capacity is weak in poor countries and strong in rich countries. It was further presumed that there are no marked differences in coping capacity inside a country.	+

Education rate	cc	soc	Measures people's ability to understand and gain information. The presumption is, that people with a low educational level do not find, seek or understand information concerning risks as well as others, and are therefore vulnerable.	—
Dependency ratio	cc	soc	Measures the proportion of strong and weak population groups. A region with a high dependency ratio is especially vulnerable for two reasons. First, elderly people and young children are physically frail and thus vulnerable to hazards. Secondly, elderly people and children may not be able to help themselves. A region with a high dependency ratio is dependent on help from the outside.	—
Risk perception	cc	soc	Indicates how people perceive a risk and what their efforts have been to mitigate the effects of a hazard.	—
Institutional preparedness	cc		Indicates the level of mitigation of a region.	—
Medical infrastructure	cc		Indicates how a region is able to respond to a hazard (e.g. number of hospital beds per 1,000 inhabitants or number doctors per 1,000 inhabitants).	—
Technical infrastructure	cc		Indicates how a region is able to respond to a hazard (e.g. number of fire brigades, fire men, helicopters etc.).	—
Alarm systems	cc		Indicates the level of mitigation of a region.	—
Share of budget spent on civil defence	cc		Indicates the level of mitigation of a region	—
Share of budget spent on research	cc		Indicates the level of mitigation of a region.	—

dp = damage potential; cc = coping capacity; econ = economic dimension; soc = social dimension; ecol = ecological dimension of vulnerability

Source: Schmidt-Thomé, 2005: 80.

219

The Delphi method is based on a structured process for collecting and synthesising knowledge from a group of experts through iterative and anonymous investigation of opinions by means of questionnaires accompanied by controlled opinion feedback. After several rounds of assigning weights, the individual scores are finally aggregated to achieve collective weights for all hazards (Table 11.1). On this basis the integration of all hazards and the production of an integrated hazard map can be achieved. For that purpose, the single range of hazard intensity (1–5) is multiplied by the Delphi weighting of a certain hazard.

Vulnerability maps

Another major component of a risk assessment is assessing a region's vulnerability to hazards. Regional vulnerability is determined by evaluating hazard exposure and coping capacity (Figure 11.1). For both hazard exposure and coping capacity a set of indicators was selected. These indicators are used to measure vulnerability at the European level and they are not necessarily applicable on a regional scale or in developing countries.

In an ideal situation it would be possible to use all the indicators introduced in Table 11.1 for measuring vulnerability. However, some of the indicators for coping capacity are in practice impossible to measure (e.g. institutional preparedness, risk perception) and problems in data availability make it impossible to use many other indicators (e.g. number of tourists, medical infrastructure). The data availability column in Table 11.1 shows the availability of data for the area of EU 27+2. In many cases there would have been data on NUTS-2 to NUTS-0 level but on the ESPON level of NUTS-3 no data was available for the whole area of EU 27+2. Therefore, a less extensive range of indicators has been used in the hazards project, which can be understood more or less as hazard independent parameters. That means these indicators are measuring the vulnerability of all hazards covered by the project.

Hazard exposure: the indicators for hazard exposure of an area's infrastructure, industrial facilities, production capacity and residential buildings are measured by the regional GDP per capita, and the area's population density stands for the probable injury to people. Finally, the fragmentation of natural areas is used as an indicator for ecological vulnerability.

Coping capacity: in contrast to hazard exposure, coping capacity reflects the response potential of an area's population. While the vulnerability of an area is defined by its population density (the maximum number of people affected by a disaster), the coping capacity is measured by the national GDP/capita, since in disaster situations the whole nation is willing to deal or cope with the consequences. Thus coping capacity reflects the financial, sociocultural and institutional potential of an area's inhabitants to prepare against and respond to hazards adequately.

As depicted in Figure 11.2, these components of vulnerability need to be aggregated to create an integrated vulnerability index. Instead of weighting all components equally, the three main components are each weighted 30 per cent and the fragmented natural areas 10 per cent, following a plausibility test of different weightings made by the ESPON project team. However, this weighting process involves a normative decision and may therefore be open to other opinions. Each vulnerability component is classified using five ordinal classes, thus facilitating the integration of the economic and social vulnerability into one vulnerability index.

Integrated risk maps

Finally, the vulnerability and hazard indices are combined. The new integrated risk index allows one to distinguish between those regions that are only hazardous and those that are at risk: that is to say they have a high degree of vulnerability as well. This methodology is derived from ecological risk analysis used in environmental impact assessments (Bachfischer, 1978; Scholles, 1997).

For the task of combining vulnerability and hazard potential a 5×5 matrix has been used (Figure 11.3). The values of a region's hazard intensity and degree of vulnerability are summed up to yield the region's integrated risk value. This aggregation procedure yields nine risk classes. The matrix shows that regions in one risk class might have the same overall risk value, but the composition of their risks may be different. For example risk class six may be reached due to high vulnerability or to high hazard intensity, or because of medium values for both items. Only after determining the risk class for each region under study is an integrated risk map produced.

Example

In the following section, key results of the EU 27+2 application are presented. The aggregated hazard map shown in Figure 11.4(c) was generated from the 15 single hazard maps.

The aggregated hazard map indicates that the greatest hazard potentials are found in parts of southern, western, central and eastern Europe. Some hotspots are located outside this area, for example in central Italy and parts of southern Scandinavia. Only a few large areas have a very low aggregated hazard, mainly in Scandinavia, the Baltic States and south-central France. The map shows that the central parts of Europe tend to be more affected by hazards than the peripheral regions. The pattern is different when comparing urban with rural regions.

Figure 11.5(c) indicates the vulnerability of different parts of the EU 27+2.

The integrated vulnerability map shows several patterns over the EU 27+2 area. Vulnerability tends to increase from west to east because of a lower coping capacity, as based on the lower GDP/capita. Less fragmented areas have a lower vulnerability because the natural environment in larger undisturbed areas can recover faster than in smaller areas. Densely populated areas with a high GDP per capita show the highest vulnerability, as the total number of people and assets per square kilometre poses a risk of higher vulnerability and greater total damage in the event of a disaster. The variance in vulnerability between western and eastern European countries is due to lower coping capacity in the latter. In consequence, the influence of the existing differences in population density and GDP per capita on the integrated vulnerability in western European countries is much greater than in eastern Europe. However, inside the single Member States the more populated central urban areas are the more vulnerable due to their higher income in combination with higher population density. Figure 11.6(c) shows an aggregated risk map based on the aggregated hazard map (50 per cent) and the integrated vulnerability map (50 per cent).

The aggregated risk map shows a similar pattern to the aggregated hazard map. The so-called "pentagon area" (which covers southern UK, Benelux, the north-east of France and western Germany, the economic heart of the European Union) displays the highest agglomeration of high risk, and the largest parts with low risk are found in northern Europe's peripheral regions. In general, urban areas seem to be more at risk than rural areas, due to the influence of the vulnerability component on the overall risk. This tendency is particularly visible if each Member State is analysed separately. In so doing the influence of the national GDP/capita (which leads for example to a higher vulnerability in eastern Europe) can be excluded.

Function and target group

Generally speaking, the approach presented here measures risk on different scales. However, the application on a Europe-wide level makes it possible to compare the different NUTS-3 regions in the EU 27+2. With the use of adjusted indicators and data, the same approach can also be utilised for regional or local spatial planning.

For the moment, the European Commission is the main target group. Risk management should be made an integral and explicit part of EU policy. This requires better coordination of policy measures at all spatial scales. Based on such a risk assessment of Europe's regions, the EC's structural funds could be used for risk management, using criteria that are relevant to risk and vulnerability to identify regions that are eligible for funding through the structural fund objectives.

Open questions and limitations

While this aggregation procedure has the advantage of being transparent and easy to perform, it does not take into account the interrelations between hazards (exacerbating or ameliorating effects). Unfortunately very little scientific work has been done so far on the effects of such cross hazards.

Developing an integrated risk index based on relative hazard intensities can be seen as a way beyond the impasse of scientific approaches from the various involved disciplines that use different risk assessment methodologies of single hazards. However, some methodological problems still remain:

Weighting problem: the Delphi method was used to weight hazards and vulnerability indicators on a regional level. Although precautions were taken to avoid such influence, events occurring during the inquiry may have influenced the attitude of participants (e. g. the South East Asia tsunami in December 2004). Nevertheless, the occurred deviation from the low estimation of this hazard (only 1.4 per cent) cannot be interpreted as distortion only. Accepting that the panel is dealing with uncertainty, each event also generates knowledge and is an impulse for reconsideration in the light of the knowledge. Thus, weighting results generated by the Delphi method may be seen as snapshots that would need regular updating.

Changes of parameters that shape risk in the future: the index presented in this chapter and used in the ESPON hazards project is based only on past data. However, to acknowledge changes of parameters, a dynamic component aiming to monitor these (changes in population density and GDP/capita on the one hand and changes in hazard intensity on the other hand) has to be integrated in a monitoring of spatially relevant trends.

Problem of data quality: when the methodology is applied it can be seen that data for the different hazards varies quite considerably. For some hazards, only the number of historic hazardous events will be obtainable while for others detailed loss data will be available. This underlines the fact that there is little comparability between the hazard intensities. On a practical level, however, the presented methodology shows a solution to this problem by using a relative scale for all data on hazard intensities. An ideal set of data would consist of reliable information for probable annual losses (PAL, for frequent hazardous events) and probable maximum losses (PML, for very unlikely events).

Limits of measurability: especially in the field of coping capacity the search for appropriate indicators and data soon reveals the limits of measurability. As the methodology has shown, some aspects can be quantified, while other aspects that might also be important in order to assess coping capacity cannot be measured quantitatively. These include social cohesion and organisational structures.

Problem of fit: this describes the problem of congruence or compatibility between hazard zones and institutional arrangements that are created to manage risks (Young, 2002). The more punctual or linear that typical hazard zones are, the more inexact the result for the whole area will be because administrative borders and hazard zones are not generally congruent.

Outlook

Before this approach can be applied for spatial planning on a very detailed scale (e. g. urban land use planning), decision makers will need a more detailed hazard and vulnerability assessment at the regional planning level to know whether risks should be tolerated or altered. In so doing, a weighing-up of trade-offs would be required to consider the appropriate level of protection in view of the different damage potentials, such as residential areas, industrial facilities or transport infrastructure. On this basis concrete designations within a regional plan or a preparatory land use plan could be made. In this context, the current European research project ARMONIA (Applied Multi Risk Mapping of Natural Hazards for Impact Assessment), aimed at the harmonisation of risk assessment methodologies for use in spatial planning, should be mentioned.

Concerning vulnerability assessment as a whole, more attention should be paid to institutional vulnerability (see e. g. ECLAC/IDB, 2000). Political and institutional vulnerability, understood as institutional weakness as a whole, and more specifically any weaknesses in the democratic system, have often been seen as one of the major causes of vulnerability where natural phenomena are concerned. The inability of traditional Government systems to involve all relevant stakeholders in decision-making from the beginning and to communicate risks has negative consequences for the efficiency of public policies, the legitimacy of Government action, and participation by citizens and the private sector in national efforts. There is a close relationship between the need to reduce vulnerability and the increase in the organisational and participatory capacity of communities, the private sector and Government. In this context, the newly emerging concept of risk governance should be highlighted (IRGC, 2005).

Notes

1. NUTS is the acronym for nomenclature of territorial units for statistics, used for statistical purposes in the European Union. NUTS-3 means a regional level.
2. A plausibility test (multiplication instead of addition) has shown the stability of the results: the ranking of the different regions is nearly the same.

REFERENCES

Bachfischer, R. (1978) *Die Ökologische Risikoanalyse*, dissertation, TU München.

Batista, M.J., L. Martins, C. Costa, A.M. Relvão, P. Schmidt-Thomé, S. Greiving, M. Fleischhauer and L. Peltonen (2004) "Preliminary Results of a Risk Assessment Study for Uranium Contamination in Central Portugal", paper presented at the International Workshop on Environmental Contamination from Uranium Production Facilities and Remediation Measures, ITN/DPRSN, Lisbon, 11–13 February.

Burby, R.J., ed. (1998) *Cooperating with Nature: Confronting Natural Hazards with Land-Use Planning for Sustainable Communities*, Washington D.C.: Joseph Henry Press.

Cooke, R. M. (1991) *Experts in Uncertainty: Opinion and Subjective Probability in Science*. New York/Oxford: Oxford University Press.

Cutter, S.L. (1996) "Vulnerability to Environmental Hazards", *Progress in Human Geography* 20: 529–539.

Cutter, S.L. and W.D. Solecki (1989) "The National Pattern of Airborne Toxic Releases", *The Professional Geographer* 41(2): 149–161.

Economic Commission for Latin America and the Caribbean, Inter-American Development Bank (ECLAC IDB) (2000) *A Matter of Development: How to Reduce Vulnerability in the Face of Natural Disasters*, Mexico City: ECLAC–IDB.

Egli, T. (1996) Hochwasserschutz und Raumplanung: Schutz vor Naturgefahren mit Instrumenten der Raumplanung: dargestellt am Beispiel von Hochwasser und Murgängen, ORL-Bericht 100, Zurich: vdf–Hochschulverlag an der ETH.

Federal Emergency Agency (FEMA) (1997) *Strategic Plan: Partnership for a Safer Future*, Washington D.C.

Fleischhauer, M. (2004) *Klimawandel, Naturgefahren und Raumplanung: Ziel- und Indikatorenkonzept zur Operationalisierung räumlicher Risiken*, Dortmund: Dortmunder Vertrieb für Bau- und Planungsliteratur.

Godschalk, D.R., Beatley, T., Berke, P., Browner, D. J., Kaiser, E. J., Bohl, C. C. and Goebel, R. M. (1999) *Natural Hazard Mitigation: Recasting Disaster Policy and Planning*, Washington D.C.: Island Press.

Greiving, S. (2002) *Räumliche Planung und Risiko*, Munich: Gerling Akademie Verlag.

Helmer, O. (1966) *The Use of the Delphi Technique in Problems of Educational Innovations*, Santa Monica: Rand Corporation.

Hewitt, K. and I. Burton (1971) *The Hazardousness of a Place: A Regional Ecology of Damaging Events*, University of Toronto, Department of Geography, Toronto: Research Publication 6.

International Risk Governance Council (IRGC) (2005) *Risk Governance: Towards an Integrative Approach*, IRGC White Paper No.1. Geneva.

Rawls, J. (1971) *A Theory of Justice*. Cambridge, Massachusetts: Belknap Press of Harvard University Press.

Schmidt-Thomé, P., ed. (2005) "ESPON Project 1.3.1 – The Spatial Effects and Management of Natural and Technological Hazards in General and in Relation to Climate Change", Geological Survey of Finland (forthcoming).

Scholles, F. (1997) Abschätzen, Einschätzen und Bewerten in der UVP: Weiterentwicklung der ökologischen Risikoanalyse vor dem Hintergrund der neueren Rechtslage und des Einsatzes rechnergestützter Werkzeuge, Dortmund: UVP-Spezial 13.

Turoff, M. and H. Linstone, H (1975) *The Delphi Method: Techniques and Applications*, Reading, Mass.: Addison-Wesley.

Young, O. R. (2002) *The Institutional Dimensions of Environmental Change: Fit, Interplay, and Scale*, Cambridge/London: MIT Press.

12

Disaster vulnerability assessment: The Tanzania experience

Lead authors: Robert B. Kiunsi, Manoris V. Meshack;
Co-authors: Guido Uhinga, Joseph Mayunga, Fanuel
Mulenge, Conrad Kabali, Nahson Sigalla and Maria Bilia

Abstract

The occurrence of hazards in Tanzania, as in many other countries, is a common phenomenon. However, the Government still does not have adequate information for drawing up appropriate plans for disaster management. Vulnerability assessment has been carried out using mainly perceptions of hazards, their causes and impacts, and coping strategies at the household, village and district levels. The common hazards, as perceived at the household level, include pests, drought and disease outbreak. The perceived causes of hazards include both natural and human factors, such as change of climate and poor farming methods, and identified hazard impact categories encompass loss of life, property and income. This study has revealed that disaster preparedness, especially at the household and village levels, is still low. The vulnerability index has indicated that the southern highlands, eastern plateau and mountain blocks are most vulnerable to pests; the central plateau is most vulnerable to drought, and the Rukwa-Ruaha rift zone to disease outbreak.

Background of the study: the need for vulnerability assessment in Tanzania

Tanzania is located in East Africa between longitudes 29° and 41° east, and latitudes 1° and 12° south. The country has an area of 945,000 km^2

227

and a population of 34.5 million people, of whom 26 per cent live in urban, and the rest in rural, areas. Administratively, Tanzania is divided into 21 regions, 113 districts and 7 agro-ecological zones (see Figure 12.1(c)) (National Bureau of Statistics, 2004; MWLD, 2003). The country is prone to a number of hazards and has a long history of their disastrous impacts. Disasters commonly occur as a result of epidemics, pests, flood, drought/famine, fire, accidents, cyclones/strong winds, refugees, conflicts, landslides, explosions, earthquakes and technological hazards (PMO, 2003). Due to the country's vast area and high diversity of geographic conditions determined by various physical, social and economic factors, each part of the country experiences different kinds of disasters. In recognition of these threats, the Tanzanian Government has made various efforts to strengthen its capacity for disaster management (e.g. preparedness, emergency and recovery plans) at different levels by introducing policies, legislation, institutions and operational guidelines, and by conducting continuous training. However, the Government's efforts are truly hampered by the lack of reliable data on vulnerabilities of communities that are exposed to these different types of disaster.

Realising that there is a great need to collect more profound data on vulnerabilities at various levels, the Tanzanian Government enlisted its Disaster Management Department (DMD), the University College of Lands and Architectural Studies (UCLAS), the University of Dar es Salaam (UDSM), assisted by the Red Cross of Tanzania, and regional disaster focal officers to conduct a comprehensive national vulnerability assessment study, which is considered a key requirement for improving disaster management in an urban area, a region or a country. The study had two phases (VA I and VA II), with the first phase conducted in 2001. This chapter deals with the results of the second phase, to develop a classification scheme and identify areas and societies that are vulnerable to different types of disaster, which was conducted by UCLAS in 2003. Specific objectives of this phase were to:

- determine the type, location and frequency of disasters at the household, village, district and national levels
- identify the current capacity and coping systems (organisational arrangement) at household, village, district and national levels
- identify direct and indirect causes of vulnerability of major hazards in Tanzania
- develop a national vulnerability index
- map out vulnerability of a given hazard at the national level
- develop a national cross-case vulnerability analysis report.

Structure and methodology

The following section discusses the theoretical and conceptual framework of the study. Subsequent sections focus on the methodology for the data collection procedure and data analysis.

Vulnerability assessment

The term vulnerability is defined in various ways. For example, de Satge et al. (2002) defines it as "the characteristics that limit any individual, a household, a community, a city, a country or even an ecosystem's capacity to anticipate, manage, resist or recover from an impact of natural or other threat (often called 'hazard' or natural 'trigger')". UNDP (1992) defines vulnerability as "the degree of loss (for example from 0 to 100 percent) resulting from a potentially damaging phenomenon". The Tanzanian vulnerability assessment study involves collecting and analysing data on four components: the hazards, elements at risk, characteristics of individuals or communities, and coping strategies or manageability.

Hazard is a natural or human-driven event that could lead to a particular level of loss, including mortality and injuries, damaged property, and disruption of economic activity and the environment. A hazard becomes a disaster when it strikes certain elements that are at risk. These can be people, resources, services or infrastructures, which are exposed to specific threats. Elements at risk are attributed by location-specific characteristics that are ruled by physical, socio-economic and political factors and that render individuals or communities defenceless against hazards. Examples of such characteristics include poverty, low levels of education, limited access to power, lack of investment, and living in dangerous locations. Thus, exposure is the degree to which people, livelihoods or property are likely to be struck or affected by a hazard (de Satge et al., 2002).

Manageability or coping strategies refer to how well households, communities and societies can anticipate, manage, resist or recover from the impact of a disaster. The degree of coping capacity is determined by the accumulation and quality of assets. These include for example physical capacities like appropriate house construction techniques or socio-economic assets.

Risk in this study is defined as the probability of a hazard occurring, and the probability that the elements of risk will be affected by a hazard, resulting in a particular level of loss, including loss of life, persons injured, property damaged, and economic activity disrupted.

Conceptual framework

The conceptual framework is based on the disaster crunch model of Wisner et al. (2004), and consists of three main components: (1) underlying causes, (2) dynamic pressures, and (3) unsafe conditions (see also Chapter 1, Figure 1.8).

Underlying causes can be characterised as a set of deep-rooted factors within a society that form and maintain vulnerability: for example, limited access to power and resources.

Dynamic pressure is defined as a translating process that turns the effects of a negative cause into unsafe conditions. This process may be due to lack of basic services or their inadequate provision, or it may result from a series of macro-forces, such as lack of appropriate skills, local markets, education and investment.

Unsafe conditions refer to the vulnerability context, where people and property are structurally exposed to the risk of a potential disaster. Factors include the fragility of the physical environment, for example living in dangerous locations, together with an unstable economy with low-income levels.

This framework assumes that communities, for example those with limited access to power or resources (i.e. underlying causes), lacking appropriate skills and education (i.e. dynamic pressure), and with low incomes, are more vulnerable to hazards than communities not exposed to such conditions. The three main conditions that make individuals or communities vulnerable to hazards are assumed to be present in Tanzania. However, due to lack of consolidated data on the physical and socio-economic conditions of the different communities, the crunch model could not be adopted. Instead, the four main parameters (hazard occurrence, effects of the last disaster occurred, hazard manageability and coping strategies) are calculated on the basis of agro-ecological zones as a spatial classification of the country. Physiographic parameters, such as precipitation patterns, dependable growing seasons and average water-holding capacity of soil, characterise these zones. They can directly reflect the physical, and indirectly the socio-economic, conditions of the different communities in the country. This is due to the fact that more than 75 per cent of the population in Tanzania still live in rural areas and mainly depend on farming to sustain their livelihoods. This means (assuming that all other parameters are equal) that areas with reliable rainfall and good soils are likely to be economically and socially better off than areas exposed to drought and with poor soils.

Table 12.1 Characteristics of agro-ecological zones

S/N	Zone	Altitude m/sea level	Precipitation pattern	Dependable growing season in months	Physiographic
1	Coastal (C)	<100 to 500	Bimodal and monomodal	3 to 10	Combination of coastal lowlands, uplands, undulating and rolling plains
2	Eastern plateau and mountain blocks (E)	200 to 2,000	Predominantly monomodal	From <2 to 7	Many physiographic types, ranging from flat areas, undulating and rolling plains, hilly mountain, plateau to mountain blocks
3	Southern highlands (H)	1200 to 2,700	Monomodal	5 to 10	Composed of flat to undulating rolling plains and plateau, hilly areas and mountains
4	Northern rift valley and volcanic highlands (N)	900 to 2,500	Monomodal	<2 to 9.5	Ranges from flat to undulating plains, hilly plateau to volcanic mountains
5	Central plateau (P)	800 to 1,800	Monomodal	2 to 6	Composed of flat plains, undulating plains, plateau and some hills
6	Rukwa-Ruaha rift zone (R)	800 to 1,400	Monomodal	3 to 9	Composed of flat terrain, rocky terrain and complex terrain
7	Inland sedimentary plateau, Ufipa plateau and western highlands (SUW)	200 to 2,300	Monomodal	3 to 9	Composed of undulating plateau, strongly dissected hills, dissected hilly plateau and undulating rolling plains.

Source: de Pauw, 1984.

Scale of the survey

The Tanzanian mainland accounts for a total of 8,811,087 households, according to a census conducted in 2002. By using a multistage sampling method, the sample size was determined to be 2,040 households at a 95 confidence interval, and a design effect of 1.3. The sample size at district level was 42 out of 113 districts, and at village level it was 84 villages. A sampling protocol was prepared to minimise biases and to include both urban and rural areas, along with all agro-ecological zones.

Methodologies

The main methodologies used for this study include questionnaire-based interviews at household, village and district levels, checklists, geographical information system (GIS) and statistical analysis.

Interviews and data collection

Three sets of questionnaires were developed: one each for the household, village and district levels. Each set of questionnaires covered the key topics: hazard occurrence, effects of the most recent disaster, hazard manageability and coping strategies, including critical facilities. Due to the differences in information available at different levels for some of the research topics, level-specific questions were developed that differed in detail and in the choice of subtopics of the main areas. For example, at the household level, the question on manageability was meant to determine levels of awareness, while at the village and district level it was meant to determine the level of preparedness.

The study is essentially based on the perceptions of the interviewees at the household and village level, and on a mixture of recorded data and insights provided by district officers. This type of approach was used in order to be able to compare the different perceptions of hazards in the country.

The household data were then used to generalise hazard and disaster occurrence in the whole country, according to agro-ecological zones. This was possible because the number of cases was statistically large enough.

Data processing and analysis

The analysis of interview data determined the different types of hazards, their effects and coping strategies in order to calculate risk levels and the vulnerability index at the household, village, district and zonal levels. The data analysis was done by using statistical packages such as S-Plus, R,

SAS and StatXact and GIS packages. Five main steps were followed when compiling and analysing data. After the first step of data cleaning, initial analysis was undertaken to determine hazard occurrence, effects and manageability at the three levels. Due to their large sample size, household data were used to obtain a broad picture of the spatial distribution of hazard occurrence in each agro-ecological zone. Subsequently, coping strategies for the three most commonly mentioned hazards were generated for each zone. This was done by matching and summarising the coping strategies identified at the household, village and district levels. The unit for processing rankings was the percentage of respondents at each level. In the next step, a risk index was calculated for each disaster by fitting the response variables of the household questionnaire linked to the impacts of the last disaster (e.g. loss of life, property and loss of income) into a statistical model.

The goal was to find the best and most parsimonious fitting, yet socially reasonable, model to describe the relationship between disaster impact (response variable) and a set of explanatory variables. Explanatory variables are characteristics or attributes of the sampling unit (e.g. a household, village or district) that influence the outcome (response) variable. Explanatory variables are sometimes referred to as predictor variables, covariates or independent variables.

In situations where one is dealing with discrete variables as responses, models are selected from a class of generalised linear models (GLM). In this particular context, logit models were chosen due to the nature of the response variables.

The logit model is a regression model that is tailored to fit a categorical response variable. In its most widely used form, the categorical response variable is a simple dichotomy, with possible values like yes/no, 0/1, present/absent, etc.

In the model selection, the variables/factors thought to influence the outcome of a disaster were added and removed in a sequential manner until a model that described the data reasonably well was obtained. Among possible approaches for model selection, the stepwise selection method was employed because it combines other approaches like backward elimination and forward selection methods. In the construction phase of the model, variables that met the criterion by Hosmer and Lemeshow were considered (i.e. with p-value of at least 0.25 in an univariate logistic regression analysis) (Neter et al., 1996: 347). The last step was to calculate the vulnerability index by using the UNDP (1992) formula:

$$\text{Vulnerability} = \frac{\text{Hazard} * \text{Risk}}{\text{Manageability/copying strategies}}$$

Example for calculating the vulnerability index in Tanzania

This section gives a more detailed example of how the vulnerability index is calculated as previously outlined. Although the vulnerability index can be calculated for each research level, the example here focuses on calculating the index on the basis of agro-ecological zones for the hazards of drought, disease outbreak and pests.

Hazard occurrences at the household, village and district levels

A total of 15 types of hazards were identified, including drought, disease outbreak, floods, landslides, pests, refugees and HIV/AIDS. It should be noted that hazard occurrences at the household and village level are mostly based on perceptions, while hazard occurrences at the district level are mostly based on records. All values are indicated as percentage of respondents. They do not sum up to 100 because multiple responses were allowed. Before aggregating values for the four most common hazards, the results were compared across all research levels, as shown in Figure 12.2.

For all levels, the study revealed that the three most commonly occurring hazards are pests, drought and disease outbreaks. Pest scored the highest for both the household and village level, followed by drought

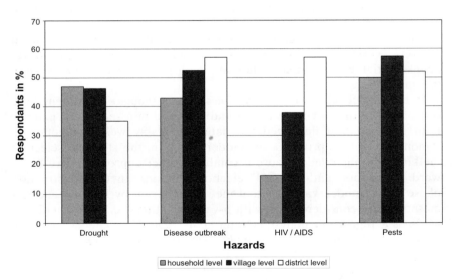

Figure 12.2 Four main hazards compared according to different levels.
Source: Authors.

and disease outbreaks at the household level, and vice versa at the village level.

The study shows that there is a difference in the order of hazard occurrences at the district level when comparing lower levels: HIV/AIDS, together with disease outbreaks, is the most common hazards at district level, followed by pests, drought and strong winds.

The differences in the ranking of the major occurring hazards between the data collected at district level and at grassroot levels (i.e. household and village) is not surprising, because the household and village data are based on perceptions while the district data are mostly based on insights and records by the administration. Another reason why there are differences in the perceptions of interviewees at different levels is because they have to deal with different issues in their daily activities. At the village level, for example, the focus of people's perceptions is more on agricultural related hazards, while at the district level, as the people questioned also live in the district capital, both agricultural and urban-related issues are significant. Furthermore, one has to take into account the different degree of openness of people dealing with sensitive questions. For instance, at the household level, people are probably less open when responding to questions about HIV/AIDS than at the village or district level. This is probably because hazards due to HIV/AIDS are comparatively lower at the household level.

Hazard occurrences at the zonal level

In order to estimate the occurrences of hazards at the level of agroecological zones, only household data were used. Figure 12.3 shows the estimated occurrences of the four major hazards in each zone. Using drought as an example, its occurrence is highest in the northern rift valley and volcanic highlands (79 per cent), followed by the central plateau (58 per cent), the Rukwa–Ruaha rift zone (43 per cent), the inland sedimentary, Ufipa plateau and western highlands (40 per cent), the eastern plateau and mountain blocks (39 per cent) and lastly the coastal zone. The estimated values for each zone were then used to produce hazard maps for the three most common hazards.

Figures 12.3 and 12.4 shows the distribution of drought occurrence across all zones. The map classifies the zones as high (Zone 4 in Figure 12.3), medium (Zone 5) and low occurrence (Zone 2, 6, and 7) levels.

Manageability/coping strategies

A number of questions were asked to assess the level of hazard manageability and coping strategies at all levels for the most common hazard types outlined above. Figure 12.5 shows the results of questions about

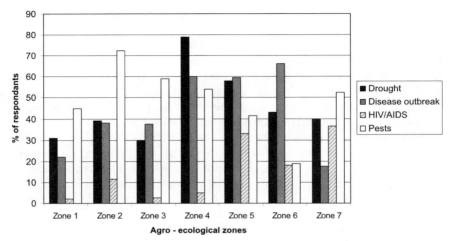

Figure 12.3 Hazard occurrence in different agro-ecological zones.
Legend: Zone 1 = Coastal; 2 = Eastern plateau and mountain blocks; 3 = Southern highlands; 4 = Northern rift valley and volcanic highlands; 5 = Central plateau; 6 = Rukwa-Ruaha rift zone; 7 = Inland sedimentary; Ufipa plateau and western highlands.
Source: Authors.

coping strategies and disaster awareness at the household level. The responses showed, for example, that the three main methods to cope with drought are the selling of assets (33 per cent), seeking employment elsewhere (29 per cent), and growing drought resistant crops (22 per cent). With regard to coping with pests, 38 per cent of the respondents stated that they used pesticides. As for information on the last disaster that had occurred (disaster awareness), responses to questions revealed that the majority of the households (32 per cent) obtained information through public meetings, followed by radio (31 per cent), newspapers (12 per cent) and posters (7 per cent).

Other questions were posed about disaster communication, for example about the number of people listening to local radio programmes, like the Jikinge Na Maafa (Protect Yourself Against Hazards) programme.

At the village level, various questions were asked to get a sense of how authorities and civil institutions were managing disasters. The data revealed that, generally, the level of disaster management is still very low. However, 65 per cent of the villages had conducted awareness-raising activities on disaster management issues within the past year. A comparison of disaster management facilities between the two levels showed better

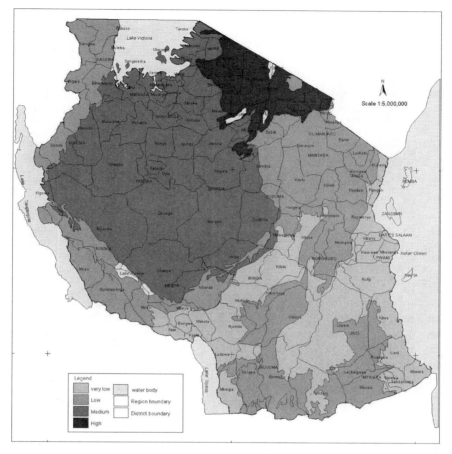

Figure 12.4 Drought occurrence in Tanzania.
Source: Authors.

service provision at the district level. Other questions were posed about the existence of critical facilities, such as hospitals and clinics.

Generalised coping strategies at zone level

After determining the coping strategies for each hazard and research level, the next step was to calculate comparable values for the three most common hazards according to agro-ecological zones. In order to obtain a cross-level value for each agro-ecological zone, the coping strategy with the highest score at the household level, together with those at the

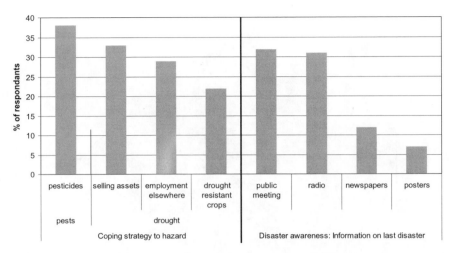

Figure 12.5 Coping strategy for drought and pests; disaster awareness and communication.
Source: Authors.

village and district level (based on the percentage of respondents at each level), were summed up and then divided by the total number of indicated coping measures. Equal weight was given to each facility found at the district level, irrespective of its capacity. In that way, an index show-

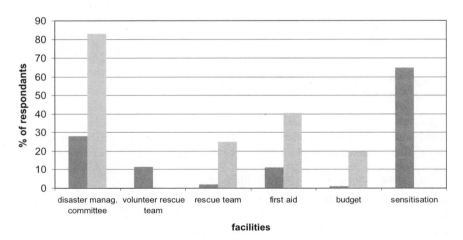

Figure 12.6 Institutional set-up for disaster manageability at the village and district level.
Source: Authors.

ing the relative strength of coping measures at a zonal level was obtained. These indices summarise the coping strength in each zone for a particular hazard, and they are also used for comparison of strengths of coping measures across zones.

Table 12.2 shows the manageability index levels for drought in each zone. The coping strategies for drought range from 70 to 78, with the eastern plateau and mountain blocks (Zone 2) having the highest values, and the central plateau (Zone 5) the lowest values.

Impacts of the last disaster

Data on impacts of the last disaster were also obtained at all levels, although in this chapter only data based on the household survey are presented. Respondents were asked to describe the impacts of the last disaster on population and property. The main impact at the household level was identified as loss of livelihood/income (48 per cent), followed by property damage (42 per cent), illness or injury (35 per cent), loss of life (28 per cent) and displacement (8 per cent).

Risk levels based on the last disaster

On the basis of responses to questions about the last disaster, three models with respect to loss of life, property and income were constructed. Each of these models met the Hosmer and Lemeshow criteria (Neter et al., 1996: 347) (i.e. all explanatory variables with a p-value of at least 0.25 in a univariate logistic regression analysis were considered for further analysis). An example for calculating disaster risk (loss of life) obtained through the selected model (see Data processing and analysis) is shown in Table 12.3; intermediate results have been omitted.

Concerning the impact variable "loss of property", hazards such as conflicts, disease outbreaks and floods contributed significantly to this effect. Factors included the level of illiteracy at household level, as well as the degree of sensitisation at village level. This effect seems to be the same among agro-ecological zones. The model also revealed that people living in the central plateau and Rukwa–Ruaha rift zone (Zone 5 and 6) are at much higher risk of loss of life when hazards occur than those living in other zones.

In the case of the impact variable "loss of income", hazards with significant impact were drought and floods. Again, as was the case with the variable "loss of property", there seem to be no differences for the "loss of income" variable across agro-ecological zones.

Using the same model as for calculating risks of loss of life, it can be seen that the hazards of disease outbreaks and HIV/AIDS had a signifi-

Table 12.2 Aggregated coping strategies for drought according to agro-ecological zones

	Manageability	Zones						
		1	2	3	4	5	6	7
Household level	Coping strategy – drought	43.86	50.62	37.63	73.27	43.11	29.41	30.38
Village level	Disaster committee	24.44	99.51	51.04	59.72	42.34	64.56	0
	Disaster budget	0	0	0	0	0	0	0
	Sensitisation	72.6	61.6	98.71	99.21	29.64	100	90.82
District level	Health centres	100	100	100	100	100	100	100
	Clinics	100	100	100	100	100	100	100
	Dispensaries	100	100	100	100	100	100	100
	Emergency plan	100	100	100	100	100	100	100
	Hospitals	100	100	100	100	100	100	100
	Food plan	100	100	100	100	100	100	100
	Disaster equipment	100	100	100	100	100	100	100
	District disaster committee	60	100	100	50	75	100	75
	Disaster budget	60	0	0	16.67	16.67	0	25
	Drought manageability index	76.99	77.82	75.95	76.83	69.75	76.45	70.86

Source: Authors.

Table 12.3 Hazards and other factors associated with loss of life

Parameter	DF	Estimate	Standard Error	Chi-Square	Wald Pr > ChiSq
Intercept	1	−32.081	0.3536	823.064	>.0001
Posters	1	0.6578	0.2648	61.705	0.0130
Disease outbreaks	1	0.6395	0.2468	67.114	0.0096
HIV/AIDS	1	13.918	0.4286	105.430	0.0012
Disaster committees	1	0.8238	0.1628	256.149	<.0001
Sensitisation	1	0.6132	0.2136	82.396	0.0041
Zone 1	1	−0.3347	0.6945	0.2322	**0.6299
Zone 2	1	0.0162	0.3218	0.0025	**0.9597
Zone 3	1	0.3683	0.2602	20.047	**0.1568
Zone 4	1	−13.701	0.3721	135.587	0.0002
Zone 5	1	0.7820	0.2634	88.114	0.0030
Zone 6	1	10.828	0.3056	125.564	0.0004
Disabled	1	0.2230	0.1047	45.345	0.0332
Distance form dispensary	1	0.2445	0.0605	163.162	<.0001

**Factors that are not significant to the loss of life (at 5% level of significance)
Source: Authors.

cant impact. Contributing factors were the number of disabled persons in the household and distance (in kilometres) from the household to the nearest dispensary. Figure 12.7 shows the strong relationship between the loss of life and the distance (in kilometres) from the household to the nearest dispensary.

Risks at the zonal level

Using the results of the fitted models displayed in Table 12.3, hazard risks for a particular effect – loss of life, loss of property and loss of income – were estimated for every individual in the study.

In order to derive risks based on three impact factors according to zones, the risk levels of individuals from the same zone were grouped and averaged. These were pooled together and weighted to obtain a single estimate across all effects. Loss of life was given the highest weight (0.7), and loss of property and loss of income were given equal weights (0.15). Table 12.4 summarises these findings. The value for risk, indicated as probabilities, ranges from 0 to 1, with 0 being an ideal desirable situation and 1 the worst case scenario. The calculated values show, for example, that even though pests rank very high as a hazard, the effects are not significant.

Table 12.4 shows that the Rukwa–Ruaha rift area (Zone 6) has the highest risk level. This means that if a hazard occurs, the possibility of

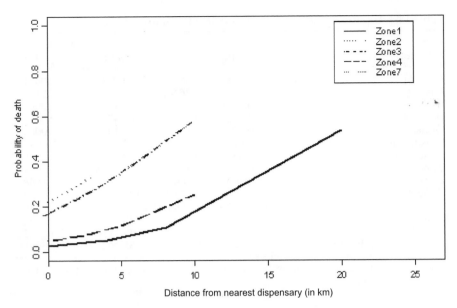

Figure 12.7 Interrelationships between distances from nearest dispensary and probability of death.
Source: Authors.

Table 12.4 Ranking of zones according to hazard risk

	Mean Probabilities				
Zone	Loss of Income	Loss of Property	Loss of Life	Pooled Probabilities	Ranking
1	0.56	0.34	0.03	0.16	7
2	0.55	0.50	0.22	0.31	4.5
3	0.57	0.45	0.32	0.38	3
4	0.75	0.60	0.16	0.31	4.5
5	0.64	0.41	0.38	0.42	2
6	0.50	0.40	0.57	0.53	1
7	0.48	0.24	0.06	0.15	6

Legend:
Zone 1 = Coastal; 2 = Eastern plateau and mountain blocks; 3 = Southern highlands; 4 = Northern rift valley and volcanic high lands; 5 = Central plateau; 6 = Rukwa-Ruaha rift zone; 7 = Inland sedimentary, Ufipa plateau and western highlands.
Source: Authors.

loss of life, property or income is higher there than in the other zones. In the Rukwa–Ruaha rift area, loss of life has the highest mean value, and this is probably compounded by death due to high levels of disease outbreak.

The vulnerability index for agro-ecological zones

The vulnerability index was determined for drought, disease outbreaks and pests by using the UNDP (1992) calculation scheme.

$$\text{Vulnerability} = \frac{\text{Hazard} * \text{Risk}}{\text{Manageability/copying strategies}}$$

The values for hazard occurrence and calculated manageability risk levels at the zonal level were merged to produce a vulnerability index for each zone, as indicated in Table 12.5. For example, the vulnerability index for drought in coastal areas (Zone 1) was calculated by multiplying the value for drought by the value for risk (see equation for calculating vulnerability in section on Methodologies), divided by manageability (Chapter 6; Table 12.2). Overall, the results show that the vulnerability index for drought is highest in the central plateau (Zone 5), for disease outbreak in the Rukwa–Ruaha rift zone (Zone 6) and for pests highest in the eastern plateau and mountain blocks (Zone 2).

Discussion of the vulnerability index results

The index will be discussed using drought as an explanatory example.

According to the vulnerability index, the central plateau (Zone 5) is the most vulnerable (0.35), closely followed by the northern rift and volcanic highlands (Zone 4) (0.33) and the Rukwa–Ruaha rift zone (Zone 6).

Even though drought occurrence is highest in the northern rift and volcanic highlands (Zone 4), this zone's vulnerability is the only second highest because it has a relatively low risk factor compared to the central plateau (Zone 5), which implies higher drought manageability capacities. The Rukwa-Ruaha rift zone (Zone 6), which is the third most vulnerable area, has the highest risk factor compared with the other zones, but has relatively low drought occurrence and the highest manageability capacities. The other zones have essentially low drought vulnerability because they have low drought occurrence and high manageability capacities.

Table 12.5 Vulnerability index parameters by zone

		Zones						
		1	2	3	4	5	6	7
Manageability of three major hazards	Drought	67.78	76.47	75.99	74.72	68.57	76.80	71.02
	Disease outbreak	71.56	69.62	68.75	65.09	64.74	74.06	62.24
	Pest	69.75	66.19	66.68	64.05	57.07	73.84	57.95
Hazard occurrence	Drought	31.06	38.94	30.00	78.91	57.69	43.04	40.21
	Disease outbreak	21.80	37.50	37.10	59.83	59.59	65.82	17.35
	Pest	44.69	72.60	58.71	53.91	41.80	18.99	52.55
Risk	Drought	0.16	0.31	0.38	0.31	0.42	0.53	0.15
Vulnerability Index	Disease outbreak	0.06	0.16	0.15	0.33	0.35	0.30	0.08
	Pests	0.05	0.17	0.21	0.28	0.39	0.47	0.04
	General	0.10	0.34	0.23	0.26	0.31	0.14	0.14

Legend:
Zones 1 = Coastal; 2 = Eastern plateau and mountain blocks; 3 = Southern highlands; 4 = Northern rift valley and volcanic highlands; 5 = Central plateau; 6 = Rukwa-Ruaha rift zone; 7 = Inland sedimentary, Ufipa plateau and western highlands.
Source: Authors.

Limitations

This investigation can be seen as a preliminary study on vulnerability assessment in Tanzania. Its main limitation is that it is based mainly on perceptions at different levels. A more reliable index can be calculated by using recorded data, but unfortunately comprehensive disaster data at all levels are lacking. The second limitation is that this study lumps together data from urban and rural areas. This makes it difficult to distinguish urban from rural vulnerabilities, which are often very different. Therefore, we recommend that a more detailed study be conducted, focusing on either rural or urban vulnerabilities and using recorded data as much as possible. This will provide more reliable and useful information on vulnerability.

Acknowledgements

The authors would like to acknowledge the support provided by the Disaster Management Department in the Prime Minister's Office and USAID, who facilitated and financed the study respectively. Special thanks go to Dr Saade Abdalah for her technical support and to Associate Professor Msafiri Jackson for his input.

REFERENCES

de Pauw, E. (1984) *Soils, Physiography and Agroecological Zones of Tanzania*, Dar es Salaam: Ministry of Agriculture and Livestock Development, Crop Monitoring and Early Warning System Project.

de Satge, R, A. Holloway, D. Mullins, L. Nchabaleng and P. Ward (2002) *Learning about Livelihoods: Insights from Southern Africa*, Cape Town: Peri peri and Oxfam.

Ministry of Water and Livestock Development (MWLD) (2003) *Tanzania Agro Ecological Zones*, Dar es Salaam: MWLD.

National Bureau of Statistics (2004) 2002 population and housing census: Regional profiles Vol. 6. Dar es Salaam: National Bureau of Statistics.

Neter, J., M.H. Kutner, C.J. Nachtshein and W. Wasserman (1996) *Applied Linear Statistical Models: Regression, Analysis of Variance, and Experimental Designs*, Boston: McGraw-Hill.

Prime Minister's Office (2003) *Disaster Vulnerability Assessment, phase II*, Dar es Salaam, Tanzania: UCLAS.

Wisner, B., P. Blaikie and T. Cannon (2004) *At Risk, Natural Hazard, Vulnerability and Disasters*, London: Routledge.

United Nations Development Programme (UNDP) (1992) *An Overview of Disaster Management*, 2nd edn, Washington, D.C.: UNDP.

13

A Human Security Index[1]

Erich J. Plate

Abstract

The chapter presents a new concept that defines human security in terms of the difference between resistance and vulnerability. Vulnerability and resistance are each defined, and the definitions compared with others reported in the literature. An index of human security is described, which could become a useful tool for setting priorities in allocating funds to disaster victims. The planning situation is also considered, illustrated by a schematic model of time development of human security. The concept supports a conceptual decision model based on risk management.

Introduction

In the report *Our Common Future* (Brundtland Report, 1987) the principle of sustainable development was proposed as a paradigm for future development. The UN Conference on Environment and Development (UNCED, 1992) adopted this principle as offering the best framework for the future of a changing world. In the words of Bruce (1992), sustainable development requires the following:

First, development must not damage or destroy the basic life supporting system of our planet earth: the air, the water and the soil, and the biological systems. Second, development must be economically sustainable to provide a continuous flow of goods and services derived from the earth's natural resources, and thirdly, it requires sustainable social systems, at international, national, local and family levels, to ensure the equitable distribution of the benefits of the goods and services produced, and of sustainable life support systems.

The principle of sustainable development is too broad to apply for direct decision-making. It must be defined more narrowly for every field of human activity. For example Jordaan et al. (1993) and ASCE (1998) have given lists of elements needed for sustainable development for water resources. Although identification of elements is necessary in planning for sustainable development, this is only the first step. Their influence on sustainable development has to be expressed in numerical quantities. Numerical quantities are needed if the concepts of sustainable development are to be integrated into decision support models.

Among the factors for determining whether development is sustainable or not is the issue of human security. It is intuitively evident that a society ravaged by extreme storms or floods or other natural and human-caused extreme events, is less capable of developing in a sustainable manner than a society that is well protected. Responsibility for increasing human security of persons or populations at risk (PAR) is the central task of Governments (Yokohama Principles, UN/ISDR, 2004a), and should be an objective of sustainable development and, for developing countries, a target for foreign aid. Because funds for improving human security are limited, it is necessary to apply the principle of maximum cost-effectiveness when allocating funds. Planning cost-effective measures to improve human security for all types of PAR requires that a model exists which is supported by a numerical decision base.

Quantification of human security is a complex subject. In this chapter, a numerical quantity – a Human Security Index – will be proposed in general terms. This is a first attempt only, and it will need to be substantiated by means of case studies. Central to the concept is the existence of a PAR that is impacted by an extreme event. Typical extreme events may be of natural origin, caused by meteorological or geologic processes; large accidents associated with chemical or other production processes, such as chemical spills or large scale pollution from agricultural sources; or events caused by failure of engineering works, such as a dam, a dyke, or a container in a chemical factory. Here we shall concentrate on natural events.

Vulnerability, resistance and human security

Definitions of vulnerability and resistance

The state of the PAR depends on many factors. We shall distinguish two types. Factors that determine the resources available to the PAR are combined into resistance, and factors describing the demand on the resources of the PAR are combined into vulnerability. In an analogy to economic terms, resistance is the supply of resources available to a person, a community or a nation, whereas vulnerability is the demand on the resources. An analogy also exists to the concept of loads and resistances, as used in structural engineering (Ang and Tang, 1984): the load is the externally imposed stress on a system (here the PAR), whereas resistance is the internal capacity of the system to withstand the load. This concept is most easily understood when interpreted in monetary or economic terms, as will be used in the examples in this chapter. However, it is also valid when the factors are social or environmental.

In international literature, the term vulnerability has been used to mean different things by different authors. All definitions agree that vulnerability must be defined with respect to a cause: an extreme event. Vulnerability to an economic impact (for example, due to a drop in commodity prices) or vulnerability to technical events, like the failure of a bridge, require different definitions and cause–effect relationships than vulnerability to natural disasters; and vulnerability to drought is different from vulnerability to floods, wildfires or landslides. Yet authors do not agree on a common definition of what vulnerability actually is.

Some authors use the term in the sense of vulnerability as a state of stress on the individuals in a society, as used in this chapter, or it has been used in the sense of resistance, in our terms. In the context of risk management for natural disaster mitigation, vulnerability is usually defined in the first sense: as "the degree of loss resulting from a potentially damaging phenomenon" (UNDHA, 1992) or, more specifically, as the consequences of the extreme natural event, expressed through monetary units. The same definition is used by the insurance industry, where the consequences of a disaster are expressed as maximum possible costs multiplied by an exposure factor (Kron, 2002). Used in this way, vulnerability is the same as the "damage cost" used in economics (Kunreuther, 2000). However, vulnerability in disaster mitigation has also been used in the second sense: that is, as a measure of resistance. Jones (1992) uses vulnerability in the context of urban management to describe: "the ability of the elements at risk in a city – i.e. the elements of the built physical environment of buildings, site improvements, and infrastructure in them to withstand the stresses imposed by natural hazards". Bohle et al. (1994) contend that vulnerability is best seen as "an aggregate measure of human welfare that integrates environmental, social, economic, and political exposure to a range of harmful perturbations". Vulnerability is seen as a multi-component, socio-economic issue, depending on many factors, such as exposure of the population, relative vulnerability of different groups of population, and people's income as affecting self-help capacity.

A very complete review of literature on the nature of risk and vulnerability of households in developing countries has been given by Vatsa and Krimgold (2000). Downing and Washington (1998) identify aspects of vulnerability, which shift the focus of vulnerability away from a single hazard to the characteristics of the social system. They state:

Vulnerability is a relative measure. The analyst, whether they are the vulnerable themselves, external aid workers, or society in broad terms, must define what is a critical level of vulnerability. Everyone is vulnerable, although their vulnerability differs in its causal structure, its evolution, and the severity of the likely consequences. Vulnerability relates to the consequences of a perturbation, rather than its agent. Thus, people are vulnerable to loss of life, livelihood, assets and income, rather than a specific agent of a disaster, such as floods, windstorms, or technological hazards. This connects vulnerability on the social system rather than on the nature of the hazard itself. The locus of vulnerability is the individual related to social structures of household, community, society and world system. Places can only be ascribed a vulnerability ranking in the context of the people who occupy them. (Downing and Washington, 1998)

A framework that encompasses these definitions has been proposed by Wisner et al. (2004), who offer what they call a "pressure and relief model" of vulnerability, which integrates extreme events, social conditions and exposure of the PAR into one common framework of vulnerability (see Chapter 1). The difficulty with all these definitions of vulnerability (with the exception of the usage in the insurance industry) is that they are descriptive and identify factors but they cannot be expressed in numerical terms. Although it seems intuitively simple to identify factors of vulnerability and resistance in a given environment, it is not easy to convert them into numbers describing human security for decision models. Here we face the problem that human security is also a poorly defined term. In general, human security is improved if either resistance is increased, or vulnerability decreased, or both.

A model is needed to determine a useful quantity to describe human security for given sets of local conditions. Furthermore, the output from the model should be so general that it permits comparison of conditions at one location with conditions elsewhere. The purpose of the model is twofold: first, to allocate resources to the most needy in the event of a disaster, and second to provide a design criterion for planning. The former deals with the impact of an extreme event that has just occurred, while the latter serves to plan protection measures against potential future threats. Ideally, the design task is to solve an optimisation problem: to define those measures which under given circumstances can improve human security to a desired degree at a minimum cost to society – this cost being understood to include not just monetary, but also social and eco-

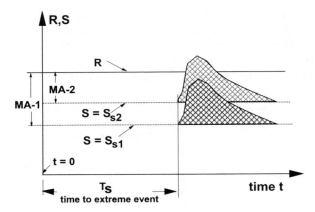

Figure 13.1 Combination of resistance and vulnerability as function of time.
Source: Author.

nomic values. This chapter provides a conceptual framework for such a model.

Indices for resistance, vulnerability and human security

Let us define resistance as an index R (i.e. as a number, possibly with a dimension, for example US$ if monetary units are used). For a given socio-economic and natural environment, R is the sum of the resources that the person or population group has available to meet the needs of the average person over a suitable time horizon: a daily average for event-based considerations, as indicated in Figure 13.1 by the straight upper line, or monthly or annual averages for planning purposes. For a nation, R is a measure of the available resources of the society and its citizens. The economic part of resistance could be measured, for example, through the index R = GNP/person as an average value, which covers the income from all economic activities over a year of a nation. However, resistance is more than an economic quantity as it also involves the capacities of both the environment and social system to absorb impacts of extreme events, so that it is a combination of many contributing factors.

Let vulnerability be identified as an index S, of the same dimension as R, which describes the demands on the resources of a person in a given population or population group. We distinguish two important components of vulnerability. The first is described by an index S_S due to continuous demands by normal living conditions, which sets the initial conditions (the vulnerability for the "normal" condition) before an extreme event strikes. This is indicated in Figure 13.1 by a horizontal dashed line. Index S_S is a measure of the effort that the individual has to exert to live in acceptable conditions, depending on a large number of indicators, such as his or her age group, income or social standing. In monetary terms "normal" vulnerability includes that part of the total available income that individuals have to invest in order to maintain their status quo and live adequately in their home community; in other words, that part of the personal income of a household that is spent on everyday living and covers the cost of food, fuel and shelter. For a household, S_S is an individual number, whereas for a nation it is a statistical average, measured in terms of fraction of GNP/person.

The second part of vulnerability index S describes the demands S_E caused by impacts of extreme events (the change in vulnerability from the "normal" to the "extreme" state due to effects of natural and other events) on the resources of the PAR. Extreme events, also termed "sudden onset" events, are defined as large, but temporary deviations from normal conditions. Typical extreme natural events are floods, drought, extreme storms and wildfires, and also earthquakes, tsunamis and volcanic eruptions. Changes, expressed by index S_E, have an impact on the normal state at a time T_s, as indicated in Figure 13.1. S_E is the area under the curve above S_S. Figure 13.1 also indicates why the resistance R is not a sufficient measure of the coping capacity of the PAR. For the same consequences of an extreme event, and the same resistance R, it is of critical importance that the normal vulnerability S_S is small enough for S_E to stay below R. If the cost of living is too high, $S_E + S_S$ exceed R, and a disaster occurs. This is illustrated in Figure 13.1 by shifting S_S from S_{S1} to S_{S2}.

We use the term disaster here in a narrow sense: a disaster is defined as a condition where the PAR cannot recover from the results of an extreme event without outside help. This corresponds to the definition of disaster given by UNDRO and others (UN/ISDR, 2004a: 17). Disasters are avoided if $MA = R - S_S$ is larger than S_E. According to this definition, a disaster is independent of the size of the PAR or the magnitude of the total consequences: if R exceeds the MA of a family, the family suffers as much and depends as much on help by the local village as when a nation's coping capacity is exceeded and help from the international community is required. Therefore, MA is a true measure of the coping capacity in the face of an extreme event. We shall call MA the margin of human security and define human security as a measure of the security condition of a person or other PARs.

Example: for illustrative purposes, economic resistance and vulnerability for an individual household are described in monetary terms. Consider the case of a homeowner who has a house and other goods valued at W \$, on which SE \$ are owed. House and goods are insured for v·W \$

against natural disasters. The homeowner also has available capital K$. These resources yield a resistance:

$$R = [(v \cdot W - SE) + K]\$ \qquad (1)$$

which is the available capital resource of the household. The normal vulnerability S_S is that part of the annual income that is used for covering all living expenses, including all interest payments on investments and possessions.

Consequently, the margin of human security is:

$$MA = R - S_S = [v \cdot W - SE + K - S_S]\$ \qquad (2)$$

Let the extreme event be a flood which destroyed a fraction rW of the possessions of the household. Furthermore, the household also suffers additional damage, i.e. to restore normal conditions, households will have to finance repairs or replacements of their homes and possessions. The resulting cost D consists of principal and interest of restoring conditions as they were before the disaster, yielding a vulnerability component S_E equal the recovery cost.

$$S_E = [r \cdot W + D]\$ \qquad (3)$$

If the cost to the household exceeds the part of the income not used for living expenses, then the household will face disaster and need outside help. For this it might be useful to define a relative quantity expressed by the relation:

$$IV = \frac{S_E}{MA} \qquad (4)$$

If IV is larger than 1, the household at risk faces disaster; if IV is smaller than 1, the family will be able to recover by their own resources. The value of Eq. 4 lies in the ability to identify personal disaster in relative terms. Clearly, a person living in an area of low cost-of-living will have a different margin of human security than a person living in a country with a high cost-of-living. To be able to compare the situation of people in different regions, it is necessary to use relative values, as expressed by Eq. 4.

The consequences of an extreme event do not affect all exposed persons in the same way. Obviously, the effect will depend on the distribution of resources over the population group making up the PAR. This is indicated in Figure 13.2, where the annual distribution of income over the population percentage n is schematically illustrated. Although the dam-

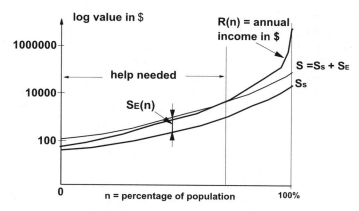

Figure 13.2 Schematic distribution of resources and resources needed as function of population income.
Source: Author.

age to the poorest group is much smaller (notice the logarithm of the value scale), the damage is so high that personal disasters occur and help is needed, whereas the wealthier group, although suffering larger losses, is not in a state of negative MA.

The indicators discussed here refer to households in a given regional environment: the groups of PAR considered are associated with the region, and the indicators have to be developed regionally. For this purpose, factors determining resistance and vulnerability have to be identified regionally, in particular if they refer to the environment. Geo-information systems (GIS) are important tools for this purpose, and are also recommended for the worldwide disaster indicator projects reviewed by the ISDR Interagency Task Force on Disaster Reduction (UN/ISDR, 2004b). For resistance and vulnerability, "normal" conditions must be identified, without regard to possible threat from natural extreme events. In an event-based assessment of what happens to a group of PAR (as illustrated in Figure 13.1) the impact of an extreme event must be seen as superimposed on the steady state condition, just as flood maps show levels for extreme floods that are overlays of regional maps.

The example given on the previous page is in monetary terms, with income and losses in monetary units. However, monetary factors alone are not sufficient in a social environment. It will be necessary to incorporate other factors into an index of vulnerability, for example by writing $S = g(U, \Pi)$, where Π is a vector of socio-economic parameters. This can be seen by comparing disasters that have the same financial consequences but occur in different countries, such as those that occurred in

December 1999 in France and in Venezuela. Landslides in Venezuela and subtropical storms in France both caused damage valued at roughly US$10 billion, but there were 123 casualties in France versus 30,000 in Venezuela. Furthermore, France recovered rapidly, as it has a high resilience, whereas Venezuela will need years of recovery time: its resilience is very low (World Bank, 2000). Obviously, costs and number of casualties – as well as other socio-economic and environmental factors – must be included to define the index of vulnerability. Although attempts have been made to define suitable vulnerability indices for response to natural hazards, in particular in the sociological literature (for example, Wisner et al., 2004), research is needed to identify a composite index that can be used as a decision tool for regional decision-making. A possible way of combining the factors into suitable indices is by means of linear combinations of weighted indicators.

Obtaining indices from indicators

It was stated earlier that in order to obtain meaningful indices for resistance R and vulnerability S, many factors must be incorporated, which reflect not only economic criteria, but also social, ecological, health and other quality-of-life criteria. They depend on sizes of populations, and therefore may be different for a family, a community, a city, a region or a country. The factors can be numbers. Economic factors often are monetary values. However, not all contributing factors are given in numerical terms. Non-numerical factors describe the state of health of a person, or the environmental condition of a field, as well as the social, economic or environmental condition of a PAR or a region. In order to make such factors quantifiable, they must be converted into numbers, which we call indicators. Note that the term "indicator" here is different from what is called indicators in most of the literature – we use the term "factor" for the descriptive quantity: an indicator is a factor converted into a number. For example, degradation of a field can be described in relative terms by factors ranging from severe erosion to no erosion, described on a scale of indicators ranging from 1 to 10.

As a next step, the indicators must be suitably converted into indices of resistance or vulnerability. There are different possibilities of combining indicators into indices. Here we obtain indices by forming sums of weighted indicators. The principle of index determination is illustrated in Figure 13.3. Indicators are denoted by I_i, where sub-index i refers to the i-th indicator out of a total of a possible number. In order to derive indices from a set of indicators, the indicators have to be weighted by weights W_i, with different weights assigned to each of the indicators to obtain their contribution to the individual indices.

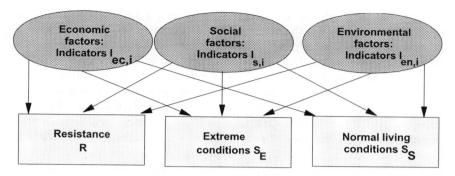

Figure 13.3 Resistance and vulnerability as indices derived from weighted indicators.
Source: Author.

It is possible that certain indicators or factors may contribute to more than one index. For example, land degradation can affect index resistance R, as degraded land reduces people's income (perhaps also changing the normal cost of living S_S), but it also may affect the runoff characteristics of the land and thus increase runoff. Increased runoff causes higher flood levels and corresponding damage as expressed through index S_E, or the land is further degraded through erosion, with the restoration costs add to S_E. As this example shows the determination of indicators involves a careful evaluation of the total vulnerability caused by an event, as the vulnerability covers not only the direct damage, but also includes the long-term consequences as well as indirect effects.

The transformation of indicators into indices is given through the following equations, where the indices *ec* refer to economic, *s* to social, and *en* to environmental conditions, and the sum is to be taken over the whole set of indicators:

$$S_S = \sum_{i1} W_{Sec,i} \cdot I_{ec,i} + \sum_{i2} W_{Ss,i} \cdot I_{s,i} + \sum_{i3} W_{Sen,i} \cdot I_{en,i}$$

$$R = \sum_{i1} W_{Rec,i} \cdot I_{ec,i} + \sum_{i2} W_{Rs,i} \cdot I_{s,i} + \sum_{i3} W_{Ren,i} \cdot I_{en,i} \qquad (5)$$

$$S_E = \sum_{i1} W_{Eec,i} \cdot I_{ec,i} + \sum_{i2} W_{Es,i} \cdot I_{s,i} + \sum_{i3} W_{Een,i} \cdot I_{en,i}$$

In Eq. 1 linear sums are taken. However, it is also possible to apply the equations to the logarithms of the indicators, in which case we obtain product formulations, with weights as power of the indicators. An example for such an index is the Human Development Index (HDI) proposed by UNDP (2000), which incorporates many different factors by giving them different weights. The HDI is, however, an index to identify the development level of a country, whereas a Human Security Index should be valid for different sizes of regions, ranging from the community level to the national level. Another index is the Disaster Risk Index (DRI) developed by UNDP–BCPR (2004), which describes the vulnerability of populations by the number of people killed as a percentage of the people exposed to extreme hazards, as a function of national population size (which in future will be applied also to smaller units, such as regions or communities). Other indices are briefly summarised and references given in UNDP–BCPR (2004) or in Villagrán de León (2005); these are useful for the purposes for which they are designed, but generally not well suited as operational tools for identifying steps to increase human security. However, they do show which data are available for use, and which factors might be significant, such as the World Development Indicators summarised by the World Bank (2000).

Development of human security over time

In this section we consider the planning stage, which must take account of the fact that neither the time nor magnitude of future extreme events can be predicted. Planning decisions, therefore, must be made on the basis of the risk: that is, of the expected value of the consequences from all types and all magnitudes of possible extreme adverse events. This can only be done by looking at future developments of resistance, vulnerability and risk.

Issues of global change

The world is continuously changing. This is reflected in resistances and in vulnerabilities of groups of people ranging from families to communities and nations, and thus also in changes of the margin of human security. A model for describing the development of human security must therefore be dynamic, reflecting these changes. Because the changes are uncertain, future developments must be expressed through probabilities, yet the deterministic component of the model outcomes must be accurate enough to form the basis for long-term decisions. These demands lead to a model for human security development which can be described without being specific of the factors that make up resistance and vulnerability.

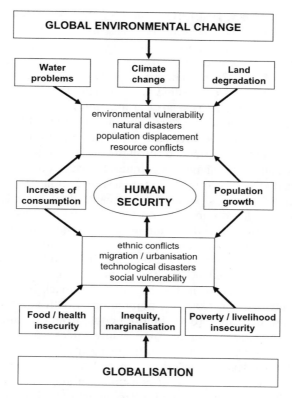

Figure 13.4 Effect of globalisation and environmental change on human security. Source: Bohle, personal communication.

In Figure 13.4 (by Bohle, personal communication) different factors affecting human security are shown. Threats to human security occur primarily through changes. If nature and social conditions remain unchanged over long periods, sustainable societies adapt to the status quo, and every person finds a niche in the fabric of a stable environment and stable society. Human security changes with the natural and/or the socioeconomic environment. The changes could come from within, or from outside of the society. Figure 13.4 shows the cause–effect chain of global change. Global change has mostly been associated with atmospheric changes. For example, predictions of future climate developments by means of mathematical models (global climate models = GCMs) yield sea level rises of dramatic consequences (Baarse, 1995), and also changes in the frequency of occurrence of weather patterns associated with floods

(Bardossy and Caspari, 1990). By combining the conditional distribution of rainfall quantities with weather patterns, it is possible to infer that if the frequency of weather patterns associated with extreme events increases, then the frequency of extreme events will also increase (Bardossy and van Mierlo, 2000). This concept was also used, in conjunction with the Drought Index of Bahlme and Mooley (1980) to predict the future incidence of droughts (Bogardi et al., 1994). Other approaches are based on forecasting floods and droughts on the basis of El Niño Southern Oscillation (ENSO) occurrences to provide the basis distribution to which rainfall events are attached through conditional probabilities.

Global environmental change is commonly discussed by looking at climate changes due to CO_2 forcing, but on a shorter timescale of a few decades other effects such as population increase, changes in land use and deforestation are globally more important. Detrimental local changes of the environment are induced by human actions. The impacts of land degradation and poor use of water may weaken the population against natural extreme events. Changes in socio-economic conditions through globalisation and social imbalances within countries and between countries also contribute. Not shown in Figure 13.4 is feedback between different contributing factors. For example, population growth and land degradation are closely correlated: to feed more people from the same land leads to overuse of the land; or population pressure forces people to move into tropical forests for short-term gains, exposing the land to sun and wind, which combine to destroy the often thin and fragile topsoil. Other connections exist between increase of consumption and water problems, or food and health insecurity and water problems; in fact, water-borne diseases, stemming from lack of quantity and/or quality of the water, are caused by many of these social factors. Furthermore, severe health problems often accompany disasters, for example, from the polluted water remaining after a flood. By these interactions human security is threatened. It is clear that complex feedback loops connect human actions, changes of environment and socio-economic conditions.

Feedback loops of this kind depend on the size of populations or population groups. The human security of individual households is threatened by the socio-economic status of householders and by local environmental factors, such as proximity to rivers, and it is therefore widely variable. The variability of the collective human security in individual communities within a country is smaller and depends on factors that may not affect individual households in the same manner, such as damage to infrastructure, like highways, railroads and other lifelines. The variability is dictated both by the natural environment and by the economic structure of the communities. The human security of a region or a country is made up of the sum of the security of all individuals and communities in it.

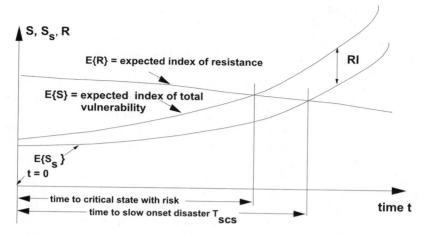

Figure 13.5 Time development of resistance, vulnerability and risk.
Source: Author.

Expected development of human security

A graphical display of the development of human security is shown in Figure 13.5. Vulnerability S_s as a function of time is indicated by the lower solid line in Figure 13.5, which describes conditions in a society whose vulnerability is increasing with time. Such a curve may be caused by a deteriorating ability of the members of a society to handle stresses, perhaps as a result of changes in economic conditions, or due to changes in the society's value system.

Now let us look at the time development of resistance indicated by the $E\{R\}$ curve in Figure 13.5. In developed countries, the resistance usually increases because of investments in protection measures. But in many developing countries just the opposite is true: due to environmental degradation and population increase the resistance per person may have decreased, as is shown in Figure 13.5. Typical causes are environmental changes – climate change and land use change – and population changes: demographic changes (numbers, age structure). These effects may be termed "creeping" or "slow-onset" effects.

Since disaster is defined as the existence of conditions where vulnerability exceeds resistance – that is, where the population or population group can no longer cope without outside help – both creeping and sudden-onset extreme events may lead to disasters. Creeping decline of resistance and simultaneously increasing vulnerability may develop into a disaster condition, so that emergency help is needed even without extreme events. This is the condition of a "creeping" or "slow-onset" disaster occurring at a time T_{scs} in Figure 13.5. Obviously, one of the most important actions to take in disaster mitigation is to ensure that the two curves always stay far apart.

The planning models used today need to build in the ability to anticipate future slow-onset disasters. They must be able to predict changes in vulnerability and resistance with time. Prediction at time t = 0 of a slow-onset disaster for some future time t > 0 requires more than a quantification of S and R. The uncertainty of these quantities also has to be considered. Resistance and vulnerability, therefore, must be considered random variables. There exists an error band around the predictions of R and S_S, each with its own probability density function (pdf) with expected values $E\{R\}$ and $E\{S_S\}$. The error band increases in width with time. The further we want to extrapolate our vulnerability estimates into the future, the wider will be the margin of error. Uncertainty of the forecast leads to uncertainty of time to disaster T_{scs}. There exists a probability density function (pdf), $f(T_{scs})$, for the time interval until disaster. As many uncertain factors contribute to this PDF, it is likely that the best approach is found from studies of scenarios typically by means of Monte Carlo simulation of the possible trajectories of "normal": vulnerability and resistance development. From such scenario calculations trajectories of possible future conditions are generated, which for every time in the future can be analysed by statistical methods.

Authorities face the challenge of preparing a population for potential disasters at household, local or regional levels. A possible decision criterion could be specifying $E\{MA\}$ in monetary terms: that is to say, the expected value of money available for disaster mitigation for a particular type of disaster, as a surrogate for human security. Then disaster mitigation could be addressed in economic terms. For potential disasters, the required expected mitigation costs could be compared with the expected available funds. The resistance $E\{R\}$ would then be the total expected amount of money available for all purposes (the GNP/person), and the "normal" vulnerability is the part of the GNP/person needed for maintaining minimum social standards. Consequently, the distance between the two curves is $E\{MA\}$, the expected monetary value of human security.

For such planning purposes, it is also necessary to predict the impact of future extreme events. For this purpose rapid onset events U_E also have to be described by a probability density function, $f(U_E)$, and the result must be a prediction of the probability that the ratio $MA/E\{S_E\}$ becomes smaller than one. Note that $E\{S_E\}$ is identical with the risk Ri from natural (and other) events; in other words, it is the expected value of the consequences arising from all events, as described below. Consequently, the ratio Eq. 4 becomes:

$$IV = \frac{Ri}{E\{MA\}} = \frac{Ri}{E\{R\} - E\{S_S\}} \tag{6}$$

This is called the expected index of human security.[2]

Eq. 6 states that a society or household will be able to cope with extreme events if, on average, IV < 1. Index IV is an interesting quantity in a planning environment; if, for instance, one determines the index of human security for all communities in a region, then it is possible to set priorities for funding remedial measures, which might be a useful approach for Governments, and in developing countries for donors. In these terms, the effect of a natural disaster for a whole country is measured in terms of the fraction of the GNP that is lost due to the natural disaster. By knowing this condition, the unit considered can accumulate enough financial reserves (for example, through insurance) to keep the average IV below 1, even if momentarily, due to a large extreme event, the index IV of Eq. 4 becomes larger than 1.

Eq. 6 indicates the direction that the improvement of a society with a low or negative index of vulnerability should take. In principle, there are two possibilities: the first is to increase the denominator: that is, to widen the gap between R and S_s. Indeed, many people contend that the best way of improving human security is by increasing R, for example by eliminating poverty. It is evident that although this would lower the index, it would do little to change the threat from a disaster. Therefore, it is generally preferred to reduce the risk.

Risk and resilience

Definition of risk

In a planning situation, the sudden-onset component is the well-known risk, expressed as the expected value of the consequence of extreme events. For each kind of extreme event, the whole family of possible events has to be considered, and expressed through the event statistics. The term hazard is used to define the combination of magnitude and probability of occurrence of harmful extreme events.

RI is calculated by means of the risk equation (for example given in Plate, 2002b), which in a simplified form can be written as a double sum:

$$RI = E\{S_E\} = \sum_i \left[\sum_j n_{ij} \cdot K_{ij} \right] \cdot P_i \qquad (7)$$

In Eq. 7 the risk RI is expressed as a function of the number n_{ij} of the objects or persons (the elements at risk) in element class j subjected to a natural or other threatening event of event class i with index i, whose probability of occurrence is P_i. The inner sum of Eq. 1 is the total consequence for all extreme events U_{ri} in class i:

$$S_{Eri} = \sum_j n_{ij} \cdot K_{ij} \qquad (8)$$

which is a measure of the vulnerability towards an extreme event in class i. K_{ij} is the vulnerability of each of the n_j elements in element class j:

$$K_{ij} = \varphi_{ij} \cdot k_j \qquad (9)$$

which consists of two parts. The first is the quantity k_j. This is the maximum effect that can occur for any event i. It can be, for example, loss of life or of health, or for material objects the replacement value of the object. The second part φ_{ij} is the relative vulnerability (also called the exposure, sometimes in combination with n_{ij}), i.e. the percentage of the maximum k_j that is affected, on average, by extreme events in class i. For example, it can be the fraction of houses that are completely destroyed, or the average damage caused to each house. Sometime the potential damage can be described as a function of a single factor, for example, in the case of a flooding event, as a function of the flood level in class i. In that case Eq. 7 applies with P_i being the probability that event i occurs. However, the vulnerability may depend on many different types of extreme events, each having a different probability function P_i. In that case P_i is the probability for the occurrence of the combination of all effects leading to vulnerability K_{ij}.

The risk equation Eq. 7 gives a means of assessing the effect of certain measures that can be introduced to reduce vulnerability to natural disasters. Coping actions for bringing the state of vulnerability back to conditions below the resistance level by reducing S_E have to be found through engineering or through planning measures. By assessing how changes due to technical or non-technical methods affect any of the quantities in Eq. 7, one can find the most effective strategy for risk reduction.

Definition of resilience

The "extreme" state due to S_E is defined by two factors: the consequences of the extreme event (i.e. the magnitude of the effect of the extreme event in terms of lives lost, property damaged, indirect costs, etc.), and the resilience (i.e. the ability to return to the "normal" state S_s).

The ability of a population to recover after an extreme event is called resilience. This definition is in agreement with the notion of recovery from the impact of a disaster, which in our terminology is a quantity that definitely involves active inputs from the PAR. A very resilient community can recover quickly, although the losses may have been very high, whereas a less resilient community may suffer for many years from the aftermath of a disaster. The higher the resilience, the more a society is capable of recovering from disaster. A good example of a resilient city is Kobe, Japan, which managed to recover (with the help of many outside donors) within a few years after the terrible earthquake of 1994.

In the framework of the definitions used here, resilience is a measure of the return to the normal state S_S. A possible schematic definition of resilience is indicated in Figure 13.6. The consequence function S_E consists of two parts: the first part is the direct impact on the resources of the population for relief and rescue, including the direct social and environmental impacts. The second part quantifies efforts needed to return the PAR to a "normal state". The first effort takes place during the direct time T_E of the event itself. It is the passive (loss) part of S_E. The second part is the active reconstruction investment during the time of recovery T_R. The time dependency of these two parts is schematically indicated in Figure 13.6 by the two curves above level S_S.

A measure of resilience should be inversely related to time of recovery T_R. This time could be expressed through the time that it takes for the

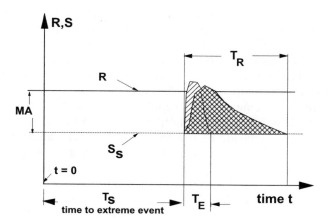

Figure 13.6 Defining resilience.
Source: Author.

PAR's state to return halfway to the original normal state (although the final state of S and R may differ from the initial states).

Risk management

As used in this chapter, the term "disaster prevention" has a very clear meaning: it implies that although we cannot prevent natural events from happening, we must strengthen our efforts to prevent them from causing a disaster: that is, from creating a negative margin of human security. By using the terminology of this chapter for defining disasters, it becomes a technical term, in contrast to the more common definition of disaster as large consequences of extreme events, involving many hundreds or thousands of people killed and many millions of US$ damage (for instance, as used in the disaster statistics of Munich Reinsurance, disaster refers to the extreme consequence of extreme events).

Prevention of a disaster is not the only purpose of risk management, nor is disaster relief. We must reduce all impacts of natural extreme events by integrating preventive measures to safeguard sustainable development. Human security is dependent on how well a population group, ranging from a household to the total population of a country, can cope with extreme events. Therefore, the ability to handle extreme events should be strengthened. The general method used for organising this task is through risk management, as described in Plate (2002b).

Conclusion

There exists a need for a tool to describe human security, for example to pre-assess the state of a population, community, population group or household, if limited funds are to be allocated to the most needy. Regardless of the constraints, the protection of human lives is a humanitarian responsibility of every country and every society. If the society cannot cope by itself, outside assistance will be needed to help prepare for the management of extreme natural events. It is evident that this is a matter of scale: outside help for a household is help from the community, while for the community, outside help is from the regional administration, and so forth. If a country cannot manage, outside help may be given by donors from other countries. But on every scale the resources available for helping are limited, and a major problem faced is to set priorities, so that the most needy get help first.

In this chapter, we have presented a model that outlines how priorities can be set. However, the model is only conceptual and requires field studies and extensive use of international and regional data before it can be used operationally. Its specific advantage lies in the fact that it clearly separates resistances and vulnerabilities, and although this is elementary in its formulation, it does shift the emphasis away from trying to integrate all negative and positive factors into one single index, and moves towards a systematic separate investigation of pertinent factors for risk, resistance and "normal" vulnerability. A prerequisite for using this model, of course, is that we are able to assign numbers to the individual terms of the equation. This requires analysis of large numbers of case studies: either of studies reported in the extensive literature (e.g. Wisner et al., 2004) or based on new cases.

Note

1. This chapter is an extensively revised version of a paper "Towards Development of a Human Security Index", presented at the Osiris workshop in Berlin, 20 March 2003.
2. In this analysis, resistance (= critical vulnerability) and load (vulnerability) are determined as random variables, depending on many factors, and second moment analysis may be the way of obtaining a safety index to be used as a decision quantity for evaluating alternative approaches to the problem of vulnerability reduction. This is similar to using the failure probability obtained by second moment analysis as a decision variable in stochastic design (Ang and Tang, 1984; Plate, 1993).

REFERENCES

American Society of Civil Engineers (ASCE) (1998) Task Committee on Sustainability Criteria, American Society of Civil Engineers and Working Group UNESCO/IHPIV Project M-4.3 "Sustainability Criteria for Water Resources Systems", Reston, USA: ASCE.

Ang, A.H. and W.H. Tang (1984) *Probability Concepts in Engineering Planning and Design*, Vol.2, New York: J. Wiley.

Baarse, G. (1995) "Development of an Operational Tool for Global Environment Assessment (GVA): Update of the Number of People at Risk Due to Sea Level Rise and Increased Flood Probabilities", CZM Centre Publications No.3, The Hague, the Netherlands: Ministry of Transport, Public Works and Water Management.

Bahlme, H.N. and B. Mooley (1980) "Large Scale Drought/Floods and Monsoon Circulation", *Monthly Weather Review* 108: 1197–1211.

Bardossy, A. and H. Caspari (1990) "Detection of Climate Change in Europe by Analyzing European Circulation Patterns from 1881 to 1989", *Journal of Theoretical and Applied Climatology* 42: 155–167.

Bardossy, A. and J.M.C. van Mierlo (2000) "Regional Precipitation and Temperature Scenarios for Climate Change", *Hydrological Sciences Journal* 48: 559–575.

Bogardi, I., I. Matyasovszky, A. Bardossy and L. Duckstein (1994) "Estimation of Local and Areal Drought Reflecting Climate Change", *Transactions* 37: 1771–1781.

Bohle, H.-G., T.E. Downing, and J.M. Watts (1994) "Climate Change and Social Vulnerability: Towards a Sociology and Geography of Food Insecurity", *Global Environmental Change* 4: 37–48.

Bruce, J. P. (1992) "Meteorology and Hydrology for Sustainable Development", World Meteorological Organization (WMO) No. 769, Geneva: WMO.

Brundtland Report/World Commission on Environment and Development (1987) *Our Common Future: Report of the World Commission on Environment and Development*, Oxford: Oxford University Press.

Downing, T.E. and R. Washington (1998) "Prediction of African Rainfall and Household Food Security", in *Proceedings of the International Conference on Early Warning*, Geoforschungszentrum (GFZ) Potsdam, Germany.

Jones, B.G. (1992) "Population Growth, Urbanization, Disaster Risk and Vulnerability in Metropolitan areas: A Conceptual Framework", in A. Kreimer and M. Munasinghe, eds, *Environmental Management and Urban Vulnerability*, World Bank Discussion Paper No. 168, Washington, D.C.: World Bank, pp. 51–76.

Jordaan, J., E.J. Plate, E. Prins and J. Veltrop (1993) *Water in Our Common Future: A Research Agenda for Sustainable Development of Water Resources*, Paris: UNESCO.

Kron, W. (2002) "Flood Risk = Hazard × Exposure × Vulnerability", in Wu, B.S. and Z.Y.Wang, Flood Defense, Vol.1, *Proceedings of the International Symposium on Flood Defences*, Beijing/New York: Science Press, pp. 82–97.

Kunreuther, H. (2000) "Incentives for Mitigation Investment and More Effective Risk Management: The Need for Public-Private Partnerships", in A. Kreimer and M. Arnold, eds, *Managing Disaster Risk in Emerging Economies*, Disaster Risk Management Series No.2, Washington D.C.: World Bank, pp. 175–186.

Munich Re (2002) Munich Reinsurance Company, *Topics: Annual Review of Natural Catastrophes 2002*, Munich: Munich Re.

Plate, E.J. (1993) *Statistik und angewandte Wahrscheinlichkeitslehre für Bauingenieure*, Berlin: Ernst und Sohn.

Plate, E.J. (2002a) "Flood Risk and Flood Management", *Journal of Hydrology* 267: 2–11.

Plate, E.J. (2002b) "Natural Disasters, Human Vulnerability and Global Change: A Framework for Analysis", in Wu, B.S. and Z.Y.Wang, *Flood Defense*, Vol.1, *Proceedings of the International Symposium on Flood Defences*, Beijing/New York: Science Press, pp. 134–145.

UN/ISDR (International Strategy for Disaster Reduction) (2004a) *Living with Risk: A Global Review of Disaster Reduction Initiatives*, Geneva: UN Publications.

UN/ISDR (International Strategy for Disaster Reduction) (2004b) *Visions of Risk: A Review of International Indicators of Disaster Risk and its Management*, New York/Geneva: UN Publications.

United Nations Department of Humanitarian Affairs (UNDHA) (1992) *Glossary:*

Internationally Agreed Glossary of Basic Terms related to Disaster Management, Geneva: UNDHA.

UNDP (2000): United Nations Development Programme *Human Development Report 2000*, Oxford/New York: Oxford University Press.

UNDP–BCPR (2004) United Nations Development Programme–Bureau for Crisis Prevention and Recovery *Reducing Disaster Risk: A Challenge for Development. A Global Report*, New York: UNDP Publications.

Vatsa, K.S. and F. Krimgold (2000) "Financing Disaster Mitigation for the Poor", in A. Kreimer and M. Arnold, eds, *Managing Disaster Risk in Emerging Economies*, Disaster Risk Management Series No.2, Washington D.C.: World Bank, pp. 129–153.

Villagrán de León, J.C. (2005) "Vulnerability Assessment in the Context of Disaster-Risk: A Conceptual and Methodological Review", to be published in UNU-EHS SOURCE series, Bonn: UN University, Institute for Human Security and Environment.

Wisner, B., P. Blaikie, T. Cannon and I. Davis (2004) *At Risk: Natural Hazards, People's Vulnerability and Disasters*, 2nd edn, London: Routledge.

World Bank (2000) *World Development Indicators*, Washington D.C.: World Bank.

Part IV

Local vulnerability assessment

14

Community-based risk index: Pilot implementation in Indonesia

Christina Bollin and Ria Hidajat

Abstract

The following chapter introduces the community-based risk index developed by GTZ and partners and tested in Indonesia. Although many activities relating to risk and vulnerability assessment are currently undertaken, only a few of these approaches actually focus on the measurement of local risk and the specific local needs of vulnerable communities. A community-based disaster risk management system would overcome this shortage. The measurement of vulnerability and risk at a local scale is an important tool for identifying the capacities of households and local communities to manage and overcome emergencies and disasters situations. A quantitative tool was therefore developed and tested in selected areas in Indonesia to assess the community-based disaster risk. The structure and methodologies as well as the results are described in this chapter. Lessons learned and future challenges are also formulated.

Introduction: Why do we need a disaster risk management index for the local level?

Traditionally disasters were viewed as isolated natural events, and few linkages were made to the circumstances of the people affected. Technical solutions prevailed, and the relief and rehabilitation measures that were normally taken were intended to restore pre-disaster conditions. Since the United Nations' International Decade for Natural Disaster Reduction (IDNDR) in the 1990s, and more recently under the UN-International Strategy for Disaster Reduction (ISDR), the paradigm has shifted towards an approach that is more development oriented. It incorporates hazard mitigation and vulnerability reduction concerns, and combines technical and scientific experiences, with special attention given to social, economic and ecological factors. The aim is to achieve comprehensive disaster risk management.

In this context and under the auspices of UNDP, a global Disaster Risk Index was developed to provide better understanding of the relationship between development and risk. The purpose of such an index is to identify a country's social and economic vulnerabilities, along with hazards caused by natural conditions and human activities that contribute to the risk. The index also makes it possible to monitor changes over years. As a first step, the index was created on the international level, comparing national data. Another shift in disaster risk management occurred over the last few years due to the growing evidence that prevailing top-down approaches in disaster risk management may lead to inequitable and unsustainable results. Many such programmes fail to address the specific local needs of vulnerable communities, ignore the potential of local resources and capacities, and in some cases even increase people's social and economic vulnerability.

The approach designed to reduce the local population's risk is called community-based disaster risk management (CBDRM). It aims to reduce vulnerabilities and increase the capacities of households and communities to withstand damaging effects of disasters. Such a system contributes to people's empowerment and participation in achieving sustainable development and sharing its benefits. According to IDNDR, the benefits of CBDRM are as follows:

- Communities are knowledgeable about their own environment. They are rich in experience of coping with emergencies. Community coping methods have evolved over time and demonstrated that they are best suited to the local economic, cultural and political environment.
- This approach has the benefit of enabling communities to be less dependent on relief during disaster periods and to increase their capacities to support their own livelihoods.
- Interventions with community participation have the potential to positively address general socio-economic concerns. Participation will empower the community with new knowledge and skills and develop the leadership capability of community members, and so strengthen their capacity to contribute to development initiatives.
- The impact of disaster situations on women, and also on women's concerns and capacity to cope and contribute, is different from that on

men. Community-based approaches, which recognise this concern, have the potential to contribute towards the social issue of gender equity.

In addition to the community itself, the local Government plays an important role in the CBDRM process. This is especially true following decentralisation, which transfers power and responsibility from the national level to lower-level Government units. They have the overall responsibility for delivering basic services for public safety and for supporting the general well-being of the community and its development. Local Government is therefore an integral part of the CBDRM process in the community. It has the responsibility for institutionalising local and community-based disaster risk management into the formal disaster management and development planning processes and system. It provides the policy and legislative environment that enables the community to become involved in disaster risk management.

To respond to the community-based approach and the increasing decentralisation of many developing countries, the Inter-American Development Bank (IDB) in 2003 requested the Deutsche Gesellschaft für Technische Zusammenarbeit (German Technical Cooperation Agency, GTZ) to conduct a study on "Comprehensive risk management by communities and local government" (Bollin et al., 2003), with the purpose of suggesting strategies and measures to strengthen local actors' capacities for disaster risk management. The study analyses institutional settings for decentralised disaster risk management systems, recognising the importance of combining strong regional and local responsibilities with an appropriate national framework.

On this basis, the authors suggest a coherent system for developing capacity and financial resources in order to make decentralised disaster risk management viable. Furthermore, the chapter presents a community-based Disaster Risk Index, which will make it possible for local Governments to manage and monitor local disaster hazards and vulnerability in a comprehensive and sustainable manner.

In this chapter we present the Disaster Risk Index. The index is based on a comprehensive indicator system, which makes it possible to gather important data on local disaster risk and to identify the main risk aspects in cooperation with the community. For this purpose a questionnaire has been developed. The indicator system provides the necessary inputs for the calculation of the index to make possible, for example, a comparison between communities. First we will present the conceptual framework and the indicator system. Afterwards, the method of calculating the index is described. As this theoretical approach was verified for its application in 2003/04 in a pilot project in Indonesia, we will describe in the conclusion some lessons learned and future challenges. In order to build support for the approach and improve the application of the method, the

GTZ Advisory Project is highly interested in an exchange of best practice and lessons learnt.

Conceptual framework of a community-based indicator system

A community-based indicator system was developed to improve the capacity of communities and local Governments to measure key elements of their current disaster risk. Using indicators at the community level in this context is a rather new and innovative approach. The purpose of the study is to propose a methodology for use at the community and local Government level that can guide decision makers in their efforts to reduce and manage risk to natural disasters. The following conceptual framework (Figure 14.1) systemises the key elements of risk management into the factors of hazard, exposure, vulnerability, and capacity and measures.

The framework helps to clarify the driving forces (factors) at work and serves to identify appropriate indicators. The resulting indicator system comprises a total of 47 individual indicators, arranged according to the identified four main factors and further broken down into factor components. Table 14.1 – Set of community-based disaster risk indicators – presents the indicators in brief, grouped according to the main factors and factor components. A more detailed "application guide and indicator description sheet", which also discusses the rationale and validity of the indicators, is available.[1]

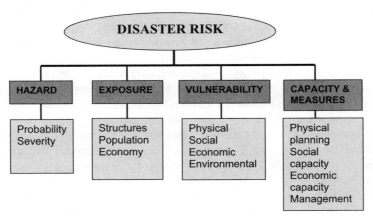

Figure 14.1 The conceptual framework to identify disaster risk.
Source: Davidson, 1997: 5; Bollin et al, 2003: 67.

For each indicator, cut-off points have to be identified that result in low/medium/high classes. This gives the local Government immediate feedback on whether their community is at the lower, middle or upper level with regard to a particular aspect of the indicator. The selection and formulation of the indicators were guided by the principle that the system needs to be applicable in data-scarce environments.

The indicator system is expected to bring benefits by:

- improving the capacity of decision makers at local and national level to measure key elements of disaster risk and vulnerabilities for communities
- providing comparative parameters for monitoring changes in disaster risk as a measure for evaluating effects of policies and investments in disaster management
- highlighting the major deficiencies in confronting natural disasters and thus indicating possible areas of intervention
- systemising and harmonising the presentation of risk information from community level.

Towards a community-based risk index

The indicator system can provide good insight into the current situation of a community with regard to the factors that determine risk, and makes it possible to track changes in those factors over time. However, in order to be able to compare different communities and to facilitate interpretation of the data, an indexing system has been proposed that will condense the technical and individual information of the indicators into summary figures.[2]

Indices are appealing because of their ability to summarise a great deal of often technical information about natural disaster risk in a way that is easy for non-experts to understand and use in making risk management decisions.

Indicator and factor scores (scaling and weighting)

In a first step, the different measurements of the individual indicators (e.g. 50,000 residents and 20 per cent poverty level) have to be made comparable through scaling. This is done by assigning a value of 1, 2 or 3, according to the category achieved (low, medium or high). A "0" is given when the indicator does not apply. (For clarification, an example will be given later from the questionnaire used in Indonesia.)

Next, since indicators have different meanings for specific hazards, a hazard-specific weight has to be found and applied. This is necessary be-

Table 14.1 Set of community-based disaster risk indicators

Main factor and factor component	Indicator name	Indicator
HAZARD Probability	(H1) Occurrence (experienced hazards) or (H2) Occurrence (possible hazards)	Frequency of events in the past 30 years Probability of possible events. Chances per year
Severity	(H3) Intensity (experienced hazards) or (H4) Intensity (possible hazards)	Intensity of the worst event in the past 30 years Expected intensity of possible events
EXPOSURE Structures	(E1) Number of housing units (E2) Lifelines	Number of housing units (living quarters) % of homes with piped drinking water
Population	(E3) Total resident population	Total resident population
Economy	(E4) Local gross domestic product (GDP)	Total locally generated GDP in constant currency
VULNERABILITY Physical/demographic	(V1) Density (V2) Demographic pressure (V3) Unsafe settlements	People per km^2 Population growth rate Homes in hazard prone areas (ravines, river banks, etc)
Social	(V4) Access to basic services (V5) Poverty level (V6) Literacy rate (V7) Attitude (V8) Decentralisation	% of homes with piped drinking water % of population below poverty level % of adult population that can read and write Priority of population to protect against a hazard Portion of self-generated revenues of the total budget
Economic	(V9) Community participation (V10) Local resource base (V11) Diversification (V12) Small businesses (V13) Accessibility	% voter turn out at last communal elections Total available local budget in US$ Economic sector mix for employment % of businesses with fewer than 20 employees Number of interruption of road access in last 30 years

Environmental	(V14) Area under forest	% of area of the commune covered with forest
	(V15) Degraded land	% of area that is degraded/eroded/desertified
	(V16) Overused land	% of agricultural land that is overused
CAPACITY & MEASURES		
Physical planning and engineering	(C1) Land use planning	Enforced land use plan or zoning regulations
	(C2) Building codes	Applied building codes
	(C3) Retrofitting/maintenance	Applied retrofitting and regular maintenance
	(C4) Preventive structures	Expected effect of impact-limiting structures
	(C5) Environmental management	Measures that promote and enforce nature conservation
Societal capacity	(C6) Public awareness programmes	Frequency of public awareness programmes
	(C7) School curricula	Scope of relevant topics taught at school
	(C8) Emergency response drills	Ongoing emergency response training and drills
	(C9) Public participation	Emergency committee with public representatives
	(C10) Local risk management/emergency groups	Grade of organisation of local groups
Economic capacity	(C11) Local emergency funds	Local emergency funds as % of local budget
	(C12) Access to national emergency funds	Release period of national emergency funds
	(C13) Access to int'l emergency funds	Access to international emergency funds
	(C14) Insurance market	Availability of insurance for buildings
		Availability of loans for disaster risk reduction measures
	(C15) Mitigation loans	Availability of reconstruction credits
	(C16) Reconstruction loans	
	(C17) Public works	Magnitude of local public works programmes
Management and institutional capacity	(C18) Risk management/emergency committee	Meeting frequency of a commune committee
	(C19) Risk map	Availability and circulation of risk maps
	(C20) Emergency plan	Availability and circulation of emergency plans
	(C21) Early warning system	Effectiveness of early warning systems
	(C22) Institutional capacity building	Frequency of training for local institutions
	(C23) Communication	Frequency of contact with national level risk institutions

Source: Authors.

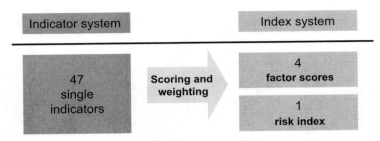

Figure 14.2 Indicator and index system.
Source: Authors.

cause some indicators are more important than others, contributing differently to each of the factors. For example, among the "capacity" factors, an early warning system is considered to be more effective than the
existence of an emergency plan. However, while this is certainly true for
"predictable" floods, in the case of "unpredictable" earthquakes early
warning is much less effective. Weighting represents the importance of
the indicator relative to other indicators. This weight has to be adjusted
for the country-specific conditions; it has been defined in Indonesia, for
example, mainly with the aid of experts from national research institutions, universities, NGOs and representatives from local Government.
A workshop was held for each hazard to discuss and define the proper
hazard-specific weight. Three different hazard-specific weights were defined during the pilot project in Indonesia for landslide, volcano eruption
and earthquake.

Separate composite indices (scores) can then be calculated for the four
main factors that contribute to the risk: hazard, exposure, vulnerability,
and capacity and measures. All the indicators that relate to hazard are
integrated into the hazard index; all those that relate to exposure are integrated into the exposure index, and so on. Depending on the scaled indicator values, the factor indices (scores) vary between 0 and 100. This
can be achieved by distributing a total of 33 weighting points (actually
33 1/3) according to the believed importance of the indicators for each
factor.

The risk index

In a final step, the "overall" composite risk index is derived from the
four factor indices, resulting again in a score that ranges between 0
and 100. As with indicator weighting, the actual relationship between
the factors cannot be determined statistically. Following the approach of

Davidson (1997), a linear relationship is assumed to be reasonable and easy to understand and implement. For the single composite risk index, the contribution of each factor is assumed to be equal. While increasing scores for the hazard, exposure and vulnerability factors represent a higher disaster risk, an increase in the capacity and measures factor reduces that risk. To use the same scale between 0 and 100 as for individual factor indices, a uniform weight of 0.33 for all factors is introduced. This way the overall risk index R can never exceed 100, and can reasonably be expected not to be negative.

Expressed as an equation:

$$R = (wHH + wEE + wVV) - wCC$$

where R is the overall risk index, H, E, V and C are the scores of the hazard, exposure, vulnerability, and capacity and measures indices, respectively, and w is the constant coefficient of 0.33 as a uniform weight for all factors.

The expected benefit is that the overall risk index tells us about the risk and the identified risk-determining factors of communities. It allows us to:

- Compare different communities across the country so as to identify and target communities with high disaster risks. This can also be done for communities that face risk from different hazards.
- Recognise the determining factors for each community behind the existing risk: that is, whether the risk stems from the hazard itself (hazard) and is due to high vulnerability levels (vulnerability) or comes from a lack of capacity (capacity and measures).
- Distinguish the different possible magnitudes of damages through the exposure score.
- Reveal deficits in risk management capacities and potential areas for interventions through a breakdown of the capacity and measures score into factor components.

Testing in Indonesia: pilot implementation in three districts

Urban Quality is an Indonesian–German bilateral technical cooperation initiative supported by the German Federal Ministry for Economic Cooperation and Development (BMZ) and conducted through the Deutsche Gesellschaft für Technische Zusammenarbeit (GTZ) GmbH. The project supports local Government in meeting the challenges and realising the opportunities of decentralisation in Indonesia. It also builds capacity, empowers decision makers and decision-making structures, strengthens institutions and enriches the policy formulation process. The section of the Urban Quality project that deals with geological hazards works in cooperation with the German Federal Institute for Geosciences and Natural Resources (BGR). The focus of the Georisk section is support for local Government in designing and implementing guidelines to improve disaster risk management associated with geological hazards like landslides, volcanoes, earthquakes and salt water intrusion, all of which are prevalent in Indonesia and hamper sustainable urban and rural development.[3]

Within the Georisk project, the "community-based Disaster Risk Index" was applied as a pilot project. Partner institutions – the Directorate of Geology and Mining Area Environment (DGMAE) and the Directorate of Volcanology and Geological Hazard Mitigation (DVGHM) – established a CBDRM working group with experts in the field of volcanology, landslide mitigation, environmental geology and regional planning. The group discussed the community-based risk index approach in the two Directorates and presented the approach also to other research institutions, universities, NGOs and local Governments in several workshops in Bandung, Jakarta and Yogyakarta.

The first visit to Yogyakarta and discussions with the local Government, NGOs anduniversities showed how necessary and important it was that the approach work at the local level. The visit supported our assumption that the set of indicators and questionnaire developed by the project's working group would provide a valuable approach for this field of intervention. With the new approach, the project will help local Governments in the selected areas to bundle activities that so far have been widely scattered, and to improve administrative effectiveness. We are conscious of the need for further application of this approach in Indonesia and other countries. The GTZ/BGR cooperation project made a contribution to the international discussion on risk indices for the local level and therefore recommends applying the set of indicators in a field operation. The CBDRM working group proposed applying the indicators and questionnaire in two different regions.

The first was in the area of Yogyakarta, Central Java, with a focus on the Sleman and Kulon Progo districts, and the second was in Flores, in eastern Indonesia, with a focus on the Sikka district. Over the past few years, many activities, covering all aspects of hazard assessment and risk evaluation, had already been carried out, especially in Yogyakarta, but the assessment of people's vulnerability (life and assets) and of their coping capacity at community level is still in its infancy. The working group was trained in the use of the new approach and qualified to apply the questionnaire. To get a representative and reliable questionnaire, it was necessary to adopt the existing (very general) indicators to the local conditions in Indonesia. The following example, using vulnerability and capacity as indicators, shows how the indicators, index system and the questionnaire are applied to obtain factor scores.

Main Factor = Vulnerability
Indicator name = Access to basic service *(V4)*
Question = How good is the access to basic health centres (e.g. community health centre, midwife centre, clinic, doctor)?
a) Health centres are available and can be reached easily by car.
   Low = 1
b) Health centres are available but not easy to reach only on foot. (X)
   Middle = 2
c) There is no health centre.
   High = 3

Hazard-specific weight for landslide = 2

If b), with value 2 applies here for landslides, the factor score is 4 (2×2).

Main Factor = Capacity and measures
Indicator name = Land use planning *(C1)*
Question = Are disaster risk reduction aspects considered in land use planning?

Yes (X)   No = 0

If yes, how are the measures being implemented?
a) Comprehensive implementation   High = 3
b) Partly implemented   Middle = 2
c) Not implemented   (X)   Low = 1
   Hazard-specific weight for volcanic eruption = 3
   If in a community a land use plan exists but is not implemented, c) with value 1 applies. As land use planning is of highest relevance to reduce risk associated with volcanic eruptions, the factor score is 3 (1×3)

Testing the questionnaire in the districts of Sleman, Kulon Progo and Sikka

The CBDRM working group conducted workshops in three districts, bringing together local stakeholders of the different communities of each district. The aim was to harmonise their perceptions of the questions in order to verify that the purpose of the questionnaire was clear and ensure that the requested data would be available, even on the local/ community level.

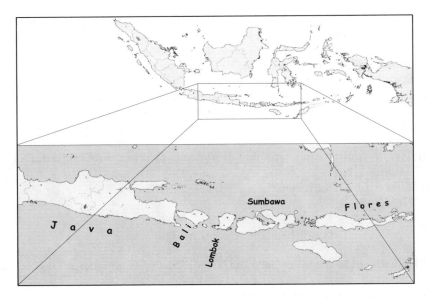

Figure 14.3 Indonesia: project location.
Source: Authors.

The index system is backed up by a database that allows for the systematic recording of the questionnaire results gathered during the workshops. The advantage of the database is that it makes it easy to create different scenarios by changing the given answers and thus identify the areas where mitigation measures could reduce risk most effectively. The results can be visualised in a chart or, if enough data are available, as a map.

Involved in the process were local Government representatives from the spatial planning, finance, infrastructure, environment, health and civil protection departments, natural hazard research institutes, local NGOs, religious and traditional leaders, mayors and community leaders.

In Central Java province, representatives of four communities from the districts of Kulon Progo and Sleman were invited to come to Yogyakarta. In east Indonesia, on Flores, six communities of the Sikka district (Nusa Tengara Timur) participated in the workshop. The CBDRM working group chose these two contrasting regions with the intention of getting representative results that were specific to the particular locations. The archipelago of Indonesia covers a huge geographical area and is very unevenly developed with regard to population density, infrastructure and economic activity, and is threatened by different types of hazards.

Picture 14.1 Discussion with local stakeholders in the district of Sleman, Yogyakarta.
Source: Authors.

Java is the "main" island, with only 7 per cent of the total land area but 80 per cent of the total population. In particular, the fertile regions around the volcanoes and coastal areas are very densely populated. In Sleman district, the average population density is around 1,000 people per square kilometre, and in some villages near the Merapi volcano it is even higher. The region has a good infrastructure, most of the roads are paved, and electricity, telecommunications and fresh drinking water are available in almost every village up the slopes of the volcano. The conditions for agriculture are very favourable and provide a good income. Health care and the provision of basic schooling and higher education are also adequate.

The Sleman district is affected by volcanic eruption (pyroclastic flow) and lahar (debris flow) from Merapi, one of the world's most active volcanoes. The Kulon Progo district is regularly and severely affected by landslides caused on the one hand by the hilly topography and geological conditions, and on the other by intense and inappropriate human activity. Many slopes have been cleared and the soil is already degraded and eroded.

In contrast to Central Java, the eastern part of Indonesia (Nusa Tenggara Timur, Nusa Tenggara Barat) is remote and less developed. The population density is on average lower but locally concentrated in the major cities, which are also important seaports. There is little diversity of economic activity; incomes are low and normally depend on few agricultural products, and manufacturing industry is lacking. Many families in rural areas have almost no income, live near the poverty line and depend on

Picture 14.2 The summit of the Merapi volcano lies only 30 km away from the capital city (left). The district of Kulon Progo is predominantly affected by landslides (right).
Source: Authors.

subsistence farming of barren land. Climate conditions are dry, and sometimes drought leads to food shortages. Basic education is available but higher education is absent or inadequate.

The project region on Flores, the district of Sikka, is prone to various natural hazards like earthquakes, tsunamis and volcanic eruptions. In 1992 an earthquake triggered a tsunami that killed 87 people, destroyed a complete village and caused much infrastructural damage. In 2003 the dormant Egon volcano became active and caused panic among the inhabitants.

Picture 14.3 In 1992 a tsunami completely destroyed the village of Wuring (left picture), but a few years later the village was rebuilt at the same site and is still extremely exposed to earthquakes and tsunamis.
Source: Authors.

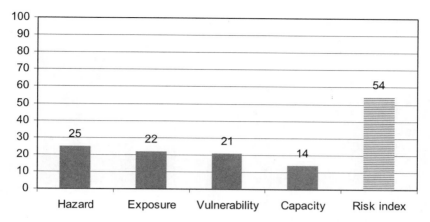

Figure 14.4 Disaster Risk Index of a community in Kulon Progo district prone to landslides.
Source: Authors.

Figure 14.4 shows the Disaster Risk Index and the factor scores of a community in Kulon Progo district, Central Java, which is prone to landslides. The hazard score is fairly high because of the high frequency of such events. Landslides happen regularly during rainy season and emergency shelters are made available for temporary evacuation. The exposure score is high because of the density of settlements and infrastructure that are threatened. The vulnerability factor score is also high, but capacity is low. It is interesting to show in more detail the breakdown of the vulnerability score (Figure 14.5) and the capacity score (Figure 14.6)

The physical and environmental vulnerability scores shows high values because many houses are in unsafe and hazard-prone areas, and much land is cleared and degraded.

A closer look at the capacity score breakdown shows deficits in the physical planning and economic capacity component. This approach could define appropriate and cost-saving intervention measures by cross-checking with the answers from the questionnaire and the results of the stakeholder discussion.

Figure 14.7 provides a direct comparison between a community in the Sikka district and a one in the Kulon Progo district. The risk index of the two communities is almost the same, and if planners take a closer look at the factor scores they can see that the hazard has almost the same value in each case. The difference between the two communities can be found essentially in their exposure and capacity scores. While both communities

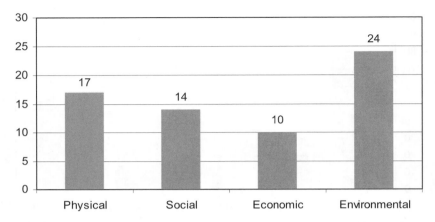

Figure 14.5 Vulnerability score breakdown of a common in Kulon Progo district. Source: Authors.

face the same value of hazard and the vulnerability of both communities is high, Kulon Progo has more property exposed to the hazard; however its capacities and measures for dealing with disasters are better than in Sikka, and the population and local Government are better prepared.

If the index were applied at regular intervals, a community risk could be observed over time and changes could be considered that might then lead to the implementation of appropriate preventive measures.

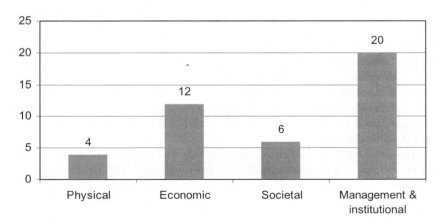

Figure 14.6 Capacity score breakdown of a community in Kulon Progo district. Source: Authors.

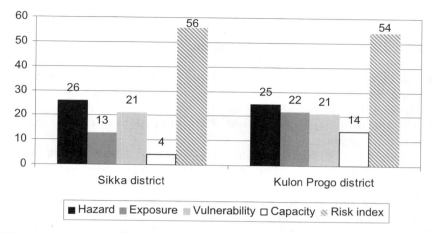

Figure 14.7 Disaster risk: Comparison of a community in Sikka district (left) and a community in Kulon Progo district (right).
Source: Authors.

Lessons learned and future challenges

It turned out that the indicator system is a good tool for sensitising decision makers and creating awareness about the complex forces driving disaster risk. It is useful to have a structured system for these different aspects of risk that helps to clarify the conceptual terms of exposure, vulnerability and capacity. During implementation, when the main task is to discuss with decision makers and the affected communities, it is better to deal with vulnerability and capacity separately. Discussing the vulnerability of a community provides the opportunity to identify deficiencies, and talking about capacity and measures shows in a positive way how to overcome vulnerability and reduce the risk. When preparing for a workshop, it is important to keep in mind that the more diverse the group of people and their knowledge is, the more representative and reliable the outcome will be. Early tests with the questionnaire showed that it is confusing to group the indicators by hazard (H), exposure (E), vulnerability (V) and capacity (C) when using the questionnaire for data collection. The questionnaire has to be made user friendly and organised more thematically, not around concepts that are familiar only to the experts. That means the indicator system and its questionnaire have to be adjusted to the political and cultural conditions of the country or region. The first field assessments backed our assumption that the index system will help in designing appropriate countermeasures for the social and management component of local disaster mitigation.

So far, however, only the first steps and experiments have been made in this pilot project. Much more effort is needed to finalise the Disaster Risk Index to make it more applicable and more user friendly. For example:

- The cut-off points of the indicator value (high, middle, low) need to be verified continuously, because the more precise the index is the more significant it will be.
- The new method needs to be tested in more regions to validate the system for weighting and scaling.
- Standardised benchmarking must be developed to interpret the index system of the Disaster Risk Index and to derive recommendations for risk reduction measures.
- The index system is still being developed. It is a process of testing and adjusting again and again.
- A pilot database to hold the raw data from the questionnaire has been created and needs to be updated continuously, like the indicator system itself.

Notes

1. Available on request at the GTZ Advisory Project "Disaster risk management in development cooperation", Email: disaster-reduction@gtz.de/.
2. Framework by Davidson (1997), adopted by Bollin et al. (2003). It is explained in detail in the manual "Towards a community Disaster Risk Index", available on request at the GTZ Advisory Project, Email: disaster-reduction@gtz.de/.
3. See the Urban Quality homepage. Indonesian–German development cooperation project: http://www.urbanquality.or.id/.
   Homepage Urban Quality component Georisk: http://www.urbanquality.or.id/Georisk_webpage/index.htm.

REFERENCES

Bollin, C., C. Cárdenas, H. Hahn and K.S. Vatsa (2003) *Natural Disasters Network: Comprehensive Risk Management by Communities and Local Governments*, Washington D.C.: Inter-American Development Bank, available from the GTZ Advisory Project's homepage www.gtz.de/disaster-reduction, English: http://www.gtz.de/disaster-reduction/english.

Davidson, R. (1997) *An Urban Earthquake Disaster Risk Index*, Report No. 121, Stanford: The John A. Blume Earthquake Engineering Center.

GTZ (Gesellschaft für Technische Zusammenarbeit) (2001) *Working Concept Disaster Risk Management*, Eschborn: GTZ.

GTZ (2003) *Community-based Disaster Risk Management*, Eschborn: GTZ.

GTZ (2004a) *Disaster Risk Management in Rural Areas of Latin America and the Caribbean: Selected Instruments*, Eschborn: GTZ.

GTZ (2004b) *Guidelines Risk Analysis: A Basis for Disaster Risk Management.* Eschborn: GTZ.

Hidajat, Ria (2002) Merapi/Java: Fluch und Segen eines Vulkans. In Geographische Rundschau 1/2002. Braunschweig.

UN/ISDR (International Strategy for Disaster Reduction) (2004) *Living with Risk: A Global Review of Disaster Reduction Initiatives*, 2004 version, Geneva: UN Publications.

UNDP-BCPR (United Nations Development Programme, Bureau for Crisis Prevention and Recovery) (2004) *Reducing Disaster Risk: A Challenge for Development*, a global report, New York: UNDP Publications.

Zentrum für Naturrisiken und Entwicklung (ZENEB) (2002) *Bericht zum deutschen Beitrag für den World Vulnerability Report des UNDP*, Bonn/Bayreuth.

15

Measuring vulnerability: The ADRC perspective for the theoretical basis and principles of indicator development

Masaru Arakida

Abstract

The ADRC (Asian Disaster Reduction Centre) employs three main indicators to identify the scale of natural disaster events and their (potential) impacts. The first indicator is a ratio of the amount of damage caused by a natural disaster to the GDP of the country concerned. The damage–GDP ratio is a good indicator for assessing events after they have occurred. A second tool developed by the ADRC is a self-assessment sheet distributed to households and local communities as well as in local Government institutions to assess resilience. The self-assessment form is easy to understand and useful for identifying future risk and to reduce vulnerability to natural disasters. A third method of measuring vulnerability to natural disasters takes account of the fact that different indicators are needed for various disaster types, depending on the type and objectives of disaster reduction measures. The ADRC defines disaster risk as a combination of hazard, exposure and vulnerability, and argues that greater emphasis should be placed on exposure in order to reduce the risk of natural disasters.

Introduction and background

The Asian Disaster Reduction Centre (ADRC) was established in 1998 in Kobe, Japan. Its aim is to promote multinational cooperation in disaster reduction by fostering the exchange of disaster reduction experts from all countries and bodies concerned, as well as by accumulating and distributing information. As natural disasters occur regularly all over Asia, the capacities of households and communities to cope with such events needs to be strengthened. Hence, it is necessary to learn from the past, to analyse present risks and thereby to reduce future dangers. The ADRC has identified a number of key indicators that are useful to describe the scale and impacts of natural disasters. It is important for each ADRC country to realise the huge amount of damage caused by past natural disasters, and the ADRC helps Asian countries to learn from the impact of past events. Trying to identify the member countries' risks in the present is far more difficult. The self-assessment and evaluation methods introduced by the ADRC offer support to community leaders and local Governments in this process through training courses.

Disaster and disaster risk

What is a disaster? Earthquakes, storms and torrential rains are natural phenomena we refer to as "hazards" and are not considered to be disasters in and of themselves. For instance, an earthquake on a desert island does not cause a disaster because there is no population or property to be affected. In addition to a hazard, there must be some "vulnerability" to the natural phenomenon for an event to constitute a natural disaster.

"Vulnerability" is defined as a condition resulting from physical, social, economic and environmental factors or processes that increases the susceptibility of a community to the impact of a hazard. "Exposure" is another component of disaster risk and refers to that which is affected by natural disasters, such as people and property. In general, "risk" is defined as the expected costs (deaths, injuries, destruction of property, and so on) that would be caused by a hazard. "Disaster risk" can be seen as a function of hazard, exposure and vulnerability as follows:

Disaster risk = function (hazard, exposure, vulnerability)

Increased exposure and delays in reducing vulnerabilities result in an increased number of natural disasters and greater levels of loss. As shown in Figure 15.1, to reduce disaster risk, it is important to reduce the level of vulnerability and to keep exposure as far away from hazards as possible by relocating populations and property. This shows how disaster risk can be reduced and indicates the area of disaster risk. The reduction of vulnerability can be achieved through such measures as mitigation and preparedness.

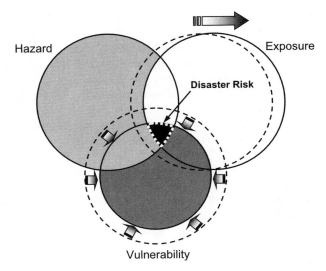

Figure 15.1 Mechanism of natural disaster reduction.
Source: ADRC, 2005.

Disaster information: selected examples

The ADRC collects and disseminates basic information about natural disasters in the Asian region. Figure 15.2 shows the percentage of the global population affected by natural disasters in Asia compared with other regions. Asia accounted for 89 per cent of the world's disaster-affected population between 1975 and 2002. But why is this? Asia is especially prone to natural disasters due to its geographical and meteorological conditions. Many of its people live in areas where disasters are prone to occur. Moreover, Asia not only has a large population in total numbers, but also an uneven distribution, with numerous densely populated urban areas, the so-called "mega-cities". It is thus a region that is highly vulnerable to natural disasters. It is the ADRS's mission to serve the people affected by these disasters.

In terms of the amount of damage, just over a third of the total is caused by floods, slightly under a third by earthquake and a quarter by windstorms. These three major types of natural disaster account for 90 per cent of total losses (Figure 15.3).

The damage/GDP ratio

Table 15.1 shows the seven worst disasters in Asia between 1975 and 2002 in terms of the amount of damage in relation to the affected region's

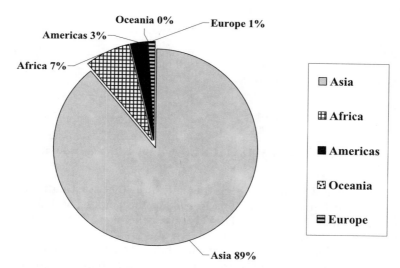

Figure 15.2 Percentage of world population affected by natural disaster in different regions (1975–2002).
Source: ADRC (2002), based on data of EM-DAT, CRED.

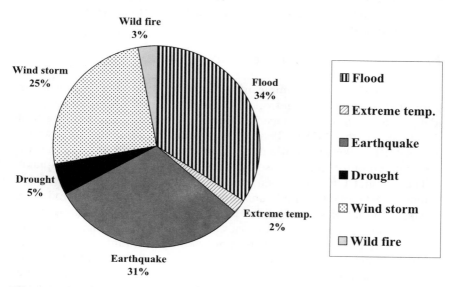

Figure 15.3 Amount of damage caused worldwide by different disaster types (1975–2002).
Source: ADRC, based on EM-DAT, CRED.

Table 15.1 Ratio of amount of damage to GDP (Asia) (1975–2002)

Country name	Year	Disaster type	Damage/GDP
Armenia	1988	Earthquake	908%
Mongolia	1996	Wild fire	192%
Mongolia	2000	Wind storm	97%
Lao, PDR	1993	Wind storm	27%
Nepal	1987	Flood	26%
Georgia	1991	Earthquake	22%
Mongolia	1990	Wild fire	21%

Source: ADRC, based on EM-DAT, CRED and WDI, The World Bank 2003.

GDP (damage/GDP ratio). Using this indicator, the worst natural disaster was an earthquake in Armenia in 1988 as the amount of damage was about nine times Armenia's GDP. This overwhelmed the national economy, and the earthquake thus constituted a national threat. For comparison, the total damage of the Great Hanshin-Awaji Earthquake in Kobe in 1995 was US$80 billion (10 trillion yen). At that time, the total GDP of Japan was US$3.5 trillion (400 trillion yen). The damage/GDP ratio was just 2.5 per cent, which was not a critical level that could affect the whole national economy and security. The damage/GDP ratio thus serves as a valuable indicator of the scale of past disasters that makes it easy to understand the impact in terms of damage caused by a disaster. However, this is not a new idea, and it is of no use for planning countermeasures against future disaster.

Household and community resilience self-assessment

Figure 15.4 shows an example of a pamphlet on disaster management published by the Tokyo metropolitan Government for non-Japanese residents. The "Quake-resilience assessment of homes" is one of the tools developed by the ADRC. People in individual households are asked to assess earthquake preparations by themselves. Thereby they can reduce their vulnerability and increase their resilience in case of an earthquake.

The Cabinet Office of Japan, one of the authorities for disaster management, developed a simple and objective self-evaluation method of assessing communities' capabilities to meet disaster risks. This approach was formed on the basis of interviews with leaders of communities that have suffered disasters, and was supported by relevant ministries. The evaluation sheets on landslide disasters and floods are available on the web. This method makes it possible to evaluate community capabilities through the results of questionnaires filled out by community leaders (Figure 15.5).

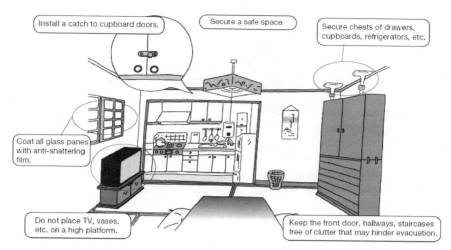

Figure 15.4 Quake-resilience assessment of homes.
Source: Tokyo metropolitan government.

Self-assessment of flood capacity: questionnaire

Q1. Who/which organisation leads the disaster prevention activities in your community? Check all that apply.

- Community leader
- Community member in charge of disaster prevention
- Fire brigade member
- Private company
- NGO of disaster prevention
- Others
- No person in charge of leading disaster prevention activities

----

Q2. Does your community always contact the fire station, fire brigade and/or flood prevention team if you find a bad business of disaster management?

- We always contact them.
- We usually contact them.
- We sometimes contact them.
- We hardly contact them.

Figure 15.5 Self-assessment of flood capacity: questionnaire.
Source: Cabinet Office, Government of Japan.

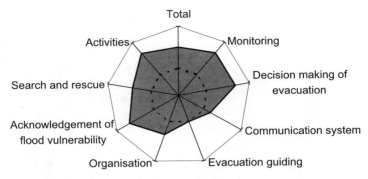

Figure 15.6 Self-assessment of flood capacity: results.
Source: Cabinet Office, Government of Japan.

Through distribution of answers and comprehensive evaluation, the strengths and weaknesses of individual communities can be identified (Figure 15.6). Through their comments and questions the respondents of the questionnaire can promote disaster prevention measures and increase the communities' capability to face disaster risks and to reduce their collective vulnerability.

The flow chart in Figure 15.7 shows the self-evaluation method in the form of a questionnaire targeting the prefectural Governments, which was conducted by the Fire and Disaster Management Agency of the Ministry of Internal Affairs and Communications.

Local disaster management plans can be graded according to nine indicators.

- Assessment, simulation.
- Mitigation, preparedness.
- Organisation.
- Communication systems.
- Resources and material.
- Countermeasures.
- Information sharing.
- Capacity building, training.
- Evaluation, review.

By answering 800 multiple-choice questions, local Governments can assess their disaster reduction efforts. The results are scored by disaster countermeasures, which in turn clarify which countermeasures need to be reinforced (Figure 15.8(c)). This evaluation, which is conducted in all prefectures, also helps to distinguish relatively well-prepared local Governments from others in terms of disaster management.

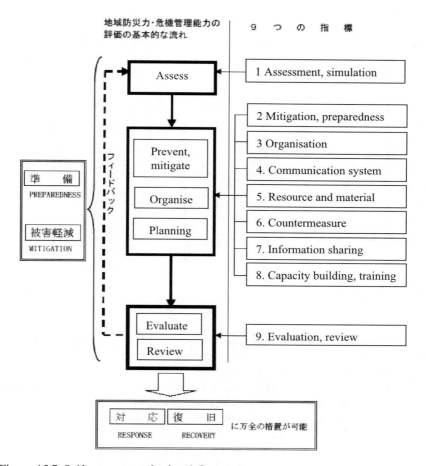

Figure 15.7 Self-assessment for local Government: assessment process.
Source: Fire and Disaster Management Agency, Government of Japan.

Disaster type indicators

As shown in Figure 15.9, identified groups such as individuals, community leaders for disaster management, and local and central Governments all have their own indicators. It should be noted that different organisations, such as those for developing human resources, producing disaster information systems or supporting economic reconstruction, require different indicators. Therefore, disaster indicators are displayed in a matrix by targets and objectives.

Targets / Objective	Persona	Community	Local government	Country	...
Capacity building					
Information system					
Renovate housing					
Evacuation					
Dyke management					
Economic support					
Social support					
...					
...					

Figure 15.9 Matrix of indicators.
Source: Authors.

Indicators for various disaster types differ from those for targets and objectives. Thus, disaster indicators can be represented in a three-dimensional matrix (Figure 15.10).

In areas where earthquakes, heavy rains, storms and other phenomena often occur, the score of the disaster indicator will be much higher than in areas with fewer hazards. While it should be recognised that high ca-

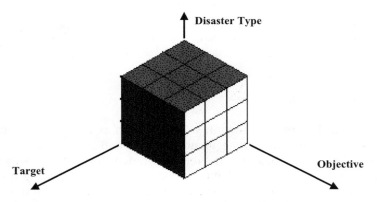

Figure 15.10 3-D matrix of indicators.
Source: Author.

pacities to manage disaster are related to high levels of vulnerability, countermeasures which are cost effective are needed in both kinds of area.

Conclusion

Disaster indicators should be used according to the group (individuals, community leaders, authorities and so on) being addressed, the objectives of the disaster countermeasures and the disaster type. Even where the assessment indicates a high level of capacity, vigilance is still needed in areas of high hazard. Disaster indicators should be used to reduce vulnerabilities and avoid exposure to risks.

The three-d-matrix of indicators has not yet been constructed because so many cells have to be standardised and not enough case studies have been undertaken so far. It will be necessary to undertake case studies of different targets, objectives and disasters and collate the results. Once the three-d matrix of indicators is available, analysis will be much easier. The proposed indicators should be used not only to assess current situations but also to indicate the right course for future improvements and suggest capacity building programmes. Bearing these objectives in mind, the ADRC aims to gather and to analyse cases in cooperation with its associates in each ADRC member country.

REFERENCES

Asian Disaster Reduction Centre (ADRC) (2002) *20th Century (1901–2000) Data Book on Asian Natural Disasters*, Vol. 2, Kobe: Asian Disaster Reduction Centre.
ADRC (2005) *Total Disaster Risk Management*, available from http://www.adrc.or.jp/publications/TDRM2005/TDRM_Good_Practices/Index.html.
Government of Japan, Cabinet Office (2004) *Self-assessment of Capacity for Flood*, March, Tokyo: Cabinet Office, available from http://www.bousai.go.jp/suigai/shindan/.
Government of Japan, Fire and Disaster Management Agency (2002) *Self-assessment for Local Government*, October, Tokyo: Fire and Disaster Management Agency, available from http://www.fdma.go.jp/en/.
Tokyo Metropolitan Government, (2001) *Protect our Tokyo from Earthquakes*, March, Tokyo: General Affairs Bureau, Tokyo Metropolitan Government, available from http://www.soumu.metro.tokyo.jp/04saigaitaisaku/14siryou/14watasitati/Protect_our1.pdf.
World Bank (2003) *Financing Rapid Onset Natural Disaster Losses in India: a Risk Management Approach*, Washington D.C.: The World Bank.

16

Vulnerability assessment: The sectoral approach

Juan Carlos Villagrán de León

Abstract

This chapter presents a quantitative approach to assessing vulnerabilities associated with various types of hazards. The approach is tailored to fit the specific characteristics of communities or societies in terms of sectors such as health, education, housing, industry and so on. In this approach vulnerability assessment is set up in a three-dimensional framework spanning the geopolitical level at which the assessment is being made, the particular sector being targeted and the component of vulnerability being assessed. The method has been developed with an eye to both policy formulation and practicality. Policy aspects are related to the institutional responsibilities regarding the reduction of existing or potential vulnerabilities and the role that national disaster management agencies must play in coordinating this effort; practicality is introduced via simple mathematical algorithms to evaluate vulnerability in a quantitative fashion. Examples of the approach are presented, as well as the links with risks assessment and risk management.

Background

Disasters have been among the many factors inhibiting sustainable development within communities in Central America and other regions of the world for many centuries. Triggered by natural phenomena, such catastrophes have been perceived until very recently as acts of nature, independent of any type of human intervention. However, in recent decades social scientists, engineers and scientists of different branches have begun to modify this view, suggesting instead that disasters are the result of a combination of natural events and the establishment of vulnerable communities, processes and services in high-hazard areas. This modern view introduces the notion of risk as a combination of hazards and vulnerabilities, but also considers risks to be processes that are generated over decades or centuries, with disasters as the end result of such processes.

The reduction of risks through the reduction of hazards, vulnerabilities, coping incapacities and deficiencies in preparedness has been identified as a target since the launching of the International Decade of Natural Disaster Reduction, IDNDR. In May 1994, the *Yokohama Strategy and Plan of Action for a Safer World* issued a set of principles that continue to be the basis for risk management in terms of risk assessment, prevention, mitigation and preparedness. In an attempt to contribute to the implementation of the Yokohama Strategy in Central America, the author has embarked since 1999 on efforts that encompass the development of various methodologies for vulnerability assessment (Villagrán de León, 2000, 2002, 2004, 2005; Hahn et al., 2003). These methodologies, which cover various types of hazards, have been applied in communities within Guatemala and Costa Rica, and expanded to the level of municipal districts and States.

While the methodologies focused initially on the housing sector, where the greatest number of fatalities occur during catastrophic events, particularly in the case of earthquakes, in recent years it has become necessary to characterise and assess vulnerability within urban centres as well. To this end, the notion of sectors as descriptors of urban centres and societies becomes useful in terms of dividing a single urban or national vulnerability into manageable segments. The framework presented in this chapter stresses the aim of making vulnerability assessments more structured, focusing on individual sectors, and quantifying vulnerability in a way that simultaneously allows for the identification of measures to reduce it.

The approach stems from the need to provide the national disaster reduction agencies of Central America with the practical tools needed to strengthen their capacities in risk management. Improving skills is a relevant issue, because the mandates emanating from summit declarations regarding disaster reduction fall directly on such agencies – at least initially – and thus it is up to these agencies to start the process of reducing vulnerability and risk. Therefore, the characterisation of vulnerabilities through sectors has an implicit policy ramification, namely that responsibility for vulnerability management is essentially removed from disaster agencies and placed instead on the agencies in charge of the different sectors.

In addition, the methodology has been designed to be easily applicable via surveys as well as simple mathematical procedures to evaluate different components of vulnerability in a quantitative fashion. This is an important consideration, especially when such assessments have to be carried out throughout the different sectors by national disaster management institutions with limited numbers of highly trained personnel to undertake the process.

Structure and methodology

In the context of natural disasters, vulnerability can be associated with the predisposition of a system, a process, an institution, a community or a country to be affected when a natural event manifests itself. A review of the literature reveals that the term has been defined in different ways by different authors. As stated by Alwang et al. (2001) and by Brooks (2001), the literature contains terms and relationships that at times are unclear, while in some cases identical terms may have altogether different meanings. A systematic analysis of the literature allows for a classification of contexts employed by various authors when defining the nature of vulnerability (Villagrán de León, 2006). These include:

• the particular state of a system before an event triggers a disaster, described in terms of particular indicators or parameters of such a system
• the probability of the outcome of the system, expressed in terms of losses, measured in terms of either fatalities or economic impact
• a combination of a particular state of the system with other factors such as the inherent capacity to resist the impact of the event (resilience) and the capacity to cope with it (coping capacities).

The International Strategy for Disaster Reduction, ISDR (UN/ISDR, 2004) defines vulnerability as the set of conditions and processes resulting from physical, social, economic and environmental factors that increase the susceptibility of a community to the impact of hazards. The physical factors encompass susceptibilities of the built environment. The social factors are related to social issues such as levels of literacy, education, the existence of peace and security, access to human rights, social equity, traditional values, beliefs and organisational systems. In contrast, economic factors are related to issues of poverty, gender, levels of debt and access to credits. Finally, environmental factors include natural resource depletion and degradation. Within the Inter-Governmental Panel on Climate Change (IPCC, 2001) vulnerability is defined as "the degree to which a system is susceptible to, or unable to cope with, adverse effects of climate change, including climate variability and extremes". It is a function of the character, magnitude and rate of climate change and variation to which a system is exposed, its sensitivity and its adaptive capacity.

Regarding various aspects or dimensions associated with vulnerability, Wilches-Chaux (1993) has proposed that vulnerability has different dimensions: physical, economical, social, educational, political, institutional, cultural, environmental and ideological. In contrast the author (Villagrán de León, 2001) identifies several components related to vulnerability: structural, functional, economic, human condition/gender, administrative and environmental.

Despite these different approaches to context, the notion of vulnerability as an essential component of risk has been fundamental in linking disasters to social processes related to development in communities throughout the world. The introduction of vulnerability in the context of disasters has allowed scientists to explain them as not arising solely as a consequence of natural events such as earthquakes or floods, or social events such as fires and explosions, but as a consequence of processes associated with development that have not taken into account the possible manifestation of such phenomena, and thus are not adapted to these phenomena. However, while the worldwide academic debate continues on the notion and the nature of vulnerability, declarations emanating from summits and international conferences are already calling on Governments and institutions to reduce it in order to promote more sustainable development.

From the policy point of view, this implies the recognition of vulnerability as a factor that contributes directly to risks, and hence to disasters. Following this line of thought, one way to start the process of vulnerability reduction is via a framework that defines it in terms that can be assessed, so that the process of reduction can be followed and assessed as well. However; at present this task is difficult as there are no standard, globally accepted methodologies to carry out such assessments. Pilot assessments have been developed at the global level by UNDP-BCPR (UNDP, 2004) and by the World Bank and other institutions through the Hotspots method (Dilley et al., 2005); more recently in the American continent the Inter-American Development Bank sponsored an assessment using novel techniques (Cardona et al., 2003). Other methods have been developed at the sub-national scale and applied in Guatemala and other countries (Villagrán de León, 2000, 2002, 2005; Hahn et al., 2003). In developing countries where disasters are frequent and resources scarce, an obvious strategy is to reduce vulnerabilities initially in those regions where vulnerability can be categorised as high. The problem then becomes one of categorising communities or geographical regions according to low, medium and high degrees of vulnerability. The next step is to develop methodologies that span the different dimensions or components of vulnerability.

In developing methodologies to assess vulnerabilities associated with natural disasters, one must understand that vulnerabilities depend on the type of hazard in question (Villagrán de León, 2000, 2002). Furthermore, a possible intrinsic relationship between vulnerability and the magnitude of the hazard should also be considered (Bogardi et al., 2005; Cardona et al., 2003). In an effort to systematise the various aspects of vulnerability, the author has proposed to consider vulnerability as a state of a particular system, excluding coping capacities, exposure and resilience. Vulnerability is considered as a dynamic quantity because there are several factors that modify it. It is a component of risk when linked to hazards and deficiencies in preparedness. In this model, coping capacities are related to the response once an event manifests itself.

The need to simplify the notion of vulnerability in terms of components becomes evident once a quantitative assessment is required. If too many elements are included within vulnerability, such as coping capacity, resilience, susceptibility and exposure, then major complications arise – first, when identifying how to assess each of these components, and then

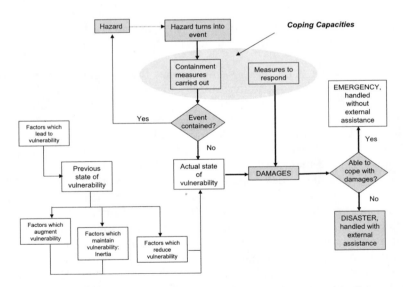

Figure 16.1 Hazards and vulnerability, and their relationship with disasters and coping capacities.
Source: Villagrán de León, 2005.

with respect to how to combine such components to obtain a final figure for the degree of vulnerability inherent in a specific system or a community. Another matter for consideration when vulnerability has to be assessed is the level at which the assessment will be carried out: is it the municipal level, the level of a single house or at the national level? Assessment at different levels encompasses different components and parameters. Integrating these previous notions regarding components of vulnerability, the level at which assessments must be carried out and the type of sectors involved, it is logical to propose that the quantitative evaluation of vulnerabilities should be set up along three dimensions (Villagrán de León, 2005):

- *The geographical level dimension*: this ranges from the human being and the single unit to the national level, and includes the local or community level, the municipal or district level, and the State or province level. The evaluation of vulnerability across this dimension is more related to public policy, as political administrations in different levels are responsible for the administration of such levels.

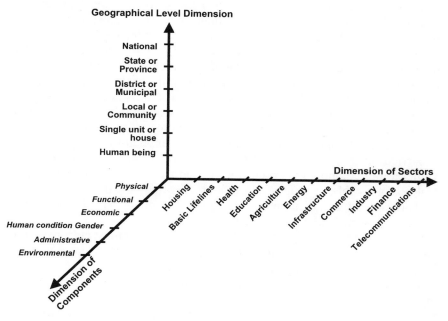

Figure 16.2 The dimensions of vulnerability.
Source: Villagrán de León, 2005.

- *The sector dimension*: this dimension is based on the typical development framework that defines society in terms of its sectors: health, housing, education, infrastructure, energy, agriculture, industry. The evaluation of vulnerability across this dimension is of interest to those institutions involved in managing such sectors, or which are already part of those sectors.
- *The components dimension*: this dimension is related to the various components that are included within the context of vulnerability. These include: structural, functional, economic, administrative, environmental and human condition/gender. Any assessment of vulnerability must start by identifying which components the definition will include or exclude.

The sectoral approach has been proposed from the policy point of view because it promotes the assigning of responsibilities for reducing vulnerabilities to those private or public institutions in charge of each sector, whether these be Government ministries or chambers of commerce, tourism, industry and the like, or bodies at other political-administrative levels. For example, at the national level it is the responsibility of the Ministry of Health to assess and reduce the vulnerability of public health centres. In contrast, it would be up to the director of a local community health clinic to manage the clinic's vulnerability by requesting whichever resources are required for this purpose from the ministry.

The assessment of vulnerability is carried out via an analysis of the components described earlier: structural, functional, economic, human condition/gender, administrative and environmental. Elements within each component are identified a priori from a classification of damage during disasters. The method then identifies options for each of these elements and assigns weights to each option according to its disposition to be affected by an event. A simple linear combination of the elements is carried out numerically to obtain a numerical output for the intrinsic vulnerability component, which can be characterised as low, medium or high using a table of ranges. All numerical values regarding options, as well as weights for combining the vulnerable elements have been deduced with the aid of expert judgment.

Vulnerability assessment using the sector approach must start by defining which sector is to be addressed and then defining the hazards and the geographical level at which the assessment is being made, and finally, the component of vulnerability being assessed. To assess the vulnerability one would then focus along the dimension of components:

- The physical component relates to the predisposition of infrastructure employed by the sector to be damaged by an event associated with a specific hazard.

- The functional component relates to the functions which are normally carried out in the sector and how prone these are to be affected.
- The human condition/gender component relates to the presence of human beings and encompasses issues related to deficiencies in mobility of human beings and to gender considerations.
- The economic components are related to income or financial issues that are inherent to the sector.
- The administrative component relates to those issues associated with the management of routine operations and the ways such administrative issues can be affected by an event.
- The environmental component continues to relate to the interrelation between the sector and the environment and the vulnerability associated with this interaction.

As stated earlier, the assessment of vulnerabilities spans the national level, the State and other lower levels. For example, the vulnerability of a particular hospital may require that the structural components of the building be analysed; a functional vulnerability would comprise those elements that are essential to the hospital's ability to function as a health facility and would include specialised medical equipment and the flow of gases, water and electricity, as well as the storage of certain chemicals and medicines in controlled environments, for example. In the case of private health institutions, economic vulnerability must be considered. In the case of a hospital, the human condition/gender is an issue, especially due to the higher vulnerability of temporarily hospitalised patients whose mobility is restricted due to injury, treatment or sickness, and to the intrinsic vulnerability of infants and incapacitated people due to their lack of mobility while remaining in the hospital. Additional issues related to administrative/organisational processes and functional relationships within different sections or departments are also important to consider (PAHO, 2000); in extreme cases there may be issues related to environmental contamination from the spill of particular chemicals, or solid and liquid waste, particularly of the biological kind.

Examples

Example 1: Structural vulnerability of a house in case of volcanic eruptions

- Sector: housing
- Geographical level: single unit or house
- Component: structural

MATRIX TO EVALUATE STRUCTURAL VULNERABILITY OF A HOUSE

		LOW	MEDIUM	HIGH
WEIGHT		→ 1	3 ←	5 ←
Walls	→15	Block, brick, metallic structure	adobe	cardboard, light wood, plastic, bamboo
Roof, materials	→10	concrete slab	galvanised sheeting, cement tiles	straw, plastic brick tiles
Roof, inclination	→ 5	very inclined	moderately inclined	low inclination
Roof, support material	→ 5	steel structure new, treated wood	old, non-treated wood	weights, stones,
Doors	→ 1	Metal, wood	small windows	large windows
Windows	→ 1	Metal, wood	small glass	large glass

$$V_{estruct} = 15\times5+10\times5+5\times5+5\times3+1\times1+1\times1$$
$$= 167$$

Degree of structural vulnerability	Numerical Range
Low	37 - 80 points
Medium	81 - 130 points
High	131 - 185 points

Figure 16.3 Matrix to evaluate structural vulnerability with respect to eruptions. Source: Villagrán de León, 2005.

Many disasters in both historical and recent times have exposed the structural vulnerability of houses which, when collapsing, provoke numerous fatalities. In Guatemala, the recent 1976 earthquake killed more than 23,000 inhabitants in urban and rural areas when adobe houses collapsed at 3:03 am. If fatalities are to be reduced, the structural vulnerability of houses must be addressed.

The following example illustrates how to calculate the structural vulnerability of a house with respect to ash deposits from volcanic eruptions. Within this framework, a house is considered at the geographic level of a structure unit, belonging to the housing sector, and the example focuses on the structural component.

In the case of volcanic eruptions, the structural vulnerability of the house is modelled using six structural elements: walls, roof materials, roof inclination, roof support material, doors and windows (Villagrán de León, 2005). The degree of low, medium and high vulnerability of each option is introduced in terms of the construction material employed and construction techniques, recognising that some are more likely to be damaged than others.

The classification of materials into the three categories has been based on an analysis of historical eruptions in Central America and the damage caused by such ash deposits on different types of houses. Numerical weights are assigned to the structural elements of the house and to the different construction materials for the various options, and are the combined in a linear fashion. The overall vulnerability is presented in terms of arbitrary units and is classified in three ranges according to such values: *low, medium* and *high*.

In this case, it is important to recognise several aspects:

- The indicators are forward-based. This is an important issue to consider especially because the vulnerability is being expressed in terms of the present condition of the house and addresses those elements likely to be damaged by the deposition of pyroclastic materials in case of an eruption.
- The method is especially well adapted to handle different hazards. Adaptation to different hazards must recognise the impact of the hazard on the various structural elements of the house, adapt the procedure, and assess the specific vulnerability of various types of construction materials and techniques for each component.
- The indicators display the vulnerability of the household in an explicit fashion through the four types of vulnerabilities. Different types of buildings and components can, it is assumed, be classified as more or less vulnerable, and the degrees of vulnerability can be computes according to the actual condition of the house.
- The indicators do not show how vulnerability depends on the magnitude of the hazard. Rather, the method is based entirely on the likelihood of a very high-magnitude event and cannot cope with small-magnitude events at this time.
- The vulnerability assessment can be employed to assess the vulnerability of a single house, but can also be aggregated at the community, municipal, province and national level. Figure 16.4(c) displays houses in the urban settlement Las Torres in Guatemala City. Lots have been classified and identified as being of low (green), medium (yellow) or high vulnerability (red) with respect to landslides (Pérez, 2002).
- The method clearly identifies options to reduce the degree of vulnerability explicitly, but has been tailored for specific regions of the world (taking account, for example, of construction materials used in a par-

ticular region). It will need to be adapted if it is to be applied in other regions of the world.

- The method requires a specific survey to gather information on the component of vulnerability being evaluated.

Thus far, the method has been applied in urban and rural communities of Guatemala and Costa Rica for a range of risks, including earthquakes, landslides, floods, high winds and volcanic eruptions.

Example 2: Functional vulnerability of a health centre in case of floods

- Sector: health
- Geographical level: single unit
- Component: functional

A similar procedure has been developed for health centres in the case of floods. In contrast to volcanic eruptions, floods are events in which

	WEIGHT	LOW 1	MEDIUM 3	HIGH 5
Level of floor first level	→10	High, > 50 cm.	Barely higher than ground, 10 cm	At ground level or lower still
Number of floors	→5	3 or more	2	1
Number of doctors and nurses attending the center	→10	More than 3, living in the community	One or two, living in the community	One, living outside the community
Potable water sources in use	→3	3 or more	2	1
Alternate power supply	→3	Yes		No
Access to facility	→3	Paved, elevated, centrally located	Paved, centrally located	Un-paved, difficult to access
Water deposit in premises	→2	Yes, elevated	Yes, underground	No
Storage of supplies, instrumentation	→2	In elevated places 2nd story	In elevated places 1st story	1

$$V_{econ} = 10 \times 5 + 5 \times 5 + 10 \times 5 + 3 \times 3 + 3 \times 5 + 3 \times 5 + 2 \times 5 + 2 \times 5$$
$$= 184$$

Figure 16.5 Matrix to calculate the functional vulnerability of a health centre with respect to floods.
Source: adapted from Villagrán de León, 2000.

infrastructure is affected by water or mud coming from the ground level. Thus, for the case of the functional component of vulnerability, the most important element is the height of the floor with respect to the ground. Experiences throughout Central America of floods in several types of hospitals and health centres indicate that the next element to be considered with respect to functional vulnerability is the number of floors that the health facility may have. Facilities with several floors are less vulnerable than facilities with a single floor.

The next element relates to the personnel in charge of caring for the sick, whether doctors, nurses or other staff. The premise is that the more staff assigned to a centre, the less vulnerable it is.

Additional elements are the accessibility of the facility, the availability of emergency electrical generators and deposits for potable water, as well as issues related to the storage of supplies and chemicals used in the health facility.

As in the case of houses exposed to volcanic eruptions, vulnerability is calculated in terms of options that make the facility more or less vulnerable. The final value obtained for a particular facility can then be used to categorise the vulnerability as low, medium or high.

Linking the components: risks and risk maps

In the context of risk management policy, risks should be conceived as composed of three measurable factors: *hazards* (the possibility of natural phenomena occurring in a certain geographical area), *vulnerabilities* (the pre-existing conditions that make infrastructure, processes, services and

Figure 16.6 The composition of risk.
Source: Villagrán de León, 2001.

productivity prone to be affected by an external event) and *deficiencies in preparedness* (those conditions that inhibit a community or a society from responding in an efficient and timely manner to minimise the impact of the event in terms of fatalities and losses).

This model assigns responsibility for hazard management to the national institutions devoted to hazard monitoring and to municipal authorities in charge of land-use norms; responsibility for vulnerabilities remains with those who generate them, and the model makes specific reference to each sector; the responsibilities related to disaster-preparedness measures are assigned to the national disaster-management agencies and those agencies devoted to response in case of disasters (such as the Red Cross).

The next step in the risk management process is the assessment of hazards by scientists from different earth science disciplines, such as geology, hydrology, meteorology and vulcanology, and assessment of vulnerabilities and deficiencies in preparedness.

Once these components have been evaluated, they can be combined to generate risk maps. The map shown in Figure 16.7 depicts risks associated with communities in the foothills of Pacaya volcano in Guatemala (Villagrán de León, 2005). The active cone is identified by a red circle on the lower right. Communities are represented by hexagons in such a way that the size of the hexagon represents the relative size of the com-

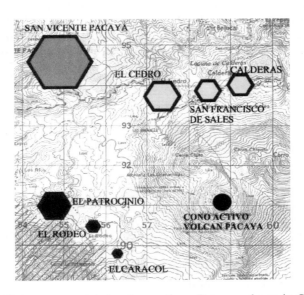

Figure 16.7 Risk map associated with eruptions, Pacaya volcano in Guatemala. Source: Villagrán de León, 2005.

munity. San Vicenta Pacaya, the largest community in the area, contains almost one thousand households. El Cedro and El Patrocinio have about 250 houses each, while El Rodeo has 17 and El Caracol only two. The colour assigned to the communities represents the level or risk, which has been classified into three ranges: low (green), medium (yellow) and high (red). In this particular case, all the communities display similar vulnerabilities, but communities to the southwest of the cone are at greater risk because the hazard is larger in this region than in regions to the north.

Open questions and limitations

As described, the sector approach is based on the notion that vulnerabilities must be reduced by those institutions in charge of the sector. To this end, each sector must recognise its responsibility and start activities to reduce vulnerability. The method used to assess vulnerabilities then relates to the identification of the hazard, the geographical level at which the assessment is being made, and the components which are to be targeted for the assessment.

In Central American nations, Hurricane Mitch was a catalyst to start such processes, and the health sector is advancing dramatically along these lines with the support of the Pan American Health Organization, the World Health Organization and the International Federation of the Red Cross. However, the current approach employed by these institutions does not cover all the components proposed in this approach.

In addition, the housing sector has received considerable attention, and in many countries the reduction of structural vulnerability is being addressed through building codes. However, lack of resources and the number of buildings constructed by individuals themselves for their own use are key factors inhibiting the implementation of building codes. Vulnerability assessments of the housing sector in various communities within different municipal districts that have been carried out by the author have analysed four of the six components (structural, functional, economic and human condition/gender). Matrices have been developed for each of these components, covering distinct type of hazards manifesting themselves in Guatemala (floods, earthquakes, landslides, volcanic eruptions, high winds and floods (Villagrán de León, 2000, 2002, 2005b;). The major drawback so far is the need to carry out specific surveys to acquire the data necessary for the assessment.

In the case of the housing sector, the method has also been adapted to the use of census data provided by the National Statistics Institute of Guatemala, but precision is lost because not all elements are considered in the census; only the structural and the human condition/gender components can be evaluated at this time with such data (Villagrán de León, 2002).

Outlook

As main conclusions regarding the methodology presented in this chapter, the following comments can be presented:
- Vulnerability assessments using this framework are easy to perform, but require a specific survey outlining the types of elements and options included in each type of component.
- The vulnerability indicators make use of"arbitrarily" set weights to combine different elements. While expert judgement has been employed, the selection of numerical weights can always be questioned.
- The indicators can deliver particular information on vulnerabilities associated with a large-magnitude event, but still lacks the capacity to handle different hazard intensities.

Regarding additional work to improve the methodology, the following comments can be made:
- The focus should be expanded to consider all sectors, encompassing all levels and components within the sectors.
- There is a need to develop models to analyse social aspects not covered within the"sector" approach.
- There is a need to develop models to analyse vulnerabilities at various levels (communities, States or provinces, and at the national level). The methodology presented in this chapter focuses on individual communities.
- There is a need to develop models to evaluate those factors that modify vulnerabilities.
- There is a need to expand the methods so that they can exhibit the level of vulnerability as a function of hazard magnitude.

REFERENCES

Alwang, J., P.B. Siegel and S.L. Jorgensen (2001) *Vulnerability, a View from Different Disciplines*, Social Protection Discussion Paper No. 115, The World Bank, Social Protection Unit, Human Development Network, available from http://www1.worldbank.org/sp/.

Brooks, N. (2001) *Vulnerability, Risk And Adaptation: A Conceptual Framework*, Tyndall Working Paper No. 38, Tyndall Centre for Climate Change Research, available from http://www.tyndall.ac.uk/publications/working_papers/wp38.pdf.

Cardona, O.D., J.E. Hurtado, G. Duque, A.M. Moreno, A.C. Chardon, L.S.

Velásquez and S.D. Prieto (2003) *Indicadores para la Medición del Riesgo, Fundamentos Metodológicos*, Manizales: Institute of Environmental Studies, University of Colombia.

Dilley, M, R.S. Chen, U. Deichmann, A.L. Lerner-Lam and M. Arnold (2005) *Natural Disaster Hotspots: A Global Risk Analysis*, Washington D.C.: The World Bank, Hazard Management Unit.

Hahn, H., J.C. Villagrán de León and R. Hidajat (2003) *Comprehensive Risk Management by Communities and Local Government, Component III: Indicators and other Risk Management Instruments for Communities and Local Governments*, publication prepared by GTZ for the Inter-American Development Bank Regional Policy Dialogue – Natural Disaster Network, Eschborn, Germany, http://www.gtz.de/de/dokumente/en-report-component-iii.pdf.

IPCC (2001) *Climate Change 2001: Impacts, Adaptation, and Vulnerability, Summary for Policy Makers*, Geneva: World Meteorological Organization.

PAHO/WHO (2000) *"Principles of Disaster Mitigation in Health Facilities"*, Disaster Mitigation Series, Washington D.C., http://www.paho.org/English/PED/fundaeng.htm, October 2004.

Pérez, I. (2002) *Determinación de Vulnerabilidades en Nueve Asentamientos Humanos del Area Metropolitana de Guatemala*, Final report of the IADB-World Bank-CEPREDENAC-CONRED project, Guatemala, Guatemala

UN/ISDR (International Strategy for Disaster Reduction) (2004) *Living with Risk: A Global Review of Disaster Reduction Initiatives*, 2004 version, Geneva: UN Publications.

United Nations Development Programme (UNDP) (2004) *Reducing Disaster Risk: A Challenge for Development. A Global Report*, New York: UNDP – Bureau for Crisis Prevention and Recovery (BRCP), available from http://www.undp.org/bcpr/disred/rdr.htm.

Villagrán de León, J.C. (2000) *Introducción a la Teoría de Riesgos*, Guatemala: CIMDEN.

Villagrán de León, J.C. (2001) "La Naturaleza de los Riesgos: Un Enfoque Conceptual", in *Aportes para el Desarrollo Sostenible*, Guatemala: CIMDEN.

Villagrán de León, J.C. (2002) *Reconocimiento Preliminar de Riesgos Asociados a Varias Amenazas en Poblados de Guatemala*, technical report prepared for SEGEPLAN, Guatemala: SEGEPLAN.

Villagrán de León, J.C. (2004) *Manual para la Estimación de Riesgos Asociados a Varias Amenazas*, Guatemala: Acción Contra el Hambre, ACH.

Villagrán de León, J.C. (2005) Quantitative Vulnerability and Risk Assessment in Communities on the Foothills of Pacaya Volcano in Guatemala, *Journal of Human Security and Development* No. 1, pp 7–27.

Villagrán de León, J.C. (2006)"Vulnerability: A Conceptual and Methodological Review", to be published in *UNU-EHS SOURCE* series, Bonn: UN University, Institute for Human Security and Environment.

Wilches-Chaux, G. (1993) in"*Los Desastres no son Naturales*", compiled by A. Maskrey, Tercer Mundo Editores, Bogotá.

17

Self-assessment of coping capacity: Participatory, proactive and qualitative engagement of communities in their own risk management

Ben Wisner

Abstract

Community-based disaster management (CBDM) is a form of highly local self-assessment, planning and action that is based on qualitative knowledge of the immediate geographical and social environment. It has evolved gradually over the past 40 years as a corrective and complementary approach to "top down" planning, which tends to rely almost exclusively on quantitative measures, emphasising the measurement of hazards such as climate variability, frequency and severity of storms or floods, and so on. CBDM, by contrast, emphasises the understanding of people's vulnerability to hazards and their capacity to cope with them. Since risk is a function of hazard, vulnerability and the capacity to cope, then the ideal approach to disaster risk reduction would be to integrate CBDM (from the "bottom up") with hazard mapping (from the "top down").

Background

This chapter describes an approach to community-based disaster management that has at its core a method of self-assessment of coping and capacity. This method has developed slowly over the past 40 years, ever since development workers first began noticing the phenomenon of differential vulnerability/capacity in the face of natural hazards. In brief, it

316

emerged as some of us began to synthesise the field observations that were coming in from different parts of the world: the Sahel famine (1967–73), the 1970 cyclone in Bangladesh and Hurricane Fifi in Honduras (1974), for example. There were several common elements in all of these observations. Chief among them were:

- Death, injury, loss and the ability to recover (that is, vulnerability) were highly associated with livelihoods (their nature and their security).
- Vulnerability was not only an economic matter, but depended also on location and access to political power.
- Vulnerability was not homogeneous in "communities", but varied widely.
- Capacity also existed. Farmers had coping strategies that relied on indigenous technical knowledge, social networks and alternative income-generating activities.
- National Government officials did not understand or trust such capacities, and national counter-disaster strategies generally came from the top down (if they existed at all in marginal, peripheral zones). On the whole these made the situation worse.

Responding to these observations, an approach was developed in the 1980s and 1990s for defining and analysing vulnerability and capacity that linked these concepts to the livelihoods, locations and ecological conditions of households, to political access and "voice", and to local knowledge and social relations. However, analysis is not the same thing as assessment. From a practical point of view, non-governmental organisations (NGOs) and other development institutions simplified the approach by creating taxonomies of "vulnerable groups" that are very familiar to us now: women, children, elderly people, people living with disabilities, ethnic and religious minorities, and the like.

Reports from Sri Lanka and India suggest that as many as a third of those killed by the tsunami of 2004 were children (Rohde, 2005). In the Great Hanshin earthquake in 1995, more than half of the 6,000 killed were over 60 years of age, and many of them elderly widows living by themselves (Wisner et al., 2004a: 293–300). There is empirical support for the use of such "check lists" of vulnerable groups, especially by hard-pressed relief personnel. NGOs like Help the Aged and Save the Children have developed sophisticated screening techniques that can pinpoint children or elderly people at risk in shelter or refugee camp situations.

While there is a lot of truth in the assertion that such groups often suffer more injury and death during disasters, and that they may have "special needs", the taxonomic approach is problematic. In between disasters, when the challenge is to work proactively with local Government, civil society and other stakeholders to assess vulnerability in advance and try to reduce it, the simple taxonomic approach fails. First, it produces too many "false positives". Not *all* women are equally vulnerable in Kenya in a drought or in the Philippines in a cyclone. Therefore, in order to deal with this weakness of the taxonomic approach, civil society organisations in many places have adopted what can be called a situational and proactive approach. Some individuals, such as Paulo Friere (1973) and Robert Chambers (1983), were influential in legitimising this kind of "bottom up", participatory approach. The fact is that the approach has evolved as civil society has evolved.

Thus one of the key goals of this approach is to empower local people so that they can understand their own daily lives and situations in a way that enables them to increase self-protection and to demand and fight for social protection. The main goal is not national or international comparison, and measurement is used here essentially as a means for providing local people more control over the conditions of their lives.

Structure, methodology and examples

A proactive and situational, dialogical approach to assessing coping and capacity

As developed and practised by a wide variety of NGOs today in many parts of the world, the approach aims to build enough trust, common purpose and motivation among a group of people so that they can use a variety of simple tools (hazard mapping, time budgets, problem trees, wealth ranking and so on) and ask key questions (e.g. what are our strengths/opportunities/weaknesses/threats?) to assess their own capacities and vulnerabilities. This is the form of self-assessment at community level that is the basis of community-based disaster management, which is very different from global assessment and measurement methodologies, for example as shown in the overview provided by Birkmann in Chapter 1.

Self-assessment is proactive because it does not focus solely on hazards and vulnerability, but also on capacities. It takes a problem-solving perspective. The approach is situational because it is place and group specific. It takes into account specificities, change and surprise. It is therefore a special case of what is more formally known as "adaptive planning". What groups in the Philippines, Bolivia and Zimbabwe (among other places) are doing is also dialogical because there is no "expert" or "teacher". The facilitator seeks to understand the reality on the ground and find the way forward together with the participants. In this way, it is also natural to begin with local knowledge (for example, of soil, weather, pests, ocean tides and storms, etc.). Outside knowledge may well be brought into the mix, but as knowledge that is added, and not as a replacement for the vernacular system of understanding.

An African example

A group in the drought-prone Chivi district in southern Zimbabwe has used wealth ranking in addition to other criteria to identify people who are most vulnerable (Murwira et al., 2000). People with fewer assets are less able to produce a surplus that can be stored against a bad rainfall year, and have fewer monetary savings or possessions (such as livestock) that can be sold to buy grain during hungry times. The self-assessment of drought vulnerability included sketch mapping of people's farms in order to identify resources and environmental constraints, and also generated participants' labour profiles. These showed the various tasks that men and women have month by month in the annual cycle of agricultural production. In this way labour constraints were identified, as well as periods when people have more free time to engage in cooperative activities to reduce drought risk.

Focus group discussions also identified the range of people's coping technology. A series of "traditional" drought-proofing measures, such as seed selection, intercropping of more than one plant and small-scale irrigation, were identified. Drought-coping mechanisms, such as the sale of livestock and of labour outside of the community, were also mentioned. However, these groups were not simply open-air seminars, but action-oriented circles. They discussed and tested various additional means for avoiding drought. These included the use of tied ridges in their maize (corn) fields. Earth is not only dug up to form ridges on the contour of the field, but also mounded to form "ties" from ridge to ridge, forming a rectangular "box". Rainwater is trapped in the box and thus can infiltrate deeper into the rooting zone of the plant. Every bit of precious rain is used in this way.

Self-assessment that is place and group specific is likely to be quite complex. In Malawi, for example, vulnerability to drought is not simply a function of agronomic practices, numbers of disposable livestock (a banking system on the hoof) or savings (de Waal, 2002, 2005). Group self-assessments there have also focused on whether an adult in the household is living with HIV/AIDS, the dependency ratio in the household and whether there is the labour capacity to carry out some of the well-known drought-avoiding practices (e.g. multiple plantings during periods of erratic rainfall, tied ridging to maximise rainfall infiltration, and earning income from casual labour).

Tools for community risk assessment

The "tool kit" for those who facilitate community risk assessment (CRA) is varied and constantly growing. Many of the tools used date from rapid rural assessment (RRA), and then later from participatory learning activities (PLA), which evolved from development practice. Since the late 1950s, with a great acceleration during the 1980s and onwards, there has been a fertile exchange back and forth between academic researchers in development studies, anthropology and geography on the one side, and their counterparts (and frequent partners) in development NGOs and other agencies on the other side. Within this academic–NGO nexus, tools were developed to elicit and systematise local knowledge. While many, if not most, of these tools were developed with uses other than CRA in mind, institutions working on disaster risk reduction have absorbed a number. An array of 20 manuals and guidebooks that contain such tools can be sampled on the website of the ProVention Consortium (ProVention, 2006). Other methods have been reviewed by Wisner et al. (2004a: 333–342).

A brief discussion of three representative examples will serve here to demonstrate their qualitative and action-oriented nature. These examples come from a manual developed for use in the Pacific Islands by UNDP (1998).

The recent historical timeline

Note that the timeline in Figure 17.1 goes back to the founding of the village some 100 years ago. Oral history is obviously the basis of some of this narrative. Note also that significant political, socio-economic and natural events are bundled together in the historical memory of this community: the establishment of a new chief, a major cyclone and an economic recovery programme. Group discussions that generate such a timeline provide the basis for planning for the future. The interconnection among different kinds of events becomes clearer, and people's potential agency in their own history becomes more than speculation.

The annual calendar and labour budget

Figure 17.2 also emerges from group discussion. It describes the cyclical rhythm of activities, both social and economic. Periods of greater or lesser vulnerability to disruptions appear, for example when credit for the harvest is more necessary in order to get by. In some cases, an annual timeline like this one is useful in identifying when surplus time is available during the year for women and men to work on cooperative projects that prepare for hazard events. Often such time/labour budgets are done separately for women and men.

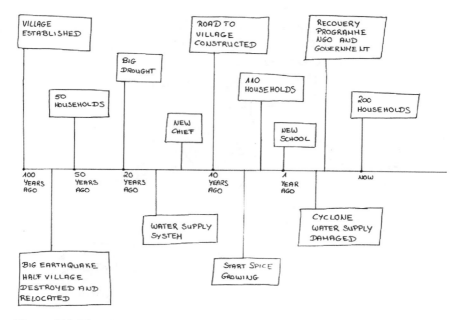

Figure 17.1 Timeline.
Source: UNDP (1998).

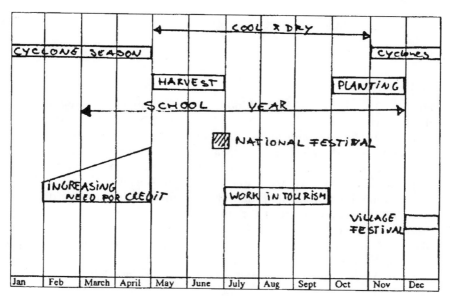

Figure 17.2 Example of a seasonal calendar.
Source: UNDP (1998).

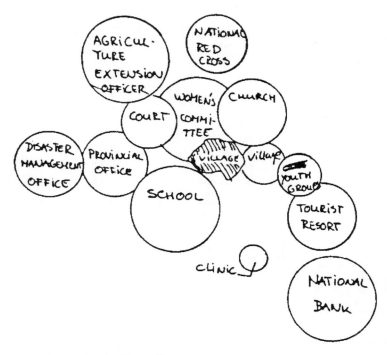

Figure 17.3 Example of a Venn diagram.
Source: UNDP (1998).

The Venn diagram

Figure 17.3 is an example of a kind of brainstorming in groups that results in clarity about the interrelationship of problems and opportunities,
strengths, weaknesses and threats. The figure is named after the logician
John Venn. Overlapping and adjacent circles "contain" processes that
are related in conceptual space. In this particular example, villagers
"mapped" the relationships among significant social, political, service
and economic institutions that might have some role in reducing disaster
risk. What became evident in the discussion and is mirrored in the diagram is that the clinic and the bank are "outliers", without strong direct
links to the village or a mediating institution, such as the women's group.
This analytical result, while qualitative, can be quite important in focusing attention on a missing link and on action that is needed to make that
link.

An American example

When one reviews the results of such group self-assessments based on these and a wider variety of other tools from many parts of the world, it is striking how complex and variable they are. In West Hollywood, part of the greater Los Angeles megacity, there are significant minorities who have special recognised needs during earthquakes and flooding. These minorities include an elderly Russian émigré population that has minimal English and little trust in authorities because of their life experiences in the former Soviet Union. After citizen-based consultations, the city has hired a Russian-speaking liaison officer and also recruited outreach volunteers among the younger elders who live in the apartment houses where this population is concentrated. Another group includes several thousand homosexual youths and male-to-female transgender individuals. They are particularly vulnerable in some cases because they inhabit derelict buildings and, like the elderly Russians, avoid and disregard authorities and their warnings. Others, who are middle class, nevertheless have special needs for privacy in common shelter situations. The city has appointed a transgender social worker to do outreach work among this population and also to give training to police and firefighters who tend to have little understanding of these people, or even act with hostility towards them (Wisner, 2004b).

Another very important point that emerges from such self-assessments is how vulnerability and capacity are treated interchangeably. Since these self-assessments are action-oriented – proactive is the term I tend to use – they focus not only on what increases the likelihood and severity of injury, loss, psychological trauma and difficult recovery, but also on what capacities can and should be developed in order to reduce these vulnerability factors. This is most striking in the case of people living with disabilities. The tendency in the past, represented by training material produced by the Federal Emergency Management Agency (FEMA) and the American Red Cross, has been to teach caregivers how to help the disabled. Little attention was given to the capacities that people living with disabilities already had or could develop. With the international "independent living movement", a more nuanced approach to disability is now emerging. The "disabled" are not simply seen as a category in a standard taxonomy; rather, each person's situation and capacities are taken into account, and the person living with a disability is seen as a partner in developing pre-disaster plans and capacities (Wisner, 2006).

Other examples

In a similar way, community-based researchers have rediscovered the large repertoire of capacities people use in Bangladesh to "live with floods" (Schmuck-Widmann, 1996, 2001) or in southern Africa to "live with drought" (von Kotze and Holloway, 1999). In my own PhD work, which I did in collaboration with the National Christian Council of Kenya (NCCK) in eastern Kenya during 1971–1976, I found that people knew of 76 different ways to cope with drought. These ranged from agronomic and ecological adjustment to social and economic and even political actions (Wisner, 1988). This pattern of indigenous coping with drought proved useful to the NCCK in designing alternatives to famine relief efforts with various groups of people in eastern and northern Kenya.

Scale, functions and target group

Scale

The approach is highly local in scale and yet very broad, including aspects of local economic, social, political, technological, ecological and geographic processes as they affect local capacity and vulnerability. However, networks of citizen-based organisations that use these self-assessment methods are beginning to develop links to national and regional hubs such as Duryog Nivaran in Sri Lanka, Peri Peri in Cape Town, South Africa, and La Red based in Colombia.

Target group of the approach

On the whole, organisers tend to approach low-income, marginalised groups of people, often in the aftermath of a hazard event such as a volcanic eruption or coastal storm. In an effort to organise people and mobilise their energy and participation in making a village or urban neighbourhood plan for the *next* hazard event, local groups, such as the one affiliated with the Center for Disaster Preparedness in Manila, build on people's experience and motivation.

For example, if one were to imagine anticipatory use of citizen-based vulnerability assessment in coastal Thailand before the 2004 tsunami, a complex and shifting mosaic of vulnerability factors would emerge. Wealth and access to resources (including information and social capital) would be important. Thus poor rural migrants, who are recent arrivals, may be more vulnerable than better-established households. However, with time, such rural migrants may become well connected. Occupation is also likely to emerge as an important factor. Those reliant on fishing were particularly vulnerable to the tsunami and in general are more vulnerable to the frequent cyclones that affect the region. Their vulnerability involves not merely their proximity to the sea, but their tendency not to want to abandon their only assets – a boat and nets – even if they have received an evacuation order. Without insurance, they will also find it very difficult to reestablish their livelihoods. Returning to the coast of Andra Pradesh eight years after a deadly cyclone hit there, Peter Winchester found that small farmers and small-scale fishermen had made least progress in recovery (Winchester, 1992). A citizen-based self-assessment of vulnerability might also have revealed that it is not customary for women and girls to learn to swim (just as in Bangladesh, gender-specific cyclone mortality is caused by the fact that women do not climb trees).

Thematic focus

The CBDM approach discussed here usually includes hazard mapping as well as vulnerability/capacity assessment. It can be focused on a single hazard or have a multi-hazard focus. Vulnerability/capacity assessment can be simple or complex. The simple version would involve a census of people and assets at risk, pinpointing some individuals or households who are at extreme risk, and a review of human, financial and technical resources available to mitigate the risk. In the simplest version, the plan that emerges would not even include mitigation, but only preparedness.

In more complex applications, the approach goes further to study the "root causes" of vulnerability and the blockage of capacity. This might involve discussion and eventual action to deal with the problem of land poverty or landlessness, exploitation by landlords, moneylenders or corrupt officials, and similar problems.

Open questions and limitations

Technical limitations

In many of the situations where this approach to community risk assessment is used, participants have a low level of formal education. A careful balance may be required between qualitative and quantitative characterisation of hazards, vulnerabilities and capacities. On the one hand, one wants results that are meaningful to local participants and that can lead to action plans and, indeed, to action. On the other, one wants to achieve a minimum acceptable standard of accuracy. However, recent experiences show that even advanced tools such as mapping with geographical information systems are accessible to untrained lay people (IAPAD, 2006).

Political limitations

"Limit situations" may be reached where participants agree that they cannot take risk reduction further without a change in policy or practices over which they have no control. In such cases, politics may come into force. In democratic, open and accountable systems of governance this should not be considered a disadvantage of the method, but rather one of its strengths. Lobbying for policy change or change in practice or implementation can result, with benefits all round. However, in non-democratic regimes, organisers and facilitators of this method may be endangered and need to be protected.

Outlook

The ProVention Consortium has already begun a process of collection, analysis and dissemination of methods used for participatory capacity and vulnerability assessment. This work will considerably expand Pro-Vention's existing "tool kit" (ProVention, 2006), where so far a compendium of some 20 sets of methods and approaches to CRA is available.

This chapter should not be misunderstood as an argument against "measurement" as quantitative assessment. The dilemma or challenge I have discussed here concerns the balance between qualitative and quantitative assessment on the one hand, and the balance between reflection and action on the other. I believe that at the local level, the balance needs to be skewed toward qualitative assessment and action, as long as that action is subject to monitoring and correction as results come in. The reality of poor, marginal and excluded people is that they have few surplus resources, time or patience for assessment without action. If they have experienced "planning" at all, it has usually been without follow-up action or beneficial results; what villagers I lived with in Tanzania in the mid-1960s called the "fruits of freedom" (*matunda ya uhuru*).[1]

Nevertheless, the debate before, during and after the World Conference on Disaster Reduction (18–22 January 2005 in Kobe, Japan) correctly identified quantitative targets as being necessary at the national level. Here I would wholly support efforts to measure vulnerability and coping capacity in terms of the investments made by national Governments in the infrastructure that supports community-based risk assessment and proactive planning. Such infrastructure logically includes the primary health care system, primary and adult/continuing education system, micro credit and micro insurance infrastructure and technical outreach in such domains as agroforestry, small scale irrigation and soil conservation.

It should be possible to quantify investments required for developing national infrastructure that provides communities with the conditions they need to implement their own risk reduction actions while simultaneously trying to meet the Millennium Development Goals.

Note

1. In the first 10 years after independence from the colonial power, the Tanzanian people expected rapid and very concrete improvements in the quality of their lives as the result of self-rule – as opposed simply to valuing the abstraction, "freedom".

REFERENCES

Chambers, R. (1983) *Rural Development: Putting the Last First*, Harlow, Essex: Longman.

de Waal, A. (2002) "What AIDS Means in a Famine", *New York Times*, 19 November, available at http://query.nytimes.com/gst/fullpage.html?sec= health&res=9B06EFD81130F93AA25752C1A9649C8B63.

de Waal, A. (2005) "HIV/AIDS and the Threat of Social Involution in Africa", in B. Wisner, C. Toulmin and R. Chitiga, eds, *Toward a New Map of Africa*, London: Earthscan.

Freire, P. (1973) *Pedagogy of the Oppressed*, London: Penguin.

IAPAD (Integrated Approaches to Participatory Development) (2006) Participatory GIS/http://www.iapad.org/participatory_gis.htm

Murwira, K., H. Wedgwood, C. Watson, E. Win and C. Tawney (2000) *Beating Hunger: The Chivi Experience: A Community-based Approach to Food Security in Zimbabwe*, London: Intermediate Technology Publications.

ProVention (2006) "Vulnerability and Risk Assessment", in *Tool Kit*, available at http://www.proventionconsortium.org/?pageid=39.

Rohde, D. (2005) "Tsunami's Cruellest Toll: Sons and Daughters Lost", *New York Times*, 7 January, p. A1.

Schmuck-Widmann, H. (1996) *Living with the Floods: Survival Strategies of Chardwellers in Bangladesh*, Berlin: FDCL.

Schmuck-Widmann, H. (2001) *Facing the Jumna River: Indigenous and Engineering Knowledge in Bangladesh*, Dhaka: Bangladesh Resource Centre for Indigenous Knowledge (BARCIK).

United Nations Development Programme (UNDP) (1998) *Guidelines for Community Vulnerability Analysis: An Approach for Pacific Island Countries*, Geneva: UNDP, available at http://www.proventionconsortium.org/files/tools_CRA/UNDP_CVA.pdf.

von Kotze, A. and A. Holloway (1999) *Living with Drought: Drought Mitigation and Sustainable Livelihoods*, Cape Town/London: David Philip Publishers/Intermediate Technology Publications.

Winchester, P. (1992) *Power, Choice, and Vulnerability: Case Study in Mismanagement in South India*, London: James and James Science Publishers.

Wisner, B. (1988) *Power and Need in Africa: Basic Human Needs and Development Policy*, London: Earthscan.

Wisner, B. (2004a) "Assessment of Capability and Vulnerability", in G. Bankoff, G. Frerks and T. Hilhorst, eds, *Vulnerability: Disasters, Development and People*, London: Earthscan, pp. 183–193.

Wisner, B. (2004b) "Urban Social Vulnerability to Disaster in Greater Los Angeles", in S. Sassen, ed., Volume on Cities for UNESCO's *Encyclopedia of the Life Support Systems*, Paris: UNESCO, access by registration at https://www.eolss.net/.

Wisner, B. (2006) "Disability and Disaster", in C. Rodrigue and E. Rovai, eds, *Earthquakes*, London: Routledge (in press). Available in draft form from http://www.radixonline.org/disability2.html.

Wisner, B., Blaikie, P., Cannon, T. and Davis, I. (2004) *At Risk: Natural Hazards, People's Vulnerability and Disasters*, 2nd edition, London: Routledge.

18

Measuring vulnerability in Sri Lanka at the local level

Jörn Birkmann, Nishara Fernando and Siri Hettige

Abstract

This chapter deals with the development and testing of different methodologies to investigate and measure the pre-existing and emergent vulnerability of coastal communities in Sri Lanka to tsunamis. It is also relevant for other coastal hazards, such as cyclones, which can induce major damages (see e.g. NDRC, 2005) and are sometimes caused by storm surges at the coastline (increased water levels and storm water waves). The chapter gives an insight into different methodologies and data sources that can be used to assess various characteristics of vulnerability. It shows the capacities and limitations of selected methodologies and also addresses the potential synergies achieved by using different techniques of vulnerability assessment at the same time in order to derive a broader picture of the past and current vulnerabilities of coastal communities in Sri Lanka. Particular emphasis is given to assessing the vulnerability of different social groups using questionnaires. Initial results of the study, especially of the questionnaire-based household survey, are shown and discussed. However, limitations and problems are also underlined. The research also tested and used approaches presented earlier in the book, such as the sector approach and the Human Security Index. The chapter is based on research undertaken within a joint project of United Nations University's Institute for Environment and Human Security (UNU-EHS), the University of Colombo, University of Ruhuna, Eastern

University and the German Aerospace Center (DLR), and the Center for Development Research (ZEF).

Background

The devastating tsunami in the Indian Ocean on 26 December 2004 hit Sri Lanka and Indonesia hardest. In Sri Lanka alone it affected more than 546,500 people or 3 per cent of the total population: about 40,000 people were killed or missing (Department of Census and Statistics, 2005).

Although the vulnerability of the coastal communities in Sri Lanka was clearly visible in the tsunami catastrophe, reconstruction, relocation and urban renewal are medium and long-term tasks, which should support and promote development of more disaster-resilient communities in coastal areas. Thus, the identification and understanding of different vulnerability patterns, coping capacities and intervention tools need to be promoted in order to be able to facilitate the reconstruction process with appropriate information to ensure sustainable development. The results presented in this chapter are based on a study currently being undertaken in Sri Lanka to measure the revealed vulnerabilities of coastal communities to tsunamis. This research is embedded in a larger project concerned with the strengthening of early-warning capacities, financially supported by the UN/ISDR-PPEW and UNU-EHS. As part of its mandate, UNU-EHS conducted research on vulnerability and human security together with local Sri Lankan universities in both the Singhalese and Tamil parts in order to strengthen local capacity and cooperation with scientific institutions as well as to facilitate the reconstruction process by providing essential scientific information.

Structure and methodology

The conceptual framework

The vulnerability assessment approach that was developed and tested aimed to explore various characteristics of vulnerability of different social groups, basic infrastructure services, economic sectors and environmental services to tsunamis and coastal hazards. At the same time, the research should also provide more in-depth knowledge regarding the capacities and limitations of the different methodologies to identify and measure vulnerability. As a theoretical framework and definition of vulnerability, the approach is based on the BBC conceptual framework (see

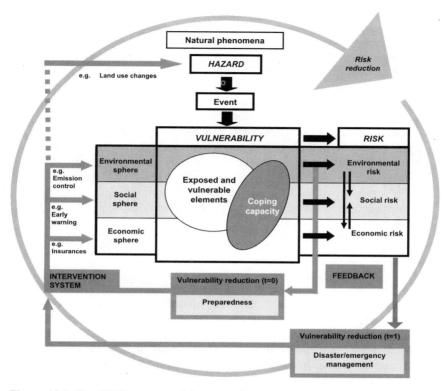

Figure 18.1 The BBC conceptual framework.
Source: Authors, based on Bogardi and Birkmann (2004) and Cardona (1999 and 2001).

Chapter 1), which stresses the fact that vulnerability is defined through exposed and susceptible elements, on the one hand, and the coping capacities of the affected entities (e.g. social groups) on the other. Moreover, the BBC conceptual framework shows that it is important to address the potential intervention tools that could help to reduce vulnerability in the social, economic and environmental spheres (Figure 18.1). In contrast to the "model for a holistic approach to disaster risk assessment and management" by Cardona (see Chapter 1) the "BBC framework" has a close link to the concept of sustainable development and therefore acknowledges three main spheres of vulnerability: social, economic and also environmental vulnerability. Conversely, current concepts often do not include the environmental sphere as potentially vulnerable to natural hazards.

Box 18.1 Overview of the four main techniques used to assess vulnerability

---

1) Assessment of the built environment with remote sensing
   Estimation of vulnerability of different urban areas
2) Critical infrastructures and sectors vulnerability
   Ground survey of the exposure and susceptibility of basic infrastructure services and their facilities, e.g. hospitals and schools
3) Vulnerability of different social groups – questionnaire based
   Interviews with households in selected locations to identify and assess the different vulnerabilities of various social groups to tsunami risk
4) Vulnerability of social groups and local communities
   Census data based assessment of vulnerability using general indicators

---

The BBC framework was used to structure the assessment of the vulnerability of different social groups to tsunamis and coastal hazards using questionnaires as a data-gathering tool. Additionally, the sustainable livelihood framework – particularly the asset pentagon – was used as another orientation for the analysis of exposed and susceptible elements, as well as coping capacities, in the social sphere (social capital, human capital, financial capital, etc.) (see in detail DFID, 1999). The BBC conceptual framework also gave guidance for the basic infrastructure and sector analysis, as well as the selection of appropriate census data to estimate the exposure of elements at risk, their susceptibility and coping capacity.

The four methodologies

Overall, the research encompassed four main techniques to identify and measure vulnerability (see Box 18.1), focusing on different data sources and different characteristics of vulnerability.

1. The first methodology is aimed at estimating the overall exposure of the settlement area, as well as examining some characteristics of the vulnerability of different city areas (Grama Niladari (GN) divisions, the smallest statistical unit in Sri Lanka) by looking at the structure and quality of the built environment. We think that the type of settlement and housing unit allows a general classification of urban areas with regard to their socio-economic status. This means we assume that, as far as an initial estimation is concerned, a higher or lower vulnerability within the community can be associated with the conditions of the built environment that different groups are living in. Thus, this approach is also intended to test the potential abilities of remote sens-

ing to estimate socio-economic conditions. However, this analysis is still ongoing and the classification and automatic analysis of different housing types especially have proved more complicated than expected. The remote sensing methodology allows for comparison of the situation before and after the tsunami, implying that one can analyse the extent to which the exposure and the structure of the buildings were major causes of revealed losses. Thus we intend to implement remote sensing techniques to measure physical vulnerability and, additionally, to test how far this information can also be used to estimate socio-economic aspects linked to vulnerability, such as poverty.

2. The second methodology explores the exposure and susceptibility of different critical infrastructures and sectors, such as education (e.g. schools), the health system (hospitals) and finance/banking (banks). The research methodology is linked to the method presented by Villagrán de León in Chapter 16 of this book. In the first phase of the research conducted in Sri Lanka, the main focus was on the degree of exposure of different units of critical infrastructures and sectors (see the section on "Selected examples and results"), although we also intend to expand the focus later to other criteria.

3. The third methodology required the most attention and included questionnaire-based interviews to explore the various vulnerabilities of different social groups in selected locations that were prone to tsunamis and coastal hazards in Galle and Batticaloa. Besides the analysis of the revealed vulnerability (retrospective focus), the in-depth questionnaire survey should also allow for a better understanding and estimation of current vulnerability after the tsunami. This methodology also addressed spatially specific features of vulnerability of coastal communities in Sri Lanka.

4. The fourth methodology focused on general indicators available in the census and local statistics to estimate the vulnerability of different social groups and economic sectors of coastal communities to tsunamis and coastal hazards. This technique also aims to use some of the data examined in the other methodologies mentioned in combination with census data, which is available for most parts of the country and its coastal areas.

Beside the analysis of the tsunami's impact, the revealed vulnerabilities of different coastal communities in Sri Lanka, the research design was intended to test and compare different techniques and methodologies to identify and measure vulnerability. This should lead to a better understanding of the benefits and advantages of the different methodologies. Additionally, the use of various methodologies provides a more comprehensive picture regarding the multifaceted vulnerability of coastal communities to tsunamis and coastal hazards. For instance, the analysis of secondary data on the tsunami disaster collected from the Department of Census and Statistics helped us to understand the macro picture of the impact of the tsunami, while the questionnaire-based analysis of households in selected GN divisions provided in-depth information regarding specific vulnerability profiles of different social groups. The questionnaire-based research was particularly important for capturing new data to identify, measure and assess human susceptibility as well as coping capacity at the local and household level. Coping is often linked to activities during a hazardous event, such as eating fewer meals in the context of famines. However, coping is also a critical issue after a tsunami since many people who may have survived have none-the-less lost important livelihood assets and are therefore forced to cope with primary and particularly secondary impacts of the tsunami, such as lack of access to water and traumatic experiences, etc. Finally, the different methodologies should also improve the quality of data and the assessment itself by applying more than one data-collecting tool, a technique also known as "triangulation of methods". The rationale behind this approach is that ideally the weaknesses of one method are offset by the strengths of the others.

Selection of the locations and study sites

As we wished to study various features and characteristics of vulnerability, especially those relating to different social groups, the built environment and basic infrastructures/sectors, we decided to focus on the city of Galle as the major study area. We also conducted similar research in the city of Batticaloa, which is located in the eastern part of Sri Lanka where intermittent violence connected with the persisting conflict is continuing. Both Batticaloa and Galle were heavily affected by the tsunami, and thus were also "excellent" areas to study the rehabilitation and reconstruction process in order to derive relevant indicators that best explain the vulnerability and unusual difficulties in recovering from a tsunami.

Galle and Batticaloa

Galle Municipal Council (MC) area spreads over 15 wards with nearly 91,000 inhabitants. Galle is a major city in the southern part of Sri Lanka with important economic and trade infrastructures, such as the Galle harbour and cement factory, and it also has an important function as a centre for medical care in the southern region (Galle Municipal Council sources). Moreover, both Galle and Batticaloa are areas where major UN and other international agencies are working; therefore it was assumed that collaboration and data sharing would be possible and beneficial for this research and the institutions involved in the reconstruction, such as Habitat and the UN/Office for the Coordination of Humanitarian Affairs (OCHA).

In contrast to Galle, Batticaloa is located on the eastern coast of Sri Lanka see Figure 18.2(c). The municipal council area (city) encompasses around 100,000 inhabitants, and more than 13,350 families were affected by the tsunami. Due to the heavy impact of the tsunami and the continuing conflict in the region, more refugee camps can be found than elsewhere and many schools, roads and hospitals have not been rebuilt so far.

Since the questionnaire-based research was a major tool to assess household vulnerability, the following section gives an insight into the structure, content and initial outcomes of the questionnaire-based research in Galle. Additionally, preliminary results of other methodologies, especially of critical infrastructure and sector assessment, will be shown.

Questionnaire-based vulnerability assessment of social groups to tsunamis

The questionnaire-based identification of vulnerability and the most vulnerable groups in Galle was executed in six GN divisions shown in Figure 18.3. In these GN divisions, which are all situated within the Galle Mu-

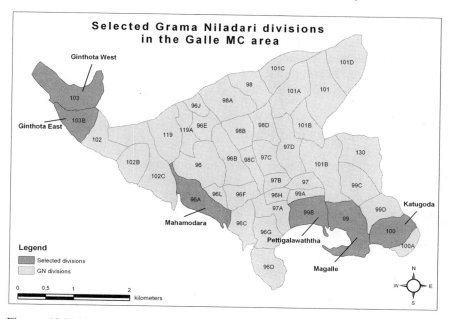

Figure 18.3 Overview of selected sites in Galle for the questionnaire-based research.
Source: Authors.

nicipal Council area, a sample of 500 households was selected by applying the stratified random sampling method to administer the interview schedule. The focus on households as an analytic unit is important within the framework of vulnerability assessment, since livelihood strategies and economic conditions can be best assessed at this basic unit level (see e.g. Green, 2004).

The criterion for selecting the GN divisions was the general degree of damage in each specific GN division; this ensured that the survey covered households in areas that had experienced heavy, medium and light damage. In other words, we acknowledged that the hazard magnitude could have an influence on the revealed vulnerability. Specific tsunami impact information (revised RF1 sheets), collected by the census after the tsunami disaster, was used as the sampling frame. In the primary stage, a structured interview schedule was developed to gather data from the selected households. The questionnaire consisted of open, coded and multiple-response questions that explored household characteristics before and after the tsunami disaster, different coping mechanisms, as well as intervention tools (see BBC framework) currently being discussed in Sri Lanka. The questionnaire also included questions about the various problems and issues that the households faced at the time of the survey (nine months after the tsunami). Five hundred interviews – comprising 73 questions with around 610 possible criteria – were conducted by the enumerators in face-to-face interaction with respondents during a four-week period in Galle. Additionally, the same household questionnaire was also conducted in selected GN divisions in Batticaloa encompassing 532 households. Thereafter the data was entered and analysed using the "Statistical Package for Social Sciences" (SPSS).

Selected examples and results

This section shows selected preliminary outcomes of the critical infrastructure and sector assessment and the vulnerability analysis of social groups based on questionnaires for the city of Galle. Some preliminary results of the findings for Batticaloa will also be shown. Initial outcomes of the comparison between Galle and Batticaloa will be outlined on the basis of the household survey.

Critical infrastructure and sector vulnerability

Although vulnerability is understood as encompassing susceptibility and exposure of the unit at risk, as well as coping capacity (see BBC framework), the analysis of the vulnerability of critical infrastructures and sectors (schools, hospitals, banks) focused in its first phase only on the degree of spatial exposure of the different units in Galle municipality.

The GIS analysis and the ground survey were designed to identify the degree of exposure of different elements and units (such as schools or banks) and assess, for example, the number of schools in the 100-metre zone (from the sea) compared to the total number of schools in the Galle municipality (see Figure 18.4(c)). Thus, as a first definition for measuring the exposure of different critical infrastructure in the high-risk zone, the governmental 100-metre zone was used as a classification. After the tsunami the Sri Lankan Government introduced a regulation under which, within a 100-metre zone from the sea in the south and southwest of Sri Lanka and a 200-metre zone in the east, reconstruction is not permitted at all, or is at least restricted. We used this 100-metre zone as a first estimation of the high-risk zone. This means that if a high concentration of facilities of a specific critical infrastructure, such as hospitals, is located within the 100-metre zone, this infrastructure or service is more vulnerable to coastal hazards and tsunamis than those whose major facilities are located further inland. In order to capture information regarding the hinterland, the research takes into account the 200-metre zone and areas 300 metres and more from the sea (see Figure 18.4(c)).

Our analysis shows that 50 per cent of the hospitals, approximately 20 per cent of the banks and 13 per cent of the schools (four schools) are located in the "high risk zone" (100-metre zone) in Galle municipality. Thus, the health infrastructure, and also the banking and schooling sector, are especially vulnerable due to their high degree of exposure in the high-risk zone compared to other infrastructures/sectors.

On the other hand, the distance from the sea is only one indicator that allows a first estimation of vulnerability regarding exposure. It is also our intention to use an elevation map to assess the exposure of different critical infrastructures and sectors in the high-risk zone. An elevation model could provide more information on how far the wave or the water could go. However, the inundation area can also serve only as a first estimation for the high-risk zone, since the likelihood of damage and loss of life does not necessarily correspond with the total inundation area; the likelihood of harm and damage also depends on the velocity of the wave and the water depth in the inundated area. Interestingly, a study by Herath (2005) shows that the inundation area often went beyond the 300-metre zone from the sea, for example in Katugoda, Magalle and Pettigalawatta (see in detail Herath, 2005; and regarding the GN divisions, Figure 18.3). Overall, the vulnerability assessment of different critical infrastructures based on the degree of exposure is only meant to give a general overview, rather than a very precise picture, which would imply also taking into account the impact of the built infrastructure, such as buildings, roads and canal systems. For a general estimation of the vulnerability of different critical infrastructures/sectors in terms of their exposure, an analysis of the proportion of exposed elements of a specific critical infrastructure and sector in the 100-metre zone compared to those outside seems to be adequate. In the recent tsunami it became evident that the proportion of damaged housing units within the 100-metre zone was around 60 per cent compared to only 3.3 per cent in the 300–400 metre zone (Department of Census and Statistics, 2005). This means the spatial distribution of critical infrastructures is an important aspect of vulnerability.

Vulnerability assessment of different social groups using questionnaires

The structure and the content of the questionnaire take into account the BBC framework; accordingly it captures vulnerability with regard to susceptibility and exposure, on the one hand, and coping capacities on the other. Moreover, it addresses vulnerability in its social, economic and environmental dimensions. It also focuses on intervention tools in order to derive information regarding potential or already implemented policy interventions to reduce vulnerability. Moreover, elements of the sustainable livelihood framework, such as the five livelihood assets (DFID, 1999), were also used to generate some of the questions and criteria regarding, for example, social capital.

The questionnaire-based research covered such topics as those shown in Box 18.2.

The household survey showed that people within the 100-metre zone from the sea suffered higher degrees of damage than those located within the 200–300-metre zone. A higher proportion of deaths (65 per cent) was reported from the households that were situated within the 100-metre zone than other areas (35 per cent). A significant proportion of housing units situated within the 100-metre zone were totally destroyed or so seriously damaged that they were uninhabitable (47 per cent). In contrast, the amount of destroyed or unusable houses outside the 100-metre zone was about 29 per cent according to the questionnaire results in the six GN divisions in Galle. Also, the degree of loss of life and damage clearly differed between the 100-metre zone and the area outside, as captured in the household survey in the selected GN divisions and shown in Figures 18.5 and 18.6.

A comparison of the category "no damage" inside and outside the 100-metre zone with the category "totally damaged" indicates clearly that the likelihood of housing damage was significantly higher in the 100-metre zone than outside. The analysis of housing damage according to housing

Box 18.2

Vulnerability
susceptibility and degree of exposure
  1) impact of tsunami on household members and their assets
  2) structure of the household
  3) housing conditions and the impact of the tsunami
  4) direct loss of possessions
  5) income before and after the tsunami, land ownership
  6) activity and occupation of the household members, such as income generating activities and sources
  7) place of the house and location of the place of work

coping capacity
  8) social networks
  9) knowledge about coastal hazards and tsunami
  10) financial support from formal and informal organizations
  11) access to information, e.g. radio

intervention tools
  12) relocation of housing and infrastructure to inland
  13) early warning system
  14) insurance preparedness
  15) 100 meter "buffer zone" (implemented by the government)

type showed that 51.9 per cent of single-storey houses were "destroyed" or "severely damaged and unusable", whereas only about 16.7 per cent of the multistorey buildings in this zone suffered these levels of damage. Outside the 100-metre zone, about 20 per cent of single-storey houses were destroyed or rendered unusable, and 11.4 per cent of the multistorey buildings. This means that both housing types suffered greater damage in the 100-metre zone than outside it, and hence the distance to the sea is an important criterion.

However, a major difference can be seen in the smaller amount of damage that was done to multistorey buildings in the 100-metre zone compared with single-storey housing units. We assume that, in general, where physical vulnerability is concerned, the multistorey houses close to the sea were constructed more robustly and to a higher standard than the average single-storey housing units. Moreover, a comparison of the damage patterns in Galle and Batticaloa underlined some interesting differences. While the damage patterns inside and outside the 100-metre zone showed clear differences in Galle, the damage in Batticaloa was

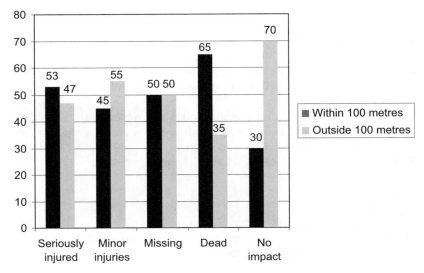

Figure 18.5 Dead and injured people in the 100-metre zone and outside in the selected GN divisions in Galle (in %).
Source: Authors.

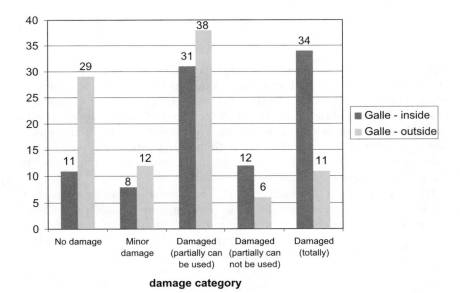

damage category

Figure 18.6 Housing damage in the 100-metre zone and outside in the selected GN divisions in Galle (in %).
Source: Authors.

also very high outside the 100-metre zone. Inside the 100-metre zone the number of houses that were destroyed or rendered unusable amounted to 70 per cent, while outside the zone 56 per cent were damaged in this way. Thus the local analysis revealed that since the tsunami also caused major destruction further inland in Batticaloa, the appropriateness of a 100-metre buffer zone has to be called into question. The different outcomes between Galle and Batticaloa underline the necessity of using different intervention tools to ensure that reconstruction promotes a reduction of vulnerability.

Indicators to measure vulnerability of different social groups to coastal hazards

The survey was intended to estimate the current susceptibility and coping capacity of different social groups and households to coastal hazards, and in particular to tsunamis. We tested different indicators and selected those listed below in order to be able to explain the revealed and to estimate the present, vulnerability and coping capacity. Among others, we selected and tested the following important indicators:

- *number of young and elderly people dead and missing in the total population* (demographic susceptibility)
- *income and employment* (economic susceptibility)
- *landownership* (socio-economic susceptibility and recovery potential)
- *social networks and membership of organisations* (coping capacity patterns)
- *loans and savings* (coping capacity).

Young and elderly people: demographic vulnerability

The statistical analysis of the questionnaire data on demographic characteristics of the dead and missing people in Galle revealed the fact that the youngest age group (from 0 to 9 years) as well as the age groups over 40 years were highly vulnerable to the tsunami, respectively suffering 25 per cent and 44 per cent of all casualties. With regard to the absolute number of dead and missing, the young age group shows the most fatalities. However, the relative number of dead and missing of specific age groups showed that elderly people were especially vulnerable to the tsunami. For example, the fatality rate among people aged 90–99 years was around 40 per cent, and in the 80–89-years age group, 13 per cent. In contrast the fatality rate among those in their thirties was only 2 per cent. In Batticaloa similar patterns can also be observed. The youngest age group (especially 0–10 years) and elderly people account for the highest relative mortality in terms of the different age groups examined.

Gender too played a role regarding the dead and missing. The questionnaire results showed that nearly twice as many females (65 per cent) as males (35 per cent) were dead or missing in Galle. Similar patterns were found in the survey in Batticaloa. The indicator shows clearly that females were – and presumably still are – more vulnerable to tsunamis than men. Women have more difficulties in getting to safety quickly (e.g. climbing the roof). Female household members might also be more exposed due to their traditional role of carrying out activities around the house. The enormous gender gap between the revealed vulnerability of females and males regarding the likelihood of being killed by the tsunami was also observed in other studies of tsunami affected areas, for example in the Aceh province in Indonesia (Oxfam, 2005 and Rofi et al., 2006) and in the Tamil Nadu province in India (Guha-Sapir et al., 2006).

Overall, one can conclude that for measuring present vulnerability, the youngest age group and especially elderly people, as well as people over 40 years of age, are more vulnerable than those age groups in between (11–40 years). The relative mortality showed that special attention has to be given to those elderly people who are 80 years and older. Moreover, households with a high number of women are generally more vulnerable than those households and areas where a mainly male population lives and works.

Income and employment

Analysis of the vulnerability of households with regard to income-earning activities showed that households earning a monthly income of 5,000 rupees or less (before the tsunami) lost a high proportion of their incomes after the tsunami as well as, in many cases, their jobs. Most of the workers were engaged in daily paid labour as mobile fish vendors and fishermen, and some were engaged in other types of small-scale self-employed activities. In contrast, the households that earned a monthly income of 21,000 rupees or more (18.4 per cent) did not suffer such negative impacts in terms of income decline. In these households there is generally more than one income earner with permanent employment in either the Government or the private sector. Small-scale businessmen, fishermen and self-employed people who earned a monthly income of 5,000 to 21,000 rupees before the tsunami are facing an income decline or have lost their jobs. However, they are better off than those households in the lowest income category, who experienced a further decline in their already low incomes.

We also analysed the unusual difficulties in recovering from tsunami impact related to the employment situation. In this context, we focused on the number of people unemployed before and after the tsunami in the selected GN divisions in Galle. On the basis of more than 2,500 valid answers, the household members were classified according to the follow-

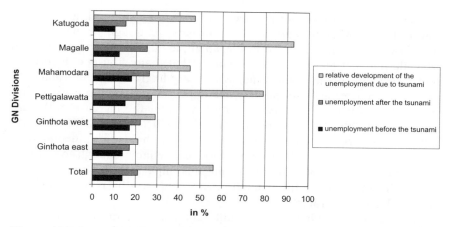

Figure 18.7 Development of unemployment before and after the tsunami in selected GN divisions in Galle.
Source: Authors.

ing activities: student, household work, unemployed, employed and unable to work. While no major changes were found in the categories "students", "household work" and "unable to work", the categories "unemployed" and "employed" indicate important changes. In all six selected GN divisions, unemployment increased due to the tsunami's impact. However, the analysis underlined important differences regarding the increase in unemployment between the different GN divisions.

Figure 18.7 shows that, especially in Pettigalawatta and Magalle, unemployment increased dramatically, by nearly 80 per cent and more than 90 per cent respectively. Katugoda and Mahamodara also faced a high increase in the unemployment rate of more than 40 per cent. By contrast, Ginthota east and Ginthota west showed only a moderate increase in relative unemployment.

The relatively low increase of the unemployment rate in Katugoda is surprising. This might be influenced by the fact that more than 25 per cent of the respondents in Katugoda were engaged in household work, which generally can be continued even with limited resources. The tsunami-induced increase in unemployment is only one indicator that illustrates the unusual difficulties faced by different groups and areas in attempting to recover from the negative impacts of hazardous events. Another important indicator is income level. Analysis of the monthly income before the tsunami shows clearly that Katugoda contained a high proportion of households with low income (5,000 rupees and less) (Figure 18.8(c)).

Thus in Katugoda the economic and financial status of households, as well as their capacity to replace the losses they suffered, is generally lower than in Magalle and in Ginthota east. Although Katugoda does not show a dramatic increase in the unemployment rate compared with Magalle and Pettigalawatta, it is evident that many households already had a very low income, which means they are likely to face unusual difficulties in recovering from the negative impact of coastal hazards. An important task for the future is to monitor whether the increase in the unemployment rate, for example in Magalle, will continue and thus will influence the socio-economic status of the households. The income data of the households after the tsunami were not reliable enough to derive a clear conclusion or evidence regarding this impact; therefore we had to use the income before the tsunami as a calculation basis to identify income-related vulnerable areas.

To measure the vulnerability of different social groups to tsunamis, we also tested the Human Security Index proposed by Plate (Chapter 13). We used and calculated a modified (simpler) version of it, based on the data we were able to capture using the household questionnaire. The modified index was used to measure the time a household will probably need to recover from property damage. Due to a lack of appropriate data, we limited the analysis to the time a household would need to repair the housing damage caused by the tsunami. This data was captured within the questionnaire, and potential reconstruction costs could also be estimated on the basis of various sources. Furthermore, housing is a human right; consequently, the reconstruction of the house is viewed as a key element within the overall recovery process.

First we calculated the disposable income of the household by taking the household income before the tsunami minus the minimum subsistence level (minimum subsistence level for Sri Lanka is calculated at 1,428 rupees per month). Thereafter we estimated the damage and the respective reconstruction costs of the specific household according to four damage categories captured in the questionnaire: "damaged totally" (250,000 rupees), "damaged partially, cannot be used" (200,000 rupees), "damaged partially, can be used" (100,000 rupees) and "minor damage" (20,000 rupees). On this basis we calculated the time (in months) that the specific household would need to repair its housing damage assuming that this household would spend all its free disposable income on this purpose, based on the income data before the tsunami (income data thereafter was not reliable).

While this assumption may not always apply, the analysis revealed that prior to the tsunami, 19 out of 500 households (3.8 per cent) in Galle would not have been able to replace any loss, while after the tsunami 31 out of 500 households (6.2 per cent) are currently not able to replace their losses using their own financial capacity.

Table 18.1 Time that households need to replace housing damage

Time to recover	GALLE Relative number	BATTICALOA Relative number
No damage	24.1%	6.0%
Up to 12 months	31.8%	12.0%
12–24 months	17.8%	12.0%
More than 2 years	26.2%	59.0%
Not able to recover at all	4.0%	11.0%
Sum and N	100% (N = 500)	100% (N = 532)

Source: Authors.

Among those households with income above the minimum subsistence level that were therefore able to replace losses they had suffered, it became evident that around 24 per cent had not suffered any damage, and thus needed zero days to replace the losses. Some 31.8 per cent of the households were able to repair or replace their housing damage within one year. In contrast to this relatively fast recovery, 26.2 per cent of the households that were able to recover would take more than two years to repair the housing damage, assuming they would spend all their freely available income on the replacement and reconstruction of their houses (see Table 18.1).

In Galle around 30 per cent of all households surveyed in the questionnaire in the six selected GN divisions needed financial support to be able to rebuild or repair their house within two years' time, based on low reconstruction costs. In contrast, in Batticaloa around 60 per cent of the households would not be able to recover unaided within two years (see Table 18.1). Thus twice as many households in Batticaloa need external financial support to repair the actual housing damage. This number might be even higher if we were to acknowledge a general decline in income after the tsunami.

We also analysed the index with regard to the different professions of household heads and the landownership for Galle. Comparing the occupations of the household heads, the calculation shows that the recovery potential of fishing households is relatively low; they need nearly 22 months to replace their housing damage, while in contrast the white-collar workers could recover much faster, requiring on average only half a year to replace the housing damage they suffered. Interestingly, the small-scale business people recover faster (7 months) than those who are self-employed (12 months). The underlying reasons for these differences may be manifold. Interestingly, not only the exposure and actual damage, but also the size of the household, the age structure and the job

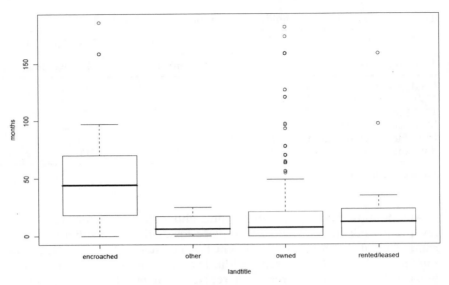

Figure 18.9 The index applied to house damage and land title.
Source: Authors.

diversity are important factors and determine the differences in the ability to recover. For example, where the household head is a clerk, the recovery potential is higher than in those households where the household head is a fisherman.

The analysis also explored the recovery process of different social groups classified according to their land title. Households that live on owned land need around seven months to replace their housing losses, while, in contrast, the group of illegal settlers and squatters needs on average about 44 months to do so (using the median). Those living in rented houses need to spend around one year of their freely available income to repair the damage to their housing (Figure 18.9).

The clear differentiation of the recovery potentials of different social groups according to their landownership already indicates that landownership itself can serve as a surrogate indicator to classify the vulnerable households and to estimate the resilience of different social groups in coastal communities in Sri Lanka.

Landownership

Landownership is a key aspect of the vulnerability of different social groups to coastal hazards and tsunamis in Sri Lanka, as poor urban households often have no legal ownership of land for housing due to excessive commercialisation of the housing and real estate market. Using landownership as an indicator allows identification of social groups that are highly vulnerable, as land serves not only as a legally recognised place to live but also as an economic and livelihood resource i.e. place for production, security for bank loans. It can even be sold in times of crisis in order to overcome the difficulties of recovering from the negative impact of a hazardous event (Farrington et al., 2002; Satterthwaite, 2000). We assume that those households that live near the sea (high exposure) and do not own land (particularly squatters and illegal settlers) are especially vulnerable because they face unusual difficulties in recovering from potential tsunamis or coastal hazard impact (see Figures 18.9 and 18.10). At present they are not allowed to rebuild their houses in the same place and do not get financial support from the Government for reconstructing houses within the buffer zone.

That said, the analysis of landownership in the six selected areas in Galle points to the fact that a significant proportion of respondents had owned land (81.2 per cent), while nearly 11.4 per cent had encroached on either Government or private land. Although, people with no land title make up around 10 per cent of the total population in this area, it is important to consider that a significantly higher proportion of squatters and illegal settlers live within the 100-metre zone (17.4 per cent) compared with the average (11.4 per cent) (Figure 18.10). The proportion of squatters in the 100-metre zone (high-risk zone) is twice as high as the

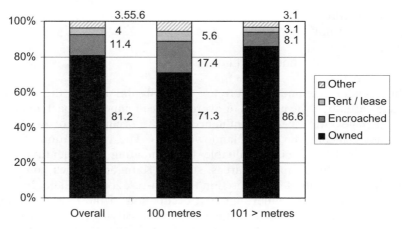

Figure 18.10 Landownership and spatial exposure in the selected GN divisions.
Source: Authors.

proportion of squatters in the areas outside the 100-metre zone in the selected divisions. This implies that a higher proportion of people within
the 100-metre zone are highly vulnerable.

It became evident that nearly 87 per cent of housing units inhabited by
squatters and situated within the 100-metre zone were totally destroyed
by the tsunami, compared to 46 per cent of those that were situated outside this zone, according to our household survey in Galle. The lack of
landownership and the low standard of squatter-occupied housing units
are root causes as to why nearly half of the squatters still live either in
relief camps or temporary shelters provided by the Government and
NGOs. When people in the buffer zone are resettled, those who had legal
title can continue to claim their property and use it for purposes other
than construction, while illegal settlers and squatters do not have this
opportunity.

According to our household survey in Galle, the pattern of landownership varies among the six field locations. The highest proportion
of squatters can be found in Katugoda (28.6 per cent), while there are
no squatters at Ginthota east (see Figure 18.11(c)). Pettigalawaththa and
Mahamodara have a higher number of squatters than Ginthota west and
Magalle. Thus, special attention regarding the problem of relocation
and rehabilitation is needed, especially in Katugoda.

Overall, it can be concluded that the following indicators do have an
important impact on the vulnerability of various households as shown
for Galle:

• the number of young and elderly people
• degree of exposure of a unit or element at risk (e.g. social group or
  infrastructure/sector)
• gender distribution
• income level
• occupation of the household head
• landownership.

While gender, age and exposure are primarily linked with the aspect of
human casualties, the abilities to overcome the financial impact and property damages are especially influenced by the occupation of the household head, income level and landownership. The combination of these
criteria can be especially valuable for estimating and identifying the
most vulnerable groups and areas, such as Katugoda, which contains a
high number of squatters exposed in the high-risk zone in Galle. Interestingly, the indicators gender, age and occupation of the household head
were also valid indicators for estimating and classifying the most vulnerable groups in Batticaloa. By contrast, the indicator landownership was
not useful, since in the selected GN divisions in Batticaloa only very few
households are squatters.

Coping capacity

In terms of how tsunami-affected people cope with the situation, social assets consisting of networks, membership in community-based organisations, relationships of trust and reciprocity, and access to wider institutions in society play important roles (Carney, 1998). Thus, it is important to consider whether there is any advantage in being a member of a local organisation. A significant proportion of household members interviewed are not members of local organisations. In fact, only 6 per cent of community members gained financial assistance from local organisations to recover from this catastrophe. Moreover, as to the question of whether community members receive any counselling or psychological support, data show that only 5 per cent have received such support. Therefore, one can conclude that a small proportion of community members receive financial assistance and counselling or psychological support from village-level organisations. By contrast, nearly 98 per cent of respondents have gained different types of aid in cash and kind from various UN agencies such as UNDP, UN-Habitat and other Government and non-governmental organisations for various purposes to help them recover from the situation. However, when the tsunami first hit, it was neighbours (55 per cent), friends (10 per cent), other family members and relatives (18 per cent) that first came to help the affected people, before the authorities, which shows the close relationship that exist among these people. It is thus clear that informal social networks play an important role in coping. In contrast, it seems that membership of a local organisation is not an adequate indicator for assessing the coping capacity of different households and household members in Galle.

Financial assets

Saving money in formal (Government, private or non-governmental organisational banks) or informal (saving in tins or *seetu* – a small group-saving system in Sri Lanka) ways is an important aspect of coping capacity, since it helps households to recover or lessens the disaster's impact. To the question of whether the interviewed respondents have bank accounts, about three-quarters (75 per cent) answered in the affirmative, and most also have a savings account. When investigating whether the affected people have taken out loans to recover from the tsunami, data shows that a significant majority of respondents have not taken out any loans (81 per cent). Of those who have taken loans, 44 per cent have borrowed from their relatives, neighbours or friends, while another significant proportion (nearly 32 per cent) have taken loans from the formal banking system. These numbers underline once more the important role of informal social networks in the immediate coping process.

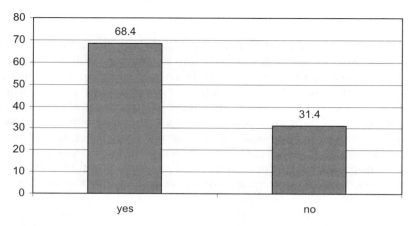

Figure 18.12 Willingness to resettle in a safer location inland-6 month after the tsunami.
Source: Authors.

Intervention tools to reduce vulnerability

100-metre buffer zone: "risk zone"

It is evident from data discussed earlier that more damage to property and life was reported within the 100-metre zone than in the area outside, especially in the case of Galle. Therefore, the Sri Lankan Government declared a strip extending 100 metres from the sea as a buffer zone of high risk for tsunami waves in the south and a 200-metre buffer zone in the north and east. In this zone the Government prohibited the reconstruction or new construction of housing units. The questionnaire survey addressed the question of whether the respondents view the 100-metre buffer zone as an appropriate or an inappropriate measure to reduce vulnerability. Although the likelihood of loss of life and property damage was significantly higher in the 100-metre zone, the prohibition has caused problems, linked, for example, to the fact that there are very few alternative relocation sites outside the prohibited zone. Moreover, squatters living in the 100-metre zone do not get financial support for the reconstruction of their houses, since they are perceived as illegal settlers. Thus the buffer zone and the relocation issue are subjects of controversial debate and are associated major problems.

Interestingly, a large majority of respondents (80 per cent) agreed that the buffer zone was an appropriate intervention tool, while only a small proportion felt it was inappropriate (19 per cent). However, the main opposition party did not agree with the idea and stated that they would remove it when they came to power. Under political pressure, the Government has reviewed the situation and has decided to reduce the buffer zone to 50 metres.

Resettlement

When we asked our respondents in Galle whether they would agree to move to a safer place and vacate their present coastal residence, more than two-thirds of respondents (68.4 per cent) would agree to do so, while nearly one-third would refuse (31.4 per cent).

When examining the relationship between socio-economic variables and the willingness to relocate, some interesting patterns emerged. For example, three quarters of the squatters who lived within the 100-metre buffer zone before the tsunami agreed to move to a safer place (75.4 per cent) compared with 67 per cent of those who owned their land or house. This shows the willingness of squatters to move out of their previous living place, not only to protect their lives and valuables from coastal hazards, but also to move out of their vulnerable situation of chronic poverty by acquiring a legally recognised and permanent place to stay. This means that in Galle, landownership is a crucial indicator for estimating the underlying vulnerability of tsunami-affected people, particularly in view of their unusual difficulties in recovering and coping.

Knowledge and early warning

The questionnaire results showed that only a negligible proportion of respondents know about tsunamis and coastal hazards. Also another study by the Asian Disaster Reduction Center (ADRC) in six coastal DS divisions, including Galle, found that over 90 per cent of the people interviewed had not heard about tsunamis before the disaster (ADRC, 2005: 27). Our study also revealed that among those people who had lived at coastal locations for more than two years, only a small proportion had any knowledge about tsunamis and coastal hazards (around 7 per cent), while people living on the coast for less than two years had no knowledge at all about them. None of the respondents interviewed had been aware that either their home or village could be hit by a tsunami prior to the events of 26 December 2004. This denotes a widespread need to increase awareness of coastal hazards and tsunamis. Although in this case, tragically, nearly all respondents have now, unwittingly, been confronted with the devastating nature of high tsunami waves. After the disaster, a significant proportion of respondents stated that an early-warning system should be installed in their areas.

Scale, functions and target group

The scale of the approach is local and is based on different analytical units encompassing single houses, households and city areas (GN divisions) as well as the whole city.

The approach aims to provide information regarding the vulnerability of coastal communities to tsunamis and coastal hazards. The identification of vulnerability of different social groups, critical infrastructures and economic sectors should provide more in-depth information about those areas and groups that need to be targeted first in emergency, rescue and reconstruction operations. The local vulnerability assessment should also function as an evaluation tool to assess the current intervention and reconstruction strategies. The information should be used in "normal" situations to reduce vulnerabilities before disaster strikes. For example, it is obvious that vulnerabilities linked to the problem of landownership need to be taken into consideration in future local and community development.

Any effective disaster and risk management requires in-depth knowledge and understanding of the respective community, particularly about the most vulnerable groups and areas, in order to be able to define priority areas for emergency response and evacuation.

The target groups of the approach are political decision makers (especially at the local level), and disaster managers, urban planners and community developers, as well as international agencies which provide financial or material support for the reconstruction. In this context an UNU-EHS workshop was held in early 2006, which brought together scientists from various disciplines, disaster managers and UN agencies, as well as NGOs engaged in the reconstruction process. Although there was a common understanding of the need to identify and measure vulnerability and risk in order to develop effective evacuation and emergency response plans, communication between social scientists and former military personnel engaged in disaster response in Sri Lanka still has to be strengthened.

Open questions and limitations

This chapter has elaborated on different approaches and indicators used to measure the revealed (pre-existing and emergent) vulnerability of coastal communities in Sri Lanka to tsunamis and coastal hazards. In the course of the research, it became evident that some indicators and criteria, such as membership in formal local organisations, do not provide adequate information regarding the intended indicandum: coping capacity.

The analysis of exposure in terms of critical infrastructures and sectors allowed an initial estimation of which infrastructures and sectors are highly exposed. However, the actual or the specific "exposure" might also be influenced by the road systems, the built infrastructure, small rivers and canals. In Galle, for example, the bus station was badly damaged by the tsunami wave, although it is located in the 300-metre zone. Therefore, the critical infrastructure analysis regarding the high-risk zone, based either on the 100-metre zone defined by the Government or on an elevation model, needs to be seen as a first overview; more in-depth studies are needed for the development of specific emergency and evacuation plans.

Analysis of the vulnerability of various social groups has provided interesting insights into the vulnerability of different professional groups and groups classified according to landownership. One can conclude that, although income-related vulnerability measures – for example the ability to replace economic and property damage that has been suffered – are often appealing and of high interest to decision makers, income data at a fine resolution is often difficult to grasp. Households might want to hide the new poverty or they might ask for external financial support. However, we were able to calculate and estimate the unusual difficulties of different social groups in recovering and repairing the actual housing damage they suffered due to the tsunami. Finally, the chapter shows the initial results of research currently in progress; more effort is needed to visualise and combine the different indicators selected and tested in an appropriate manner, as well as in an appropriate visualisation tool. The measurement of revealed vulnerabilities allows us to identify and measure past and current vulnerabilities. Some vulnerabilities were directly revealed to the event while other underlying vulnerabilities became evident during the coping and reconstruction process. Examples of the latter include the lack of access to land and other resources for squatters during the recovery phase. However, this approach also revealed limitations, such as the need to examine vulnerability on the basis of actual and potential damage patterns and scenarios.

Outlook

The combining of different methodologies and data sources seems to be an important step forward in overcoming the specific limitations of a single methodology. For example, it would be very helpful if remote sensing were able to identify more precisely the extent of squatter occupation and illegal settlement of a specific local unit (e.g. GN division) in order to estimate the most vulnerable areas. Given the nature of such settlements, they may not only incur higher losses but may even impede speedy evacuation of people following an early warning. This issue needs further investigation.

As for resettlement, further investigation is needed of those who are unwilling to move as well as of those who have already moved. It is necessary to explore what alternative settlement options are available for them and how, or whether, the new locations meet the expectations of the migrant families. These questions are currently being analysed within the project by the Center of Development Research (ZEF). Overall, reconstruction and relocation are long-term issues that need further investigation, including closer consultation with affected people and communities. Little attention seems to have been paid so far to diverse settlement and housing options in the recovery process. The continuously changing information and rules about the so-called buffer zone, and the lack of transparency regarding relocation options, sites and timeframes, are major problems within the general framework of the recovery process in Sri Lanka.

The estimation and assessment of vulnerability will be a key issue also for the future, especially with regard to the reconstruction process and the implementation of an early-warning system. Since the buffer zone has been reduced to 50 metres, allowing the proliferation of settlements and other structures close to the sea, it will be important to learn more about the specific vulnerabilities of different social groups or critical infrastructure facilities in order to be prepared for emergency situations and future coastal hazards.

This chapter has shown that single indicators, as well as highly aggregated measures, at the household and local scale are important in order to be able to compare the vulnerability of different social groups and different locations.

Although the research on the four different tools and methodologies tested in Sri Lanka is still ongoing, the time and costs also have to be taken into account when considering the appropriateness of different methodologies to measure vulnerability in different regions. While, for example, census data can often make possible an initial and general estimation of demographic and social vulnerability, the more in-depth questionnaire methodology allows a better understanding of specific vulnerability patterns among the different social groups. This is also important in order to estimate and evaluate the different intervention tools and their impacts. The available census data can often be analysed within one or two months, while the development, testing and implementation of a household questionnaire survey takes at least four to six months. The remote sensing analysis, although not fully presented here, makes it possible to estimate the impact of disasters on physical structures almost all over the world. However, although a satellite can provide actual information for almost any part of the world, the approach is costly, since one satellite image at the high resolution required to assess the structure of a single building could cost around US$ 5,000 to 10,000 and analysis requires special software and trained personal. Furthermore, the remote sensing analysis depends on ground truth data to verify the classification methodology. This means a combination of different methods is needed. An important advantage of using different methodologies is that they can be combined and these synergies exploited. For example, the in-depth questionnaire research conducted by the enumerators also captured information about house types and the damage they suffered, as well as the socio-economic condition of the household living in them. This is important information for the remote sensing analysis and for the verification of the classified settlement-structure types.

This research in Sri Lanka, which started five months before the time of writing, is only a first step. Future investigations need to strengthen our understanding of how to combine different methodologies effectively. Moreover, one has to explore how to integrate this information into development and emergency preparedness plans in order to ensure that vulnerability assessment supports activities that progress towards disaster-resilient communities in a practical manner. Finally, it will be important for vulnerability assessment to be transformed into a continuous monitoring system. This would allow better and continuous information on the medium and long-term impact of the December 2004 tsunami, the great difficulties in recovering, and the consequences of different intervention measures undertaken to reduce the vulnerability of coastal communities.

Note

1. The project received financial support from the International Strategy for Disaster Reduction (UN/ISDR) through the UN Flash Appeal for the Indian Ocean Earthquake, which was supported by the European Commission and the Governments of Finland, Germany, Japan, Netherlands, Norway and Sweden.

REFERENCES

Asian Disaster Reduction Center (ADRC) (2005) *Report of the Survey on Tsunami Awareness in Sri Lanka, Urgent Study of the Great Sumatra Earthquake and Tsunami Disaster*, Kobe: ADRC.

Carney, D. (1998) "Implementing the Sustainable Rural Livelihoods Approach", in D. Carney, ed., *Sustainable Rural Livelihoods: What Contribution Can We Make*, London: Department for International Development (DFID).

Department of Census and Statistics (2005) *Post Tsunami Census Data*, Colombo, Sri Lanka: Department of Census and Statistics.

Department for International Development (DFID) (1999) *Sustainable Livelihood Guidance Sheets*, London: DFID, available at http://www.livelihoods.org/info/info_guidancesheets.html.

Farrington, J., T. Ramasut and J. Walker (2002) *Sustainable Livelihood Approaches in Urban Areas: General Lessons with Illustrations from Indian Cases*, Working Paper, No. 162, London: ODI.

Green, C. (2004) "The Evaluation of Vulnerability to Flooding", *Disaster Prevention and Management* 13(4): 323–329.

Guha-Sapir, D., L.V. Parry, O. Degomme, P.C. Joshi and J.P. Saulina Arnold (2006) *Risk Factors for Mortality and Injury: Post-Tsunami Epidemiological Findings from Tamil Nadu*, Draft Report, Brussels: Centre for Research on the Epidemiology of Disasters (CRED).

Herath, S. (2005) "December 24 Tsunami Disaster in Sri Lanka and Recovery Challenges", Presentation at the Workshop Tsunami Impacts, Colombo.

Kumar, R. (1996) *Research Methodology: A Step by Step Guide for Beginners*, Melbourne: Longman Press.

National Disaster Reduction Center (NDRC) (2005) *International Natural Disaster Reduction Day Commemoration Report*, Colombo.

Oxfam (2005) *The Tsunami's Impact on Women*, Oxfam Briefing Note, Oxford, available at http://www.oxfam.org.uk/what_we_do/issues/conflict_disasters/downloads/bn_tsunami_women.pdf.

Rofi, A., S. Doocy and R. C. Robinson (2006) "Tsunami Mortality and Displacement in Aceh Province", Indonesia, *Disasters* (forthcoming in early 2006)

Satterthwaite, D. (2000) "Seeking an Understanding of Poverty that Recognizes Rural and Urban Differences and Rural–Urban linkages", Paper presented at the World Bank's Urban Forum for Urban Poverty Reduction in the 21st Century, London: International Institute for Environment and Development.

Part V

Institutional vulnerability, coping and lessons learned

19

Assessing institutionalised capacities and practices to reduce the risks of flood disaster

Louis Lebel, Elena Nikitina, Vladimir Kotov and Jesse Manuta

Abstract

In this chapter we propose a framework and some methods for an institutionally oriented analysis of the capacity of societies to reduce the risks of flood disasters. It is intended to complement other approaches to vulnerability assessment that characterise flood hazards and their impacts. We focus on the formal institutions created by States to deal with flood-related disasters and how these interact with local, often informal, institutions. The interplay of institutions not only defines what and who will be at risk, but also shapes the way flood disasters are defined, perceived and acted upon. The framework is most useful in situations recently affected by major floods because it requires investigation of practices and performance that often depend on obtaining primary data.

Our initial application of the framework revealed several important issues which have been hinted at before but which can now be more systematically exposed. Four stand out. First is the misplaced emphasis on emergency relief to the detriment of crafting institutions to reduce vulnerabilities and prevent disasters. Second is the self-serving belief that disaster management is a technical problem needing expert judgments that systematically exclude the interests of the most socially vulnerable groups. Third is the over-emphasis on structural measures, which again and again have been revealed to be more about redistributing risks in time and place rather than reducing them. Fourth is the failure to integrate flood disasters as inevitable challenges into normal development planning in flood-prone regions. Our empirical studies demonstrate that a systematic approach to diagnosis of institutionalised capacities and practices in flood disaster management is feasible and can yield practical insights.

Reducing risks of flood disasters

The role of social institutions in altering the vulnerability of households and communities to extreme floods is increasingly well understood (Adger, 2000; Chan, 1997; Few, 2003). We know that it is often the poor, the elderly, women-headed households, ethnic minorities and other social groups with the least access to critical resources for coping and adapting who often have to bear the largest involuntary risks (Blaikie et al., 1994; Dixit, 2003; Morrow, 1999). Very often it is concurrent social and economic changes associated with modern development that amplify some of these vulnerabilities, at least temporarily, by disrupting traditional institutions that in the past provided social safety nets (Adger, 1999). We also know that much of what passes for institutional reform at the basin or State level to reduce risks of disaster might really be about redistributing risk away from central business districts and valuable property, rather than reducing risks to livelihoods of the poorest (Manuta et al., 2006). We know that the flood problem is often not just one of "high water", but also one of rate of onset, duration, sedimentation, debris flows and poor water quality with their impacts on ecosystems, infrastructure, health and livelihoods. Finally, we also know that where authorities work closely with the public, from negotiating risks and preparing for flood events, to designing institutional responses for relief and recovery, that the risks of disaster can be greatly reduced (Takeuchi, 2001). After all, people in many parts of the world have been living with recurrent floods for thousands of years and have often learnt useful ways of coping with and living with floods, even the more extraordinary ones (Wong and Zhao, 2001).

The overall global picture, however, is worrisome on two fronts. First, despite better understanding of disasters, losses of life and property from flood disasters remain unacceptably high and are increasing (Vorobiev et al., 2003; White et al., 2001). Second, climate change is likely to result in significantly more intense rainfall events which, depending on trends in other factors affecting runoff and river flows, will result in more extreme flood events in some places (Adger et al., 2005; Kundzewicz and Schellnhuber, 2004). Clearly, it would be highly desirable to have more systematic methods for assessing institutional influences on key vulnerabilities, and consequently, on the risks of flood-related disasters.

In this chapter we propose a framework and some methods for an institutionally oriented analysis of the capacity of societies to reduce the risks of flood disasters. We focus on the formal institutions created by States to deal with flood-related disasters and how these interact with local, often informal, institutions. The interplay of institutions not only defines what and who is to be at risk but also shape the way flood disasters are defined, perceived and acted upon. The language of "disasters" itself, for example, creates stories of uniform and large negative impacts of singular events which may or may not reflect realities very accurately (Bankoff, 2004).

The framework presented here is in its second iteration. It began life as part of a pilot comparative study of flood disaster risk management in Vietnam, Thailand, Japan and Russia (Nikitina, 2005). It was developed to complement emerging frameworks on vulnerability and disaster management with a more biophysical focus. We were inspired by some of the earlier work on vulnerability, risk and natural disasters, especially the book by Piers Blaikie and colleagues (Blaikie et al., 1994) and the work of Neil Adger and colleagues on vulnerability to climate change in coastal areas (Adger, 1999, 2000; Tompkins and Adger, 2004). We have also been influenced by writings on human security (Kotov and Nikitina, 2002), vulnerability (Turner et al., 2003) and resilience of social-ecological systems (Carpenter et al., 2001; Gunderson and Holling, 2002). Each of these bodies of theoretical work, we believe, has important contributions to make to the practical challenge of assessing and measuring vulnerabilities and risks.

Our starting point is the need to link insights about social institutions important to livelihood security and social justice with the increasingly sophisticated institutional frameworks being proposed by States to *manage* disasters. To make the link requires careful attention to matters of scale and cross-scale interaction. Thus, even with a focus on the vulnerability of individuals, households, and communities we still need to consider the institutions operating at the scale of basins or regions, and, invariably, the State.

Our earlier research also suggests that matters of governance, both the institutional structures and the process by which they come about, are crucial for both reducing and redistributing involuntary risks to flood disasters (Manuta et al., 2006; Nikitina, 2005). Here we propose that the presence of institutionalised capacities and practices to deal with flood-related disasters are themselves important indicators and criteria of vulnerability and coping capacity.

This chapter is primarily about methods. In it we justify our approach and illustrate various measures with short examples. The chapter is organised as follows. In the section on the "Assessment framework" we describe our overall framework, defining key terms, measures and sources of data. In the next section we illustrate the core parts of the framework. The final section outlines some of the main strengths and weaknesses of our approach and the prospects for refining the diagnostic framework further for practitioners involved in reducing the risks of flood-related disasters.

Assessment framework

Institutional influences on vulnerabilities

Institutions, whether purpose-built to address floods or flood-related disaster risks or not, may influence the vulnerabilities of households and communities through several pathways (Figure 19.1). In our conceptualisation the influence of institutionalised capacities and practices (inner

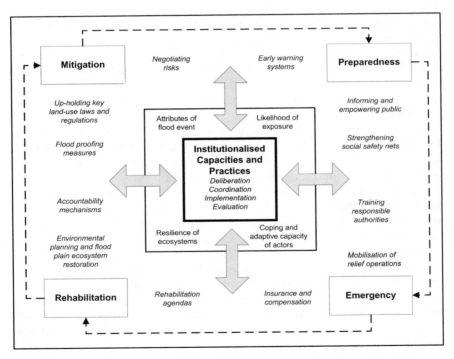

Figure 19.1 Institutions modify vulnerabilities and hence risks of flood-related disasters through several pathways.
Source: Authors.

box) on the disaster cycle (outer ring) are mediated by ecological and social resilience as well as attributes of the flood event itself (middle box). Some examples of typical institutions are shown (outer box). The pathways themselves may be complex. For instance loans for investments in structural measures and regulatory practices with respect to land-uses in the basin, will alter the attributes of floods in terms of onsets, durations and peak flows by altering runoff, retention times and river-flow regimes, Other pathways alter how involuntary risks are distributed, either by modifying likelihoods of exposure or the capacities of different actors to avoid, cope with or adapt to floods.

The pathways most important in particular places depend on socio-economic development, political systems and the attributes of the flood event itself. Understanding these at least partially is important for the context in which institutional performance can be assessed. The hydro-climatic triggering events, for example, may be a cloudburst, a period of prolonged rainfall, snowmelt, glacial lake outburst, or dam failure (Dixit, 2003). Interactions with landslides in mountainous areas are common, causing temporary blockages and breaks, scouring, deposition and massive debris flows. Although many of the institutional issues are similar, the actors involved, the preventative measures needed and technical difficulties can be quite different.

Institutional analyses can be very powerful, but also confusing if terms are used loosely. In our framework institutions are rules or norms which define the roles, rights and responsibilities of actors (Young, 2002). Institutions, by definition, are relational. An organisation, on the other hand, is a type of actor. Like a household, community, firm or State, an organisation may host many kinds of institutions that guide the behaviour of their members. Institutions can be formal, with supporting written legal documentation, or informal, like social norms or customs which nobody articulates but most people follow. Informal institutions are often much harder to identify and assess than formal ones, but nevertheless could be crucial to social responses. The emergence of informal "shadow" institutions to perform certain functions at times of crisis, like flood-related disasters, may reflect inflexibilities or gaps in the formal institutional framework. Finally, institutions are systems of rules, and rules invariably get broken. We need to explore the main reasons for gaps and deviation from norms as these may provide guides to both adaptive and mal-adaptive practices from which society can learn.

Capacities and practices

Significant capacities to reduce the risks of flood disasters lie both within actors and in the relationships among actors. Our focus in this chapter is on assessing capacities that are relational. We call relations that regularly define roles and responsibilities and rules of engagement in ways that enhance the capacities of actors, *institutionalised capacities*. Relationships among actors have different functions that may be institutionalised. Our assessment framework focuses on four classes of institutionalised capacities and practices (Table 6.1). The capacity for *deliberation* and negotiation is important for ensuring that interests of socially vulnerable groups are represented, that different knowledge can be put on the table for discussion and that, ultimately, fair goals are set. The capacity to mobilise and then *coordinate* resources is often critical to prevention and response actions. The capacity to use those resources skilfully to carry out actions transforms potential into implementation. Finally, the capacity for *evaluation* is important because it can be the basis for continual improvement, adaptive course corrections and learning by key actors. We illustrate each of these capacities in detail, including issues of measurement, in the section on "Institutionalised capacities and practices".

We also ask questions about each kind of relationship across four conventionally designated phases of the disaster cycle (Table 19.1). We intentionally developed our framework around the conventional idea of a disaster cycle so that we could introduce an approach to institutional analyses in a context that would be familiar for practitioners in disaster management. The phase or cycle idea is also useful because the institutional issues at times of crisis and during re-organisation may be qualitatively different from periods between such events (Gunderson and Holling, 2002). Of course in any particular setting not all boxes are equally important or modifiable so the analyst will need to prioritise measurement and assessment efforts.

Finally, gaps between stated policy goals and practice, or between design and action, contribute to increased vulnerabilities. A broad *variety of factors* influence institutionalised practices. External factors that may affect implementation include financial deficiencies, administrative barriers and conflicts between organisations, corruption, poverty, lack of economic incentives, low public participation and awareness. Situational factors might block or alter the performance of institutions or modify the designed pathways for implementation of policies and tools. In our work we sought out flood disasters (of modest proportions) to try and understand which institutions come into play in practice and which remain paper-bound.

Data sources

Our research methodology is scale-dependent (Table 19.2). We assess information about institutions at the national and basin or regional scale mostly through review of documents and interviews, but we evaluate performance and practices at this and more local scales through analyses of particular flood events. Our original comparison included two-level case studies in Vietnam, Thailand, Russia and Japan (Nikitina, 2005).

We begin with the general features and structure of domestic institutional frameworks, policies and measures to reduce risk of floods. We suggest starting with an assessment of the presence or absence of key institutional relations and then moving on where possible to analyses of their comprehensiveness. Information that may need to be scrutinised includes legislation, programmes, strategies, action plans, task forces, administrative organisation, financial mechanisms and tools and insurance schemes.

Exploring specific cases of severe floods that have recently taken place is often crucial for understanding institutionalised practices, the divergence between rules on paper and in use, and underlying diversity of actor behaviours. Our approach, therefore, is most appropriate for areas that have recently experienced major floods, whether or not they resulted in disasters, as it requires asking actors to recall information about actions taken by themselves or others (Kitamoto et al., 2005; Kotov and Nikitina, 2005; Manuta et al., 2006). Although secondary information such as newspaper and agency reports is also important, good primary data is crucial for validation. In local-community-level studies of flood events in urban, rural and remote rural locations in Thailand, we used household questionnaires to: characterise flood events, identify prevention and mitigation measures; assess effectiveness of relief, compensation and rehabilitation actions; explore household and village-level coping and adapting strategies; and assess channels for public participation and accountability of decisions.

Institutionalised capacities and practices

Deliberation and empowerment

Who and what should be at risk? This is the central unasked question in disaster management. The only way the sharing of involuntary risks can be negotiated is if the interests of marginalised and vulnerable groups are represented, the quality of evidence is debated and challenged, and authority is held accountable for its decisions. Alternative dialogues, the mass media and acts of civil disobedience may be critical to incorporate issues of unfair distribution of involuntary risks into the design of flood and disaster programmes. Without opportunities for deliberation, women-headed households, the elderly, ethnic minorities and other mar-

Table 19.1 Framework for assessing institutionalised capacities and practices with regard to flood-related disasters

Functions	Phase of disaster cycle (Timing)			
	Mitigation (Well before)	Preparedness (Before)	Emergency (During)	Rehabilitation (After)
Deliberation *What should be done?*	How were decisions made about what and who would be at risk? Whose knowledge was considered, whose interests were represented?	Was the public consulted about disaster preparations? How were decisions to give special powers to particular authorities made?	How were decisions made about what and who should be saved or protected first? What special directives or resolutions were invoked?	How were decisions made about what is to be on the rehabilitation agenda? Whose knowledge was considered, whose interests were represented?
Coordination *Who was responsible?*	What national and basin-level policies, strategies or legislation were in place to reduce risks of disaster?	How were responsibilities divided among authorities and public? Was an appropriate early warning system implemented?	How were specific policies targeting emergency operations implemented? Were there gaps between stated responsibilities and performance of key actors? Who was in charge?	Were the resources mobilised for recovery adequate? Were they allocated and deployed effectively? How was rehabilitation integrated into community, basin or national development?
Implementation *How was it done?*	What structural measures were undertaken to reduce likelihood of severe flood events?	Were public authorities well prepared? Was the public well informed?	How were emergency rescue and evacuation operations performed?	Did the groups who most needed public assistance get it? Who benefited from reconstruction projects?

Evaluation *Was it done well?*	To what extent were laws and regulations regarding land-use in flood prone areas implemented? What measures were taken to improve coping and adaptive capacities of vulnerable groups?	How were specific national or basin-level policies targeting disaster preparedness implemented?	Were special efforts made to assist socially vulnerable groups? Were any measures taken to prevent looting?	Was insurance available and used and if so how were claims processed? Was the compensation process equitable and transparent?
	How is the effectiveness of risk reduction measures assessed?	How is the adequacy of preparedness monitored?	How is the quality of emergency relief operations evaluated?	How is the effectiveness of the rehabilitation programs evaluated?

To whom and how are authorities accountable?
Were institutional changes made to address capacity and practice issues learnt about in the previous disaster cycle?

Source: Authors.

Table 19.2 Illustrations of scale-dependent actors, institutions and perceptions with regard to flood-related disasters

Scale of interest	Key Actors	Examples of institutional responses	Common perceptions of disaster
Nation	National governments, multilateral banks	Funding mechanisms, loans, debt relief, regional cooperation agreements	Infrastructure losses and re-building costs; losses of investments, debt-burden
Basin, coastal region	Local governments, river basin organisations, sector associations	State laws, policies and programmes, Insurance, state of emergency legislation	Destruction of infrastructure, disruption of regional economy
Community	Households, firms, local government authorities	Local government by laws, social safety nets, revolving loans, micro-credit schemes	Loss of social control and safety nets (e.g. looting), Displacement-induced breaking of social networks
Household	Individual	Family, marriage, kinship networks	Loss of home, crops and family members, livelihood disruption and insecurity

Source: Authors.

ginalised groups are unlikely to benefit and may even be disadvantaged by programmes and policies aimed at reducing risks of flood disasters. For example, minority households affected by landslides and floods in one of our studies were ineligible for most kinds of post-disaster assistance because they were poorly informed about correct reporting procedures or did not hold citizenship documents (which the State had failed to provide for them) (Manuta et al., 2006). Small fishers in southern Thailand had similar difficulties navigating bureaucratic barriers and corruption in compensation programmes after the Indian Ocean tsunami (Lebel et al., 2006).

Debate, consultation and planning procedures for floods and disaster management need to be assessed by criteria similar to those used to analyse "good governance" (Table 19.1). In particular, focus is needed on issues of participation, representation and sources of knowledge. In most countries such an assessment would highlight how, at least until fairly recently, the public has been treated as irrelevant to the technical exercise of assessing and managing risks and designing institutional responses.

Things may be changing. A return to community-based flood disaster management is being widely promoted by international agencies, but only cautiously adopted by national ones (Few, 2003; Morrow, 1999). The key idea is that greater involvement of the public in decisions about all stages of the disaster cycle will make better use of local knowledge and capacities and help identify both risks and pragmatic opportunities to address them. Early results of community-based flood management strategy (CFMS) pilot areas in Bangladesh suggest huge dividends in reducing vulnerability of affected communities during the 2004 flood (Ahmed et al., 2004).

The area requiring the most profound engagement with wider group of stakeholders is in assessing and addressing the underlying causes of vulnerability. State agencies usually find these tasks very difficult because fundamental issues of governance and social justice have to be addressed, and this may undermine positions of authority. Extremely low asset levels, poor access to natural resources, and insufficient rights to public goods and services are often at the core of these vulnerabilities (Blaikie et al., 1994; Dixit, 2003).

In contrast to the neglect of questions about *"Who will be at risk?"*, issues of *"Who will pay?"* are intensely debated from day one. The main debate is often between levels in the administrative hierarchy: should funds come from local, regional or central budgets? Local Governments often find they need to obtain additional sources to fund recovery and rehabilitation operations. Thailand, for example, has a fairly clear set of rules for passing budget requests up the hierarchy depending on levels of damage. The problems are with accountability and the timeliness of available funds. In Russia, the vertical division of responsibilities is institutionally fixed by national rules, but in crisis and emergency situations the provinces and locales tend to do their best to bargain with the national administration for extra resource allocations (Kotov and Nikitina, 2005). Constant debates and controversies between the "centre" and the regions requesting increased involvement and support from the central authorities, especially at recovery stages where mobilisation of significant funds is essential, can turn into conflicts and gridlocks that weaken institutional performance.

In many places there is a need to go beyond participation being defined as simply informing the public or being seen as an opportunity to shift onto communities the burden for actions that should have been the responsibility of public authorities. Participation should result in empowerment of marginalised and vulnerable groups in decision-making around who and what should be at risk.

Coordination and cooperation

Who is or should be responsible? Being able to count on institutionalised capacities to mobilise and coordinate resources when and where they are needed is crucial in all phases of the disaster cycle, sometimes with very little scope for delay or errors of judgment. Because there are many uncertainties about knowing where disasters will occur and exactly how they will unfold, it is important that this "institutionalising" aspect fosters flexible and adaptive responses that rely on coordinated, as opposed to uni-dimensional, assignment of responsibilities.

Assessment requires attention to bureaucratic procedures for re-allocating resources and the existence of coordination mechanisms. The effectiveness of public mobilisation can be assessed at primary level by looking at the extent to which it is "better prepared, but not scared". The best insights, however, are usually obtained from observing actual efforts at preparedness and emergency and recovery responses to major flood events, as these provide a genuine test of the flexibility inherent within disaster management systems that may otherwise be hard to ascertain. Issues of trust in public institutions also arise. It is also useful to analyse how well activities are coordinated across Government agencies and between authorities and the public in order to understand both institutionalised operations and their practice (Table 19.1). Effective mobilisation and coordination means that societies' response is appropriate to the risk and that the most vulnerable groups are being taken into account.

Because most river basins cut across administrative jurisdictions (within and among nation-states) they create special challenges for coordination and assigning responsibilities in disaster management. In Thailand, the notion of organising water management through river basin organisations is only now being introduced (Thomas 2005). In 2001, the Mekong River Commission began to more systematically address international coordination issues for the Lower Mekong countries through the creation of a strategy-oriented Flood Management and Mitigation Programme (Asian Development Bank, 2003; Fox and Schmit, 2003). In Russia, river basin management administrations have been in place for a number of years, but coordination problems between State and federal agencies persist.

A lack of clearly defined roles and responsibilities among State agencies is an indication of poor institutional capacity. In Thailand the problem has been acute, so much so that nothing happens unless the

Prime Minister personally commands and directs the response effort (Tingsanchali et al., 2003). More complex and integrated systems of disaster management, however, may have trade-offs in terms of responsiveness and reach. Thailand has re-oriented its approaches to a more proactive integration of mitigation and preparedness in the overall scheme of disaster risk management. This was initially done by establishing in 2002 the Department of Disaster Prevention and Mitigation (DDPM), which consolidated several different agencies into a one-stop-shop for disaster coordination. Unfortunately its mandate far exceeds the actual capacities of the organisation and its relationships (Manuta et al., 2006).

The State may fail to deliver an appropriate response to marginalised people living in remote areas. Our own fieldwork in the mountains of northern Thailand confirms both the challenge and failure of Thai State to deliver reasonable service to a remote area (Manuta et al., 2006). Floods accompanied by severe landslides in and around several villages in Om Koi district of Chiang Mai province in 2004 did not generate a relief/emergency response until three or four days after the event. It is noteworthy that nearby villages were able to self-organise food and shelter for affected people quickly and this was sustained for several weeks. An upland Karen village where crops, livestock and homes were devastated faced starvation because its livelihoods had been destroyed and no follow-up assistance after the initial emergency relief was provided by the State. The legal requirement for people to hold Thai citizenship cards before receiving compensation was an important practical constraint, and was compounded by the fact that in at least some cases the State itself has been at fault as a result of discriminatory practices for not issuing such cards to long-term residents in the first place.

Lack of trust in public institutions can also hinder the ability to prepare for emergency operations. The catastrophic loss of life in the Lena River flood in Sakha Republic of Russia was in part caused by the combination of local cultural norms that were dismissive of future threats and mistrustful of authority (Nikitina, 2005). Warnings to prepare and evacuate went largely unheeded by both local authorities and populace, because people were afraid that if they abandoned their homes they would be looted. The response of the State disaster agency was to propose compulsory evacuation measures.

Coordination among agencies and stakeholder groups is important for flood mitigation, in particular the design and execution of programmes and policies to help address underlying causes of extreme vulnerability. In urban areas of Asia, the problems of flooding can be severe and almost chronic for slum dwellers forced into high-risk zones because of the lack of low-cost housing in more desirable areas. Insecurity over settlement rights combined with poor or non-existent access to drinking water, waste disposal or drainage services compound the risks of flood-related disasters. These are not voluntary risks, but rather a structural outcome of urban development that is focused on serving the wealthy (Manuta et al., 2006).

Mobilising adequate funds, both for protection measures before an event and for recovery and rehabilitation of affected areas and livelihoods afterwards, is the core "coordination" and "cooperation" issue for local authorities because it has a large bearing on their ability to implement plans. What will be the major sources of funding? Who will benefit most from their deployment? In Russia, Vietnam and Thailand, flood insurance schemes are at a very rudimentary stage so there is a strong reliance on the State to come to the rescue. In more wealthy countries like Japan, State guarantees have allowed significant entry by the private sector into insuring against flood disasters (Kitamoto et al., 2005). Here damage is compensated for by the comprehensive insurance provided to households by the private insurance companies. Insurance is optional, but people who take out loans to build or buy houses are obliged to buy comprehensive insurance.

If local authorities have the capacity and legal framework that enables them to seek loans and private sector cooperation, then they may be able to secure more, and more diverse, funds for disaster risk management. For example, after the 2001 Lena river flood the Sakha Republic administration applied for central bank credit for housing renovation; it also formed a partnership with the Alrossa company, a leading diamond producer based in Sakha, to help rehabilitate and restore livelihoods (Kotov and Nikitina, 2005). Elsewhere there are examples of non-governmental organisations venturing into micro-finance, training and mobilisation in intervention programmes to reduce disaster risk. For example, in the aftermath of the Indian Ocean tsunami in 2004 that caused severe coastal flooding in southern Thailand, fishing communities established "community shipyards" with the support of a private firm (the Siam Cement Group) and an NGO (Save Andaman Network) (Lebel et al., 2006). A community banking and revolving fund system were established for recovering people's livelihoods (Achakulwisut, 2005).

Coordination of activities across phases of the disaster cycle is necessary because there is often a need to link or transfer responsibilities and budgets for programmes over time. One approach is through cross-agency and multi-stakeholder taskforces, set up for a limited period with clear objectives, that can help guide these transitions.

Implementation and stewardship

How was it done? Wonderful planning and coordination mean nothing when it comes to reducing the risks of disaster if there is no follow-through because of corruption or other institutionalised and *ad hoc* incapacities that prevent appropriate use and allocation of the resources available.

Assessing institutionalised capacities to effectively use resources and execute critical actions requires several different kinds of measures corresponding to different kinds of resources and actions. At the simplest and most conventional level we need to look at actual structural and non-structural measures undertaken in preparing for, and responding to, flood disasters.

Forecasting and early warning systems are often the weakest element in the chain of purpose-built institutions for reducing risks of flood disasters. First, there are the technical challenges of obtaining critical information and sharing it in a timely fashion. Second, there are organisational and individual behaviours that undermine otherwise sound information-sharing arrangements. For example, in Russia in 2001, the Hydromet service provided early warning forecasts of dangerous spring thaw conditions in the Lena River basin. Local and provincial administrations in the Sakha Republic were slow in responding. As a result, the population was not well informed and losses were much higher than they needed to be (Kotov and Nikitina, 2005).

In most countries a national-level institutional framework for *emergency* response is well established. Normally, such frameworks incorporate a set of administrative structures, governmental programmes and legal frameworks defining the necessary conduct and interactions between specialised task forces, which are usually well trained and able to perform skilfully in extreme situations. Often the military is involved.

States differ greatly in how they view their own involvement in recovery. In centrally planned economies like Russia and Vietnam, the State's role remains dominant in all aspects. Thus, in the case of the Lena flood in Russia, a combination of tools was applied, including (1) introduction of a programme to resettle populations from the affected areas, (2) subventions from the federal to the provincial budget for this purpose, (3) allocation of housing certificates from the State Reserve Emergency Fund for the population affected by flood, and (4) material compensation for the affected livelihoods (although too modest to restore them).

For the most part, implementation always lags far behind promises and ideals when it comes to addressing the underlying causes of disasters. Consider, for example, issues related to housing and road construction both in mountain areas and in floodplains. Economic imperatives would argue for taking structural measures to protect these investments before disaster strikes, rather than exploring their role as contributing causes of disasters after the fact. Poorly constructed roads destabilise slopes or act as channels for debris in mountain areas, while in deltas and wetland areas they can prevent or alter natural drainage, thus increasing the duration and height of floods.

During post-disaster periods there is often a flurry of programmes, investments and rule changes. All such actions are far more likely to be followed-up and implemented if there is a significant group of stakeholders involved, who have a sense of ownership and responsibility for them. This means going beyond the project-bounded logic that "implementation" ends when the final budget item of the initial action has been completed, and rather moving towards integrating projects and programmes into local development. In a real sense it is about creating a sense of stewardship for disaster risk management. This is most likely to be fostered when there is significant decentralisation to local authorities, who are in turn accountable to local affected communities.

Evaluation and learning

Was it done well? The performance of institutions and organisations should be monitored and evaluated. This has to be done with a degree of independence or the opportunities for organisations to learn, for authorities to be held accountable, and for success at reducing the risks of the next disaster will themselves be reduced.

The presence of institutionalised evaluation and monitoring procedures for the disaster management system must be present. Otherwise, there can be no improvements in performance or adjustments to take account of changing contexts like altered flood regimes resulting from climate change. A more thorough assessment would also need to take a historical perspective to review the extent to which learning had actually taken place (Krausmann and Mushtaq, 2006), above and beyond factors simply reflecting technological change or increasing wealth. Apart from social learning, conventional learning by key individuals about risks, vulnerable groups and places, or about experiences from other places and times may be important in reducing risks of disaster too. The capacity for current arrangements to foster these kinds of learning should be also assessed.

In our studies of upland flash flood events in northern Thailand, conflicts arose with respect to irregularities, and a lack of transparency or accountability in compensation payouts involving the village heads (Manuta et al., 2006). A mobilisation by villagers was able to oust corrupt officials, but delayed compensation. Similar problems have plagued recovery processes in small fisher villages in southern Thailand after the tsunami of December 2004 (Lebel et al., 2006; Manuta et al., 2005).

An assessment framework like the one we are now discussing could itself be part of an institutionalised learning process for key disaster organisations. Regular assessment exercises by particular publics and bureaucracies could consult expert advice as needed. Thorough and well-communicated research could contribute to such evaluations.

Prior to reforms in October 2002, the Thai approach to disaster was explicitly reactive, focusing on readiness and response. Since then a more proactive rhetoric has been adopted, which aims to minimise the risks and impacts by using both structural and non-structural measures that include preparedness by mobilising the resources of the Government offices, private sector and community (Tingsanchali et al., 2003). This development might be evidence of nascent learning. The huge problems with the still technocratic institutional response to the Indian Ocean Tsunami (Lebel et al., 2006; Manuta et al., 2005) underlines just how many more lessons still need to be learned.

Limitations and prospects

The preliminary framework proposed here (Figure 19.1) for assessing institutionalised capacities is intended to complement, not replace, other approaches to vulnerability assessment that characterise flood hazards and impacts more thoroughly. The initial application of the framework revealed several important issues which have been hinted at before but which can now be more systematically exposed. Four stand out. First is the misplaced emphasis on emergency relief to the detriment of building up institutions to reduce vulnerabilities and prevent disasters. Second is the self-serving belief that disaster management is a technical problem that calls for expert judgments that systematically exclude the interests of the most socially vulnerable groups. Third is the overemphasis on structural measures, which again and again, have shown themselves to be more about redistributing risks in time and place than reducing them. Fourth is the failure to integrate flood disasters as inevitable challenges into normal development planning in flood-prone regions.

Our experiences in four countries confirm that a systematic approach to diagnosis of institutionalised capacities and practices in flood disaster management is feasible and can yield practical insights. At the same time our empirical investigations have revealed several challenges that limit the situations in which our framework can be realistically applied to help reduce risks of flood disasters.

First, there is often a lack of relevant and available data and documentation about the process through which various flood policies, programmes and laws were set up. Much of this has taken place behind the closed doors of technical bureaucracies. Deliberation has not been viewed as an important aspect of disaster risk management, because the initial assumption has been that it is primarily about emergency relief operations and this is clearly a time when authoritarian measures are needed.

Second, although direct observations and interviews at critical times during an event or in the immediate emergency response provide superb data on practice, such behaviour may be unethical and put lives at stake. Researchers and assessors caught in such events will, like other people, be anxious to help and act, and leave most reflection to later.

Third, the presence of institutionalised capacities is not on its own a reliable indicator or criterion of a capacity to reduce the risks of disaster. Some forms of bureaucratisation, for example, may actually result in loss of flexibility or reduce opportunities for self-organisation that could help avert the worst of a disaster – what we have called, "*institutionalised incapacities*" (Manuta et al., 2006). Assessment, therefore, cannot stop at documenting capacities on paper but must also delve into relationships and practices on the ground. Major flood events provide the right kind of challenge to learn about these.

Some of these limitations could be overcome through joint design and implementation of assessment exercises, especially following major events. Institutional analysts can use theory and reasoning to help provide more logical frameworks of analysis and synthesis, but at-risk communities and authorities with operational and planning responsibilities can identify more sensitive measures and estimates of institutionalised capacities. If such exercises were treated as important learning opportunities, then we think the large knowledge-to-action gaps in much disaster risk management could be narrowed.

In this chapter we intentionally developed our framework around the conventional idea of a disaster cycle (Figure 19.1 and Table 19.1) so that we could introduce an approach to institutional analyses in a context that would be familiar to practitioners of disaster management. The language of "disaster", because it focuses attention on events with large singular and negative impacts, constrains thinking about the full diversity of possible institutional responses to flood risks and events. Looking ahead, we see value in going further and treating management of flood-associated risks, together with other disturbances that can have large impacts on society, as a *normal* rather than *extraordinary* part of development. In such a re-conceptualisation, the language of discrete "disasters" might be replaced with an understanding of the unfolding of cycles of change across scales in a language of shifting vulnerabilities, capacities to cope and underlying changes to system resilience.

In most flood-affected and flood-dependent regions, especially in the developing world, institutionalised capacities and practices to reduce the risks of flood disasters remain weak. This is especially true in the fast-developing regions where the entire livelihood and socio-economic context is in flux; traditional institutions may no longer be relevant or functioning well, and new relationships among firms, communities and State agencies have not emerged or kept pace with shifting risks. The mature industrial and service economies have fewer institutional gaps, but still face daunting challenges of escalating costs as the legacy of controlling, rather than living with, floods. The prospects of climate change further altering flood regimes, which society has already struggled to respond to, suggest that the institutional challenges are going to become more important and tougher. A systematic approach to diagnosis of institutionalised capacities and practices in flood disaster management could help societies identify critical gaps beforehand, and thus learn more from experience.

REFERENCES

Achakulwisut, A. (2005) "All in the Same Boat", *Bangkok Post*, 21 February.

Adger, N.W. (1999) "Social Vulnerability to Climate Change and Extremes in Coastal Vietnam", *World Development* 27(2): 249–269.

Adger, N. W. (2000) "Institutional Adaptation to Environmental Risk under the Transition in Vietnam", *Annals of the Association of American Geographers* 90: 738–758.

Adger, N.W., N.W. Arnell and E.L. Tompkins (2005) "Successful Adaptation to Climate Change across Scales", *Global Environmental Change* 15: 77–86.

Ahmed, A.U., K. Zahurul, K. Prasad, S.N. Poudel and S.K. Sharma (2004) "Synthesis of Manuals on Community Flood Management in Bangladesh, India and Nepal", *Asia Pacific Journal of Environment and Development* 11(1 and 2): 1–39.

Asian Development Bank (2003) *Technical Assistance for Support for the Mekong River Commission Flood Management and Mitigation Program*, Manila: Asian Development Bank.

Bankoff, G. (2004) "In the Eye of the Storm: The Social Construction of the Forces of Nature and the Climatic and Seismic Construction of God in the Philippines", *Journal of Southeast Asian Studies* 35: 91–111.

Blaikie, P., T. Cannon, I. Davis and B. Wisner (1994) *At Risk: Natural Hazards, People's Vulnerability, and Disasters*, London: Routledge.

Carpenter, S., B. Walker, J.M. Anderies and N. Abel (2001) "From Metaphor to Measurement: Resilience of What to What?", *Ecosystems* 4: 765–781.

Chan, N.W. (1997) "Institutional Arrangements for Flood Hazard Management in Malaysia: An Evaluation using the Criteria Approach", *Disasters* 21: 206–222.

Dixit, A. (2003) "Floods and Vulnerability: Need to Rethink Flood Management", *Natural Hazards* 28: 155–179.

Few, R. (2003) "Flooding, Vulnerability and Coping Strategies: Local Responses to a Global Threat", *Progress in Development Studies* 3: 43–58.

Fox, I.B. and P.M. Smidt (2003) *Technical Assistance for Support for the Mekong River Commission Flood Management and Mitigation Program*, Manila: Asian Development Bank.

Gunderson, L.H. and C.S. Holling, eds (2002) *Panarchy: Understanding Transformations in Human and Natural Systems*, Washington D.C.: Island Press.

Kitamoto, M., E. Tsunozaki and A. Teranishi (2005) "Institutional Capacity for Natural Disaster Reduction in Japan", USER Working Paper No. 2005/11, Chiang Mai: Unit for Social and Environmental Research, Chiang Mai University.

Kotov, V. and E. Nikitina (2002) "Mechanisms of Environmental and Human Security", in E. Petzold, ed., *Responding to Environmental Conflicts: Implications for Theory and Practice*, Amsterdam: Kluwer Academic.

Kotov, V. and E. Nikitina (2005) *Institutions, Policies and Measures towards the Lena River Flood Risk Reduction*, Chiang Mai: Unit for Social and Environmental Research, Chiang Mai University.

Krausmann, E. and F. Mushtaq (2006) "Methodology for Lessons Learning: Experiences at the European Level", in J. Birkmann, ed., *Measuring Vulnerability to Hazards of Natural Origin*, Tokyo: UNU Press.

Kundzewicz, Z.W. and H.-J. Schellnhuber (2004) "Floods in the IPCC TAR Perspective", *Natural Hazards* 31: 111–128.

Lebel, L., J. Manuta and S. Khrutmuang (2006) "Tales from the Margins: The Implications of Post-tsunami Reorganization and Politics on the Livelihood Security of Small-Scale Fisher Communities", *Disaster Prevention and Management* 15(1): 124–134.

Manuta, J., S. Khrutmuang, D. Huaisai, and L. Lebel (2006) "Institutionalized Incapacities and Practice in Flood Disaster Management in Thailand". *Science and Culture* 72(1–2): 10–22.

Manuta, J., S. Khrutmuang and L. Lebel (2005) "The Politics of Recovery: Post-Asian Tsunami Reconstruction in Southern Thailand", *Tropical Coasts* July: 30–39.

Morrow, B.H. (1999) "Identifying and Mapping Community Vulnerability", *Disasters* 23: 1–18.

Nikitina, E. (2005) "Institutional Capacity in Natural Disasters Risk Reduction: A Comparative Analysis of Institutions, National Policies and Cooperative Responses to Floods in Asia", Progress Report for APN project 2004-11-NMY-Nikitina, Moscow: Eco-Policy Research and Consultancy.

Takeuchi, K. (2001) "Increasing Vulnerability to Extreme Floods and Societal Needs of Hydrological Forecasting", *Hydrological Sciences Journal* 46: 869–881.

Tingsanchali, T., S. Supharatid and L. Rewtrakulpaiboon (2003) "Institutional Arrangement for Flood Disaster Management in Thailand", Chiang Mai: 1st Southeast Asia Water Forum.

Thomas, D. (2005) "Developing Watershed Management Organizations in Pilot Sub-basins of the Ping River Basin". Office of the Natural Resources and Environmental Policy and Planning, Ministry of Natural Resources and Environment. December 2005.

Tompkins, E.L. and N.W. Adger (2004) "Does Adaptive Management of Natural

Resources Enhance Resilience to Climate Change?" *Ecology and Society* 9(10), available at http://www.ecologyandsociety.org/vol9/iss12/art10/.

Turner, B.L., R. Kasperson, P. Matson, J.J. McCarthy, R. Corell, L. Christensehn, N. Eckley, J. Kasperson, A. Luers, M. Martello, C. Polsky, A. Pulsipher and A. Schiller (2003) "A Framework for Vulnerability Analysis in Sustainability Science", *PNAS*, 100(14) 8074–8079.

Vorobiev, U., V. Akimov and U. Sokolov (2003) *Catastrophic Floods of the Beginning of the XXI Century* [in Russian], Moscow: Deks-Press.

White, G.F., R.W. Kates and I. Burton (2001 "Knowing Better and Losing More: The Use of Knowledge in Hazards Management", *Environmental Hazards* 3: 81–92.

Wong, K. and X. Zhao (2001) "Living with Floods: Victim's Perceptions in Beijiang, Guangdong, China", *Area* 33: 190–201.

Young, O.R. (2002) *The Institutional Dimensions of Environmental Change: Fit, Interplay and Scale*, Cambridge: MIT Press.

20

Public sector financial vulnerability to disasters: The IIASA CATSIM model

Reinhard Mechler, Stefan Hochrainer,
Joanne Linnerooth-Bayer and Georg Pflug

Abstract

This paper addresses the financial vulnerability of developing country Governments to disasters of natural origin. A framework of public sector financial vulnerability and its components of economic risk and financial resilience have been developed. The International Institute for Applied Systems Analysis Catastrophe Simulation (CATSIM) model, which is an interactive simulation tool for building the capacity of policy makers to assess and reduce public sector financial vulnerability by employing pre-disaster financial instruments, is presented. As a case study, the tool is applied to Honduras. We conclude with some observations on the opportunities and limitations of vulnerability indicators, such as those employed in the CATSIM tool.

Introduction

The public sector plays a major role in reducing the long-term economic repercussions of disasters by repairing damaged infrastructure and providing financial assistance to households and businesses. If critical infrastructure is not repaired in a timely manner, there can be serious effects on the economy and the livelihoods of the population. The repair of public infrastructure, however, can be a significant drain on public budgets, especially in developing and transition countries. In Poland, for example, public infrastructure damage from the 1997 floods amounted to 41 per cent of the reported direct losses (Kunreuther and Linnerooth-Bayer, 2003). The Polish Government absorbed close to half of these losses, which increased its budget deficit substantially. Governments of disaster-prone developing countries, such as Honduras, the Philippines, Mexico and regions in China, face such large liabilities in repairing their critical infrastructure and providing subsistence to disaster victims that without international assistance they can be set back years in their development. After Hurricane Mitch devastated Honduras in 1998, GDP growth in the following year (despite the growth impetus from reconstruction) dropped from an estimated 3.3 per cent to −1.9 per cent (Mechler, 2004). Typically, disasters affect Government budgets by reducing tax revenue, increasing fiscal deficits and worsening trade balances (Otero and Marti, 1995). Governmental support of relief and reconstruction is critically important for economic recovery and ultimately for preventing the long-term hidden deaths and suffering caused by disasters.

The State can be physically and financially vulnerable to natural disasters (what we refer to as *public sector financial vulnerability*), especially in highly exposed developing countries. Developing country Governments frequently lack the liquidity, even with international aid and loans, to repair damaged public infrastructure fully or provide sufficient support to households and businesses for their recovery. For example, following the 2001 earthquake in the State of Gujarat, India, recovery funds from the central Government and other sources fell far short of promises, and actual funding only covered around 30 per cent of the State Government's post-disaster reconstruction needs (World Bank, 2003). Gujarat and other recent cases of Government post-disaster liquidity crises have sounded an alarm, prompting financial development organisations, such as the World Bank among others, to call for greater attention to reducing financial vulnerability and increasing the resilience of the public sector (Pollner et al., 2001; Gurenko, 2004). In this context, *resilience* refers to the capacity of a social system to absorb economic disturbance and reorganise, or to "bounce back" so as to retain essentially the same function, structure and identity (Walker et al., 2002).

This chapter addresses the financial vulnerability of developing country Governments to disasters of natural origin, and examines pre-disaster (ex ante) financial measures for increasing the coping capacity and resilience of the public sector. In the next section, a framework of public sector financial vulnerability and its components of economic risk and financial resilience are discussed, along with measurable indicators of these concepts. The IIASA CATSIM model, which is an interactive model for increasing the capacity of policy makers to assess and reduce public sector financial vulnerability, builds on these indicators and is discussed in the next section. Later on, the tool is applied in a case study of Honduras. We conclude with some observations on the opportunities and limitations of vulnerability indicators such as those employed in the CATSIM tool.

Public sector financial vulnerability

Turner et al. (2003) define vulnerability as the degree to which a system or sub-system is likely to experience harm due to exposure to a hazard, either as a perturbation or stressor. Some communities suffer less harm than others from hurricanes, fires, floods and other extreme events because they can mitigate the damage and recover more rapidly and completely. As a case in point, Bangladesh has become less physically vulnerable to cyclones. Over the past four decades deaths from cyclones have decreased by two orders of magnitude as people have learned to respond to warnings and use storm shelters. Moreover, the people in Bangladesh may become less economically vulnerable to long-term economic losses from cyclones and other disasters as affordable micro-insurance and other financial hedging instruments become available (Linnerooth-Bayer and Mechler, 2005).

In the literature, work on economic vulnerability to external shocks (often of small island developing States) has focused on the structure of an economy (e.g. commodity-based versus high-technology), the prevailing economic conditions (e.g. degree of inflation, economic recession) and the general stage of technical, scientific and economic development (Benson and Clay, 2000). Economic vulnerability is assessed by a set or a composite index of indicators such as the degree of export dependence, lack of diversification, export concentration, export volatility, share of modern services and products in GDP, trade openness or simply GDP (Briguglio, 1995; Commonwealth Secretariat, 2000).

This chapter focuses on the financial vulnerability of the public sector as a subset of economic vulnerability. Public sector financial vulnerability is defined as the degree to which a public authority or Government is likely to experience a lack of funds for financing post-disaster reconstruction investment and relief. As illustrated in Figure 20.1, financial vulnerability depends on the asset risks the country is facing from natural hazards, which can be measured by the *hazard* frequency and intensity, the public and private capital *exposure* and the *sensitivity* of the public and private assets to the hazard.

A second important component of public sector financial vulnerability is the *resilience* or financial capacity of the public authorities to cope with the losses. This can be measured by calculating the available financial resources for meeting unexpected liabilities of the public sector. If the Gov-

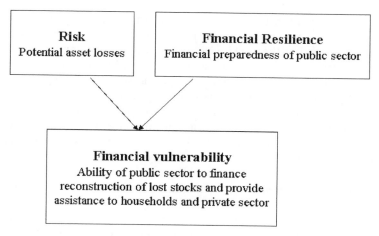

Figure 20.1 Public sector financial vulnerability to natural hazards.
Source: Authors.

ernment has sufficient reserves or insurance cover to finance its post-disaster liabilities, or can easily raise capital through its budget or borrowing, then it is financially resilient to the disaster shock. However, if the asset risks are high and the Government cannot cover the anticipated losses, then a *financing gap* may occur. The potential for a financing gap is an indicator of financial vulnerability. The term financing gap has been coined in the economic growth modelling literature as the difference between required investments in an economy and the actual available resources. In consequence, the main policy recommendation has been to fill this gap with foreign aid (Easterly, 1999).[1] In this report, this tradition is followed and the financing gap is understood as the lack of financial resources to restore assets lost due to natural disasters and continue with development as planned.

An assessment of public sector financial vulnerability, or the potential financing gap therefore considers the following two questions:

- Given the country's current exposure to hazards and changes in future conditions, what are the Government's capital asset *risks* over the planning period?
- Given the Government's financial situation and history of external assistance, is it financially *resilient* to these disasters in the sense of being able to access sufficient post-disaster funding opportunities to cope with losses and liabilities?

The risk of direct economic losses and financial resilience are thus essential concepts for addressing public sector financial vulnerability to natural disasters. Public policy measures can focus on reducing risks by reducing asset exposure, for instance with structural measures or land-use planning, or by reducing the sensitivity of structures, for example by seismically retrofitting the public infrastructure. In addition, policies can improve the resilience of the private or public sectors, through measures such as developing appropriate systems for insuring or transferring the risks. To reduce their financial vulnerability, public authorities can consider investing both in risk reduction and in financial instruments for assuring financial resilience. In what follows, we discuss these concepts with reference to how they can be assessed and measured.

Direct asset risk: hazard, exposure and sensitivity

Risk is generally defined as the probability and magnitude of an adverse outcome, and includes the uncertainty over its occurrence, timing and consequences (Covello and Merkhofer, 1993). Risks of extreme events can be characterised by the frequency and intensity of such events, as well as the exposure and sensitivity of physical assets. A common measure is the probabilistic loss exceedance curve, which indicates the probability of certain losses exceeding a certain amount; for example, if there is a 1 per cent probability (called a 100-year event) that losses may exceed US$1 billion.

Financial resilience

Originating in the field of ecology, a key concept in vulnerability research is *resilience*, which refers to the capacity of a system to absorb disturbances and reorganise so as to "bounce back" to essentially the same function and structure (Walker et al., 2002). A resilient ecosystem can withstand shocks and rebuild itself when necessary. Similarly, a resilient social system, in our case the public sector, can absorb shocks and rebuild the economy so that the country or region stays on a similar economic trajectory. Systems with high resilience are able to reconfigure themselves without significant declines in crucial functions in relation to primary productivity and economic prosperity. Resilience in social systems has the added capacity of humans to anticipate and plan for the future.

Due to the role of the public sector in financing reconstruction, financial preparedness is essential for countries or regions to "bounce back" from major shocks. The preparedness of the public authorities for financing disasters depends on their access to capital after a disaster, which, in turn, depends on the Government's tax base, budget deficit, and internal and external debt, among other fiscal indicators. In addition, regional

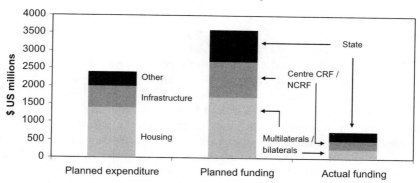

Figure 20.2 Financing gap in India after the Gujarat earthquake.
Source: Modified based on World Bank 2003: 22.

Governments of developing countries rely extensively on national and international loans and aid. Despite often generous international support, developing countries frequently encounter shortfalls in financing reconstruction and relief following disasters. One example, as mentioned above, is the earthquake of 2001 in the State of Gujarat in India, where planned funding from Government relief funds, bi- and multilateral sources and budget diversions would have exceeded planned expenditure; however actual funding disbursed amounted to only 32 per cent of the planned amount (World Bank, 2003). As shown in Figure 20.2, the Gujarat Government experienced a severe *financing gap* with regard to the planned expenditures for repairing the housing stock and public infrastructure as well as providing relief to the affected population.

Financial preparedness can be enhanced by pre-disaster planning. The public authorities can set aside reserves in a catastrophe fund (such funds exist in India), or alternatively they can purchase instruments that transfer their risk to a third party. Insurance is the most common pre-disaster instrument, but recently other types of novel risk-transfer instruments have emerged. These instruments and their costs will be discussed in more detail in the next section. The important message is that pre-disaster measures exist to improve sovereign financial resilience for highly exposed countries. Given that these measures are costly, it is important to ask which countries need them (which countries are financially vulnerable?) and what are their costs and benefits? These questions are addressed by the CATSIM model as described in the following section.

Assessing financial vulnerability with CATSIM

The experience of India and many other disaster-prone developing countries raises the question of how policy makers can reduce public sector financial vulnerability. The IIASA CATSIM tool was developed to provide insights on this question (for a detailed discussion of CATSIM see Hochrainer, Mechler and Pflug, 2004; Freeman et al., 2002a). CATSIM uses Monte Carlo simulation of disaster risks in a specified region and examines the ability of the Government to finance relief and recovery. It is interactive in the sense that the user can change the parameters and test different assumptions about the hazards, exposure, sensitivity, general economic conditions and the Government's ability to respond. CATSIM can provide an estimate of a country or region's public sector financial vulnerability. As a capacity building tool, it can illustrate the trade-offs and choices the authorities confront in increasing their resilience to the risks of catastrophic disasters.

The CATSIM methodology consists of five stages or modules as described below and illustrated in Figure 20.3.

- *Stage 1:* The risk of direct asset losses expressed in terms of their probability of occurrence and destruction in monetary terms is modelled as a function of hazard (frequency and intensity), the elements exposed to those hazards and their physical sensitivity.
- *Stage 2:* The financial preparedness of the public sector to meet the direct losses is assessed. Financial preparedness is a measure of financial resilience and can be defined as the access of the State or central Government to funds for financing the reconstruction of public infrastructure and the provision of relief to households and the private sector. Financial preparedness will, in turn, depend on the general economic conditions of the country.
- *Stage 3:* Financial vulnerability, measured in terms of the potential financing gap, is assessed by simulating the risks to public infrastructure and the financial resilience of the Government to cover its post-disaster liabilities following disasters of different magnitudes.
- *Stage 4:* The consequences of a financing gap on the macroeconomic development of the country are characterised by using indicators such as economic growth or the country's external debt situation. These indicators represent consequences to economic flows as compared to consequences to stocks addressed by the asset risk estimation in Stage 1.
- *Stage 5:* Strategies are developed and illustrated that build the financial resilience of the public sector. The development of risk financing strategies has to be understood as an adaptive process, where measures are

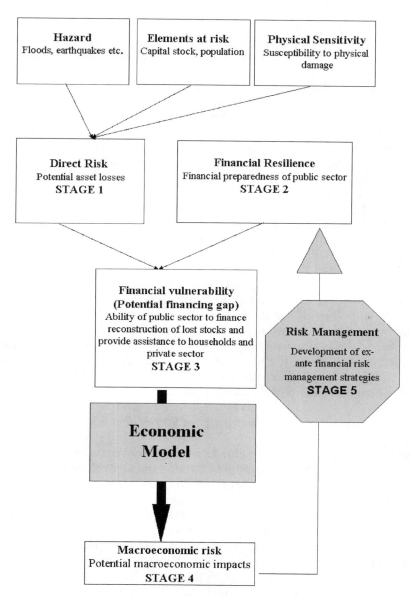

Figure 20.3 Financial vulnerability and the CATSIM methodology.
Source: Authors.

continuously revised after their impact on reducing financial vulnerability and risk has been assessed within the modelling framework.

Stage 1: Assessing public sector risk

The stage 1 CATSIM module assesses the risk of direct losses in terms of
the probability of asset losses in the relevant country or region. Consistent with general practices, risk is modelled as a function of hazard (frequency and intensity), the elements exposed to those hazards and their
physical sensitivity (Burby, 1991; Swiss Re, 2000).[2] In other words:

- Natural hazards, such as earthquakes, hurricanes or floods, are described by their intensity (e.g. peak flows for floods) and recurrency
  (such as a one in 100 year event, i.e. with a probability of 1 per cent).
- Exposure of elements at risk: Total private and public capital stock is
  estimated.
- Physical sensitivity describes the degree of damage to the capital stock
  due to a natural hazard event. Fragility curves, which set the degree of
  damage in relation to the intensity of a hazard, are commonly used for
  this purpose.

Using data on the return period and losses in per cent of capital stock,
CATSIM generates loss frequency distributions describing the probability of specified losses occurring, such as a 100-year event causing a loss
of US$200 million of public assets, a 50-year event causing a US$40 million loss, and so on.[3] It should be kept in mind that top-down estimates at
this broad scale are necessarily rough. Since most disasters are rare
events, there is often little historical data available; furthermore it is difficult to include dynamic changes in the system, for example, population
and capital movements and climate change.

Stage 2: Assessing public sector financial resilience

Using the information on direct risks to the Government portfolio, financial resilience can be evaluated by assessing the Government's ability to
finance its obligations for the specified disaster scenarios. Financial resilience is directly affected by the general conditions prevailing in an economy; changes in tax revenue have important implications on a country's
financial capacity to deal with disaster losses.

The specific question underlying the CATSIM tool is whether a Government is financially prepared to repair damaged infrastructure and provide adequate relief and support to the private sector for the estimated
damages of 10-, 50-, 100- and 1,000-year events? For this assessment, it
is necessary to examine the Government's resources, both those that will
be relied on (probably in an ad hoc manner) after the disaster and those

Table 20.1 Ex-post financing sources for relief and reconstruction

Type	Source	Considered in model
Decreasing government expenditures	Diversion from budget	Yes
Raising government revenues	Taxation	No
Deficit financing	Central Bank credit	No
Domestic	Foreign reserves	No
	Domestic bonds and credit	Yes
Deficit financing	Multilateral borrowing	Yes
External	International borrowing	Yes
	Aid	Yes

Source: Authors.

put into place before the disaster (ex-ante financing). These sources are described below.

Ex-post financing resources

The Government can raise funds after a disaster by accessing international assistance, diverting funds from other budget items, imposing or raising taxes, taking a credit from the Central Bank (which either prints money or depletes its foreign currency reserves), borrowing by issuing domestic bonds, borrowing from international financial institutions (IFIs) and issuing bonds on the international market (Benson, 1997; Fischer and Easterly, 1990). Each of these financing sources can be characterised by costs to the Government as well as factors that constrain availability, which are assessed by this CATSIM module. Sources not considered feasible are not included in the module.

As shown in Table 20.1, ex-post financing can be constrained. As an example, disaster taxes are expensive to administer and generally not part of the public sector financing portfolio. As a second example, borrowing can also be constrained by the existing country debt. CATSIM assumes that the sum of all loans cannot exceed the so-called *credit buffer* for the country. In the Highly Indebted Poor Countries Initiative (HIPC) the credit buffer is defined as 150 per cent of the typical export value of this country minus the present value of existing loans (World Bank, 2002). These ex-post instruments have (sometimes high) associated costs; even budgetary diversions have associated opportunity costs in terms of other Government investments like building highways or schools.

Ex-ante financing sources

In addition to accessing ex-post sources, a Government can arrange for financing before a disaster occurs. Ex-ante financing options include reserve funds, traditional insurance instruments (public or private), alternative insurance instruments, such as catastrophe bonds, or arranging a contingent credit. The Government can create a reserve fund, which accumulates in years without catastrophes. In the case of an event, the accumulated funds can be used to finance reconstruction and relief. A catastrophe bond (cat bond) is an instrument whereby the investor receives an above-market return when a specific catastrophe does not occur, but shares the insurer or Government's losses by sacrificing interest or principal following the event. Contingent credit arrangements call for the payment of a fee for the option of securing a loan with prearranged conditions after a disaster. Insurance and other risk-transfer arrangements provide indemnification against losses in exchange for a premium payment. Risk is transferred from an individual to a (large) pool of risks. These ex-ante options can involve substantial annual payments and opportunity costs; statistically the purchasing Government will pay more with a hedging instrument than if it absorbs the loss directly.

Given the costs, many developing country Governments are asking whether public sector insurance is desirable for improving financial preparedness. According to an early discussion by Arrow and Lind (1970), Governments should generally not purchase insurance. Due to the large number of public assets in different locations, the Government is sufficiently diversified, and post-disaster expenses can be spread over a large base of taxpayers. This means that the public authorities are not risk averse and therefore do not need to purchase insurance or other financial hedging instruments. Disaster risks and other stochastic shocks to public budgets can thus be ignored in public planning and budgeting decisions. Recent research undertaken by IIASA, however, has shown that the Arrow-Lind theorem does not hold for hazard-prone developing countries if they are facing high risks, if the pool of publicly owned assets is too narrow for sufficient diversification, and if they cannot raise sufficient funds after a disaster to finance the recovery process (Freeman et al., 2002a; Mechler, 2004). Whether insurance is desirable for a developing country's Government will thus depend on the Government's financial vulnerability and the cost of insurance instruments compared to the cost of other financing options.

The Government's portfolio of ex-ante and ex-post financial measures is critically important for the recovery of the economy should a disaster occur. For this reason, an assessment of the Government's asset risk and financial resilience is an essential part of disaster risk management. An IIASA study has carried out such an assessment for four highly at-risk Latin American countries: Bolivia, Colombia, the Dominican Republic and El Salvador (Freeman et al., 2002b). The study revealed differences in their financial preparedness for disasters. At the time of the study, none of the four countries had ex-ante instruments like reserve funds or insurance in place. Bolivia and Colombia were, however, better prepared than the Dominican Republic and El Salvador to meet their liabilities. The reason was that they could more readily divert funds within their current budget, although Colombia was far more constrained with respect to other ex-post options, such as borrowing domestically and internationally. These indicators of financial resilience can be combined with the risk each country is facing to yield an indicator of potential financial vulnerability. The results are discussed below.

Stage 3: Measuring financial vulnerability by the "financing gap"

Comparing available financing with the Government's post-disaster financial obligations yields an estimation of the potential *financing gap*. In the IIASA study, the potential financing gap for Bolivia, Colombia, the Dominican Republic and El Salvador was assessed for a range of probabilistic disaster losses. Figure 20.4 illustrates this gap only for the 100-year event in each country. In this figure, financing sources available to the Governments of the four countries are compared with the Governments' potential financial obligations calculated for the 100-year disaster. The shortfall between financial sources and obligations is the financing gap.

Estimates show, for example, that the losses to the Bolivian Govern-

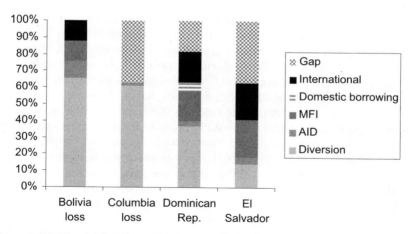

Figure 20.4 Financial vulnerability to 100-year event in four Latin American countries.
Source: Authors.

ment due to a 100-year event would have amounted to US$500 million (from damaged public infrastructure and obligations for relief). If this event had occurred in the 2002 budget period, Bolivia could have financed all but about one per cent of its obligations by accessing the following: international and domestic capital markets, support from international financial institutions, international donor aid, and, most importantly, diversions from its domestic budget. Colombia, the Dominican Republic and El Salvador can expect far larger financing gaps mainly because of less slack in their domestic budgets. Because of their lack of resilience and the risks they are facing, in 2002 these Governments were highly financially vulnerable to the 100-year disaster event.

Stage 4: Illustrating the developmental consequences of a financing gap

Financial vulnerability can have serious repercussions on the national or regional economy and the population. If the Government cannot replace or repair damaged infrastructure, such as roads and hospitals, nor provide assistance to those in need after a disaster, this will have long-term consequences. The consequences on long-term economic development can be illustrated by the CATSIM model. For example, Figure 20.5(c) shows the results of the simulations of growth paths in El Salvador with and without the purchase of insurance for public assets as an ex-ante financial tool.

As seen in Figure 20.5(c), El Salvador is expected to grow over time (with the current year as the base year) as investment adds to the capital stock. However, the country can experience disasters, which can be thought of as stochastic shocks to the growth trajectory. CATSIM simulates 5,000 trajectories, although in this figure only 100 are summarised for illustrative purposes. The trajectories do not have equal probability. The trajectories in the upper part of the figure, which show economic growth proceeding in the absence of shocks, have a higher probability of occurrence than the catastrophic cases in the bottom of the figure. Economic growth in El Salvador is higher on average if the Government does not allocate its resources to catastrophe insurance (upper figure), but the economy has fewer extremes and is more stable with public sector insurance (lower figure). Investing in the risk financing instruments can thus be viewed as a trade-off between economic growth and stability. Budgetary resources allocated to catastrophe reserve funds, insurance and contingent credit (as well as to preventive loss-reduction measures) reduce the potential financing gap, and thus can ensure a more stable development path. On the other hand, ex-ante financing and prevention measures come at a price in terms of other investments foregone and will inevitably have an adverse impact on the growth path of an economy. The IIASA model assesses this trade-off by comparing the costs of selected ex-ante measures with their benefits in terms of decreasing the possibility of encountering a financing gap.

Stage 5: Reducing financial vulnerability and building resilience

Vulnerability and resilience must be understood as dynamic. In contrast to ecological systems, social systems can learn, manage and actively influence their situation. There are two types of policy interventions for reducing public sector financial vulnerability: those that reduce the risks of disasters by reducing exposure and sensitivity, and those that build the financial resilience of the responding agencies. On the basis of an assessment of the financing gap and potential economic consequences, CATSIM illustrates the pros and cons of strategies for building financial resilience using ex-ante financial instruments. Four ex-ante financing policy measures are currently considered in the CATSIM model: insurance, contingent credit, reserve funds and cat bonds. Also, one generic option for loss reduction measures has been implemented in the model in order to analyse the linkage with risk financing. More detail on the model can be found in Hochrainer, Mechler and Pflug, 2004.

Example: the case of Honduras

Honduras illustrates the case of a country with a potential financing gap. Over the last decade it has experienced a number of hurricanes and other weather disasters. With over half of its 6.5 million people living in poverty, Honduras is socially and economically vulnerable to extremes in weather. Recent IIASA studies examined the conditions under which the Government can expect to be short of funds to finance disaster relief and reconstruction, and the effectiveness of ex-ante financial measures for building financial resilience (Mechler and Pflug, 2002; Mechler, 2004). Relying on historical data the CATSIM simulation model provided insights on the overall risks of flood and storm events in the country, and the ensuing liabilities for the Government. The analysis looked closely at the capacity for the Government to raise funds through borrowing, raising taxes and diverting from other budgeted items. In addition, the likely availability of external aid and assistance was examined. As shown in Figure 20.6(c), the main hazards in Honduras are hurricanes and other windstorms originating from the northern coast that cause flooding and landslides.

Information on the intensity and frequency of hazards as well as the

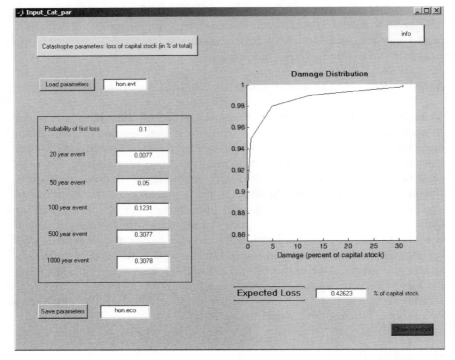

Figure 20.7 Cumulative probability distribution of direct asset damages for storm and flood for Honduras.
Source: Authors.

sensitivity of the exposed assets to these hazards was obtained from Swiss Re. Capital stock was estimated at US$13.9 billion for 2004. It was assumed that about 30 per cent of capital stock is public and that Government will finance another 20 per cent of total capital losses due to its political commitment to relief for private victims after disasters (Freeman et al., 2002b). These assumptions are consistent with country data and past experience.

From this information, direct asset losses were estimated. Figure 20.7 shows a screen shot of the CATSIM model illustrating the cumulative loss exceedance curve for public sector assets plus anticipated relief to the private sector in 2004.

As shown in Figure 20.7, for very rare storm and flood events (once in 1,000 years) the capital stock losses could approach 30 per cent of the total capital stock in Honduras. Lower frequency events, for example, the

100-year storm and flood, are estimated as likely to destroy around 12 per cent of total capital stock. The expected losses due to storm and flood risk are 0.43 per cent.

Figure 20.8(c) displays the CATSIM screen shot illustrating the financial vulnerability of the Honduran Government to floods and storms. As shown in this figure, the Government could (in 2002) depend on traditional sources to finance the losses from moderate flood and storm disasters (with a recurrence period of less than about 100 years) and thus should not consider any form of risk transfer covering these events. But for very rare, high-consequence events – one-in-109 years or worse – there is a sizable financing gap. This means that Honduras will not be able to provide sufficient relief to private victims nor repair its infrastructure in a timely way, which can set the country back significantly in its economic development.

On the basis of an assessment of financial vulnerability and its economic consequences, a case for increasing financial resilience using ex-ante instruments may be justified. The IIASA CATSIM model illustrates the cost efficiency and economic consequences of selected ex-ante instruments, including their consequences for public sector indebtedness and economic growth. More details on the development and illustration of ex-ante risk financing strategies can be found in Hochrainer, Mechler and Pflug (2004).

Beyond indicators: building capacity for reducing vulnerability

The financial vulnerability of the public sector represents only one aspect, albeit an important one, of vulnerability to natural hazards. Other indicators are necessary to give a more complete picture of vulnerability. For example Cardona et al. for the Information and Indicators Program for Disaster Risk Management of the Inter-American Development Bank (IADB), Economic Commission for Latin America and the Caribbean (ECLAC) and Instituto De Estudios Ambientales (IDEA) have complemented the IIASA methodology of financial vulnerability (termed *Disaster Deficit Index* in their report) with other vulnerability indicators, such as the *Prevalent Vulnerability Index* that accounts for social vulnerability in terms of exposure in hazard-prone areas, socio-economic fragility and social resilience (Cardona et al., 2005).

These and other indicators of vulnerability generally rely on quantitative indicators and thus communicate a degree of objectivity, which can be misleading if not handled with great care. Since the numbers often rely on incomplete data and numerous assumptions, there can be large uncertainties and subjective choices. Because of these uncertainties and subjective judgments, indicators may work best if they are created and applied within a participatory approach that includes the key stake-holders (Morse, 2004).

CATSIM has been created as a participatory, interactive tool for build-ing the capacity of policy makers by sensitising them to the tradeoffs in-herent in planning for disasters. By means of a graphical user interface the user can explore financing issues in the probabilistic context of natu-ral disasters, can change important parameters and test the sensitivity of outcomes to those changes. In addition, the user is cautioned that the model does not yield "optimal" strategies, but gives insights into the pros and cons of different policy options.

The model underlying CATSIM was originally designed for the Re-gional Policy Dialogue of the IADB, where it was applied to Latin Amer-ican case studies (Freeman et al., 2002b). The CATSIM simulation model was developed from the IADB model and has been successfully em-ployed by economists, financial experts and policy makers in workshops for stakeholders who are interested in taking account of disaster risk in public finance theory, and in the financial management of disaster risk. The first multi-country workshop sponsored by the ProVention Consor-tium and the World Bank was held at IIASA in 2004 with participants from Mexico, Colombia, Turkey, India and the Philippines. Several follow-up efforts are underway and more national or regional workshops are envisaged.

IIASA will continue developing and extending the CATSIM modelling framework. Work is under way to improve the evaluation of mixed ex-ante and ex-post financial instruments and to make the model more dy-namic by taking account of future changes in the risks (including climate change) and financing capacity. Furthermore, the representation of the private sector and its vulnerability to natural hazards needs to be mod-elled more explicitly. The tool will be tested further in participatory stakeholder workshops involving policy makers who are intent upon re-ducing the vulnerability of their countries or regions to the long-term consequences of natural disasters.

Notes

1. This approach has been criticised by Easterly (1999) among others as generally failing to account for the role of incentives and institutions in economic growth. Nevertheless, there is no doubt that capital investment plays an important role in economic growth.
2. In the hazards and risk community, "sensitivity" is referred to as "vulnerability", and ex-posure is often included in the sensitivity component; thus, risk is defined by hazard and vulnerability. In catastrophe models carried out for insurance purposes, the contract specifications of the underwritten and exposed portfolios are added as a fourth compo-nent (e.g. Swiss Re, 2000).
3. It is standard practice to refer to 20-, 50-, 100-, 500- and 1,000-year events.

REFERENCES

Arrow, K.J. and R.C. Lind (1970) "Uncertainty and the Evaluation of Public Investment Decisions", *The American Economic Review* 60: 364–378.

Benson, C. (1997) *The Economic Impact of Natural Disasters in Fiji*, London: Overseas Development Institute.

Benson, C. and E. Clay (2000) "Developing Countries and the Economic Impacts of Catastrophes", in A. Kreimer and M. Arnold, *Managing Disaster Risk in Emerging Economies*, Washington D.C.: The World Bank, pp. 11–21.

Briguglio, L. (1995) "Small Island Developing States and their Economic Vulnerabilities", *World Development* 23(9): 1615–1632.

Burby, R., ed. (1991) *Sharing Environmental Risks: How to Control Governments' Losses in Natural Disasters*. Boulder, Colorado: Westview Press.

Cardona, O.D. et al. (2005) Results of Application of the System of Indicators on Twelve Countries of the Americas. IDB/IDEA Program of Indicators for Disaster Risk Management, Manizales: National University of Colombia, available at http://idea.unalmzl.edu.co.

Commonwealth Secretariat (2000) *Small States: Meeting Challenges in the Global Economy*, Washington D.C./London: World Bank Joint Task Force on Small States/Commonwealth Secretariat.

Covello, V.T. and M.W. Merkhofer (1993) *Risk Assessment Methods: Approaches for Assessing Health and Environmental Risks*, New York: Plenum.

Easterly, W. (1999) "The Ghost of Financing Gap: Testing the Growth Model Used in the International Financial Institutions", *Journal of Development Economics* 60(2): 424–438.

Fischer, S., and W. Easterly (1990) "The Economics of the Government Budget Constraint", *The World Bank Research Observer* 5(2): 127–142.

Freeman, P.K., L.A. Martin, R. Mechler, K. Warner and P. Hausman (2002a) "Catastrophes and Development: Integrating Natural Catastrophes into Development Planning", Disaster Risk Management Working Paper Series No. 4, Washington D.C.: The World Bank.

Freeman, P.K., L.A. Martin, J. Linnerooth-Bayer, R. Mechler, S. Saldana, K. Warner and G. Pflug (2002b) *Financing Reconstruction: Phase II Background Study for the Inter-American Development Bank Regional Policy Dialogue on National Systems for Comprehensive Disaster Management*, Washington D.C.: Inter-American Development Bank.

Gurenko, E. (2004) *Catastrophe Risk and Reinsurance: A Country Risk Management Perspective*, London: Risk Books.

Hochrainer, S., R. Mechler and G. Pflug (2004) *Financial Natural Disaster Risk Management for Developing Countries*, Proceedings of the XIIIth Annual Conference of European Association of Environmental and Resource Economics, Budapest, June 2004.

Kunreuther, H. and J. Linnerooth-Bayer (2003) "The Financial Management of Catastrophic Flood Risks in Emerging Economy Countries", in J. Linnerooth-Bayer and A. Amendola, Special Edition on Flood Risks in Europe, *Risk Analysis* 23: 627–639.

Linnerooth-Bayer, J. and R. Mechler (2005) *Financing Disaster Risks in Develop-*

ing and Emerging-Economy Countries, Proceedings of OECD Conference on Catastrophic Risks and Insurance, Paris, 22–23 November 2004.

Mechler, R. (2004) *Natural Disaster Risk Management and Financing Disaster Losses in Developing Countries*, Karlsruhe: Verlag für Versicherungswissenschaft.

Mechler, R. and G. Pflug (2002) *The IIASA Model for Evaluating Ex-ante Risk Management: Case Study Honduras. Report to IDB*, Washington D.C.: IADB.

Morse, S. (2004) *Indices and Indicators in Development*, London: Earthscan.

Otero, R.C. and R.Z. Marti (1995) "The Impacts of Natural Disasters on Developing Economies: Implications for the International Development and Disaster Community", in M. Munasinghe and C. Clarke, eds, *Disaster Prevention for Sustainable Development: Economic and Policy Issues*, Washington D.C.: World Bank, pp. 11–40.

Pollner, J., M. Camara and L. Martin (2001) *Honduras: Catastrophe Risk Exposure of Public Assets. An Analysis of Financing Instruments for Smoothing Fiscal Volatility*, Washington D.C.: The World Bank.

Swiss Re (2000) *Storm over Europe: An Underestimated Risk*, Zurich: Swiss Reinsurance Company.

Turner, B.L., R. Kasperson, P. Matson, J.J. McCarthy, R. Corell, L. Christensen, N. Eckley, J. Kasperson, A. Luers, M. Martello, C. Polsky, A. Pulsipher and A. Schiller (2003) "A Framework for Vulnerability Analysis in Sustainability Science", *PNAS*, 100(14): 8074–8079.

Walker, B., S. Carpenter, J. Anderes, N. Abel, G. Cumming, M. Jansen, L. Lebel, J. Norberg, G. Perereson and R. Pichard (2002) "Resilience Management in Social-ecological Systems: A Working Hypothesis for a Participatory Approach", *Conservation Ecology* 6(1): 14, available at http://www.ecologyandsociety.org/vol6/iss1/art14/print.pdf.

World Bank (2002) *About the HIPC Debt Initiative*, Washington D.C., available at http://www.worldbank.org/hipc/about/hipcbr/hipcbr.htm.

World Bank (2003) *Financing Rapid Onset Natural Disaster Losses in India: A Risk Management Approach*, Washington D.C.: The World Bank.

Box 21.1 Effective measurement of vulnerability is essential to help those most in harm's way

Simon Horner

For the European Union, showing solidarity with the victims of natural and man-made disasters is an important principle. Since 1992, the European Commission has provided relief to people caught up in crises through its humanitarian aid department (ECHO). Today, it is one of the largest relief donors, channelling funds through United Nations agencies, NGOs and the Red Cross/Red Crescent movement. The EU as a whole (Commission plus Member States) supplies more than half of the world's publicly financed humanitarian aid.

Much of the assistance goes to help victims of tragedies of human origin in conflict zones around the world. A substantial proportion, however, is provided for relief in disasters involving the power of nature: hurricanes, droughts, floods, volcanic eruptions and earthquakes.

The Commission's humanitarian aid is guided above all by needs. Support is provided impartially to people suffering most in crises, irrespective of their nationality, ethnic origin, religion or gender.

One implication of the Commission's needs-based approach is its continuing focus on crises that gain little media attention and have difficulty in attracting relief funds. In 2001, the Humanitarian Aid department established basic principles and a methodology for identifying forgotten crises. The approach, which has subsequently been refined, combines "bottom-up" and "top-down" assessments – reports by field experts and desk officers together with aggregated data analysis of 130 countries (the "Global Needs Assessment") – to determine the areas of greatest need. Measuring the vulnerability of different populations and communities in the face of existing humanitarian challenges is a necessary part of this analysis and any new method for achieving this more effectively is to be welcomed.

The expression "prevention is better than cure" is well-worn but nonetheless valid. While we must be ready to help the victims of conflict, in an ideal world it would be better to prevent them. The same is true of natural disasters. And where we cannot prevent nature from following its awesome course, lives can at least be saved, and suffering limited, through effective preparedness.

This is where measuring vulnerability is even more relevant. Natural disasters affect more than 300 million people every year and an essential element is the identification of geographical zones most at risk: the coastlines where a tsunami may strike, the population centres located close to major geological faults, the regions likely to suffer

Box 21.1 (cont.)

from droughts or floods, the communities that lie in the potential path of tropical storms.

DIPECHO programmes

In 1996, following the United Nations' adoption of the Yokohama Strategy for a Safer World, the Commission's Humanitarian Aid department launched its own regional disaster preparedness programme, DIPECHO. By the time of the World Conference on Natural Disaster Reduction in Kobe in January 2005, DIPECHO activities were taking place in six of the world's most vulnerable regions. Typically, the projects cover training, capacity building, awareness raising, early warning, planning and forecasting.

A key lesson that the Commission has drawn from its experience, both in responding to disasters and in trying to prepare for them, is the importance of local capacity. This is a key variable that makes measuring vulnerability such a complex task. Within a country, and even within a region, capacity is often far from uniform. The most effective vulnerability assessments are those made at community level, integrating local knowledge with expert observation and localised scientific data.

Thus, for example, while an entire region or country may be exposed to hurricanes, some communities are particularly vulnerable to their effects: where poverty levels are high, homes are less well built, protective infrastructures are non-existent or badly maintained and there are few safe refuges available. Where education levels are low, it is less likely that community members will be trained in preparedness and first-aid. In some places, people are excluded from local government preparedness measures because of language barriers or ethnic discrimination. In others, the problem is simply geographic isolation.

Complex local realities require a highly participatory disaster preparedness approach. The fastest life-saving support will almost always come from volunteers within the affected communities. Local rescue teams may be energetic and enthusiastic, but too often they lack resources, equipment and training. DIPECHO programmes therefore focus on local reaction capacity, to enable people to prepare for future disasters.

With finite resources, it is clearly not possible for the global humanitarian community to offer a programme for every village and district

Box 21.1 (cont.)

Box 21.1 Water Supply at Kasab camp.
Source: EC/ECHO/Greta Hopkins.

under threat from natural disasters. The Commission recognises this reality, which is why it is increasingly stressing disaster prevention as an issue for longer-term development policy.

In DIPECHO, the emphasis is on pilot projects that can have a multiplier effect. While each community has its own specificities there are more general lessons that can also be learned, and communicated more widely. The actions with the greatest impact are those that inspire similar projects in neighbouring communities. This is why DIPECHO also supports national consultative meetings that bring together the essential actors in disaster prevention, including local. and national authorities, international agencies and NGO partners. The dialogue is designed to ensure that projects directly target real needs and that key local actors participate from the outset.

The impact of any prevention initiative is likely to be hard to quantify. However, when seasonal floods struck Bangladesh hard and early in 2004, they affected some areas where community-based preparedness had been undertaken, and others where they had not. Although

Box 21.1 (cont.)

coping capacities were stretched to the limit in all the worst affected districts, it was reported that significantly fewer people needed relief assistance in areas where preparedness had taken place. Communities in these areas were generally better organised and able to rehabilitate more quickly.

For many, 2005 will be remembered as the year of natural disasters. More attention is now being paid to ways of mitigating the effects of these terrible events by being more ready to deal with them in advance.

Simon Horner, European Commission, Humanitarian Aid Directorate-General (ECHO)

21

Overcoming the black hole: Outline for a quantitative model to compare coping capacities across countries

Peter Billing and Ulrike Madengruber

Abstract

Coping capacity, defined as the level of resources and the manner in which people or organisations use these resources to face the adverse consequences of natural disasters, is a key concept in vulnerability assessments. However, very few, if any, datasets and methodologies actually permit a quantitative comparison of coping capacity across countries. Such a methodology would offer a valuable support tool for strategic planning of disaster reduction measures. As a first step in this direction, this chapter develops a simple and pragmatic coping capacity model to rank countries in four categories, from high to very low coping capacity. It uses a combination of selected proxy indicators from UN-Habitat (Global Urban Indicators), the World Bank (mitigation projects), the Red Cross (volunteers) and UNDP's Disaster Reduction Index. By clearly outlining the limitations of the approach, the chapter also provides orientation for further research.

Background and rationale

The notion that populations affected by natural disasters have a coping capacity is a key concept in vulnerability assessment (e.g. Schneiderbauer and Ehrlich, 2004; Bogardi and Birkmann, 2004; see also Chapter 1).

Coping capacity is also often referred to in the context of climate change. The Intergovernmental Panel for Climate Change (IPCC, 2001), for instance, concludes that countries with limited economic resources, poor infrastructure and weak institutions have little capacity to adapt and cope with climate change and are more vulnerable to its effects, including natural disasters. Paradoxically, however, there seem to be few systematic methodological approaches to the subject. In particular, very few, if any, datasets and methodologies actually permit a *quantitative* approach to comparing coping capacity across countries. Such a methodology would offer a valuable support tool for the design of natural disaster reduction strategies by humanitarian or other donor organisations.

This chapter attempts to bridge this gap by developing a simple and pragmatic coping capacity model based on quantitative methods. It uses a combination of selected proxy indicators taken from UN-Habitat, the World Bank, the International Federation of the Red Cross (IFRC), UNDP and others.

Coping capacity can be measured at both the local and other societal levels. Different levels require different approaches. The main focus of this chapter is the societal assessment of coping capacities as a fundamental element for the thorough understanding of a country's overall risk to disaster. A systematic assessment of what enables communities and countries to cope with, recover from and adapt to various risks and adversities provides emergency planners, disaster risk managers and development actors with a clearer understanding of the foundations on which to build interventions that support coping capacity. It also allows a better targeting of external assistance, a main concern for international donors of emergency assistance, such as the European Commission. Up to now, coping strategies of at-risk populations have been poorly understood. Knowledge of what makes up coping capacity, how it can be measured and, above all, how it can be strengthened, is still limited compared to our understanding of what constitutes need, hazard, risk or vulnerability.

This chapter will address that shortcoming by developing a preliminary coping capacity model for populations affected by natural disasters. It proposes a methodology and a tentative country ranking based on an empirical analysis of several datasets using pertinent proxy indicators.

As regards the scale of the model, it deals primarily with the international level (priority setting among donors) but also looks at the level of disaster-affected countries to see where they stand in international comparison.

The Coping Capacity Index (CCI) is created by combining the Global Urban Indicators (GUI) dataset (UN-Habitat, 2003), figures on Red Cross volunteers (IFRC, 2003) and disaster mitigation projects (World

Bank, 2005). It is complemented by components from UNDP's Disaster Risk Index (DRI) (UNDP, 2004). The CCI provides a list of countries divided into four categories of coping capacity: *very low, low, medium* and *high*. Again, the main focus is to facilitate the setting of priorities in the decision-making process of the international donor community, and as a complement to in-depth case studies. Unlike other approaches, the model is not just based on economic factors but also includes a societal dimension (Red Cross volunteers).

By clearly outlining the limitations of the approach ("Open questions and limitations" section), the chapter also provides orientation for further research ("Conclusion and outlook" section).

Methodology: Coping Capacity Index

Definitions and terminology

Both "vulnerability" and "coping capacity" have been defined and used in a variety of ways and contexts. They are not independent of each other. In a sense, they can be considered two sides of the same coin. According to Bohle (2001), coping covers the "internal side of vulnerability" referring to the intrinsic capacity to anticipate, resist or recover from the impact of a hazard. For instance, a community that is unorganised for disaster response has inadequate coping capacity (low capacity) and therefore is likely to suffer more from the exposure to risks and shocks (high vulnerability).[1]

In order to overcome this terminological ambiguity, the following generic definitions will be used for the purpose of this chapter.

- *Vulnerability:* the conditions determined by physical, social, economic and environmental factors, which increase the susceptibility of a community to the impact of hazards (UN-ISDR, 2004).
- *Coping capacity:* the level of resources and the manner in which people or organisations use these resources and abilities to face adverse consequences of disasters (UN-ISDR, 2004).

It should be mentioned that a distinction between *individual* (referring to an individual's strategy and capacity to deal with natural disaster risks) and *institutional* coping capacity (referring to the coping capacity provided for by the society, the Government, etc.) can be made. However, as this chapter is not concerned with funding decisions for individual humanitarian disasters but rather responds to the planning needs of a donor when establishing its global strategy, we will deal solely with coping capacity at national level.

Selection of the country sample

The countries covered by this methodology were selected from the countries listed in the UN Office for the Coordination of Humanitarian Affairs' (OCHA) Relief Web, assuming that any country or territory listed in Relief Web would be in a situation of some humanitarian relevance. The Humanitarian Aid Department of the European Commission's (ECHO) 2005 Global Needs Assessment was also used as a reference base (ECHO, 2005).

Highly industrialised countries (mainly OECD members and their dependent territories) as well as all EU and recent EU accession countries were deleted from the selection. Wealthy developing countries that can be assumed to be able to cope with humanitarian disasters themselves, such as Kuwait or Brunei, were also removed. We also omitted very big countries like China or India because pockets of low coping capacities cannot adequately be addressed. Finally, countries in protracted crisis (e.g. Afghanistan, Burundi and Somalia) as well as those for which only two data points were available (e.g. the Former Yugoslav Republic of Macedonia) were also discarded. Some 100 countries were ultimately retained in the list.

Selection of indicators

Several attempts have been made in the recent past to develop models measuring coping capacity. Indicators attempting to measure coping capacity include human and environmental resources, economic capacity, indigenous knowledge, macro-trends (GDP/capita), and tools and processes of disaster management (Bogardi and Birkmann, 2004).

In our view, these models suffer from several shortcomings: the data may be too highly aggregated, or datasets may not be focused on specific aspects of coping capacity and the data used may be either old or incomplete. Even though this is partly true for the model presented here as well, the novelty of this model is that the disaster management instruments measured by our indicators have a direct impact on coping capacity.

For the purpose of this CCI four main indicators were selected: the level of *institutional preparedness* (e.g. existence of disaster management plans and building codes) was chosen as a starting point to assess the coping capacity of a country. In order to achieve a more refined assessment, the level of *mitigation measures taken* by a country and the *number of IFRC volunteers per inhabitants* in the same country were also taken into account, as well as the DRI developed by UNDP.

For the purpose of the model we assumed that the coping capacity of a country is higher:

- if institutional disaster management measures have been established by the Government (e.g. building codes, hazard mapping, disaster insurances for cities)
- if the country has a high "density" of trained Red Cross/IFRC volunteers in relation to the total population (IFRC national society profiles)
- if the level of investments in mitigation measures per inhabitant is high (World Bank Disaster Management Facility).

Although those indicators do not cover all resources available to reduce the level of risk – as required by the UN-ISDR definition – they represent a significant, relevant and important proportion of a nation's capabilities to cope with a disaster.

Furthermore, we assumed that the exposure of a country to natural hazards and risks is likely to have an influence on its coping capacity, especially in cases of recurrent disasters which use up resources, slow down recovery and gradually erode national coping capacities. However, the fact that no preparedness instruments have been installed in a country does not automatically mean that it has a low coping capacity. It could simply mean that there is a low level of disaster in that country and therefore disaster preparedness instruments are not needed in the first place.

In order to avoid "penalising" these countries in the analysis, we added a *fourth indicator* based on a somewhat adapted and simplified DRI from UNDP. This enables us to adequately distinguish low-hazard countries from disaster-prone countries in relation to institutional coping mechanisms. The lack of such mechanisms is obviously more significant in the latter. We therefore assume that an accurate picture of the coping capacity of a country can be obtained by combining the abovementioned four elements: the degree of preparedness of a country, the amount spent on mitigation projects per inhabitant, the number of IFRC volunteers in a country and the modified UNDP DRI (see in detail Chapter 8).

Description of indicators used

Red Cross volunteers

In the profiles of national Red Cross and Red Crescent societies from 2002 to 2003 (IFRC, 2003), the numbers of all volunteers and permanent staff in all existing International Federation of Red Cross and Red Crescent societies (except Mali and Comoros) are available. This can give an approximate idea of the response capacities available in case a disaster strikes.

The population was divided by the number of volunteers in the relevant country. The list was then ranked and divided into four even sections, for which a value of 1 to 4 was granted, with "1" indicating the category of countries with the highest number of volunteers per capita.

Mitigation projects

Since 1980, the World Bank has approved more than 500 operations related to disaster management, amounting to more than US$40 billion (World Bank, 2005). These include post-disaster reconstruction projects, as well as projects with components aimed at preventing and mitigating disaster impacts. Common areas of focus for prevention and mitigation projects include forest fire prevention measures (early warning measures and education campaigns to discourage farmers from slash and burn agriculture) or flood prevention mechanisms (e.g. shore protection and terracing in rural areas).

The financial volume of all mitigation projects in a country since 1980 (in US$) was added up and then divided by the size of the population, resulting in the amount spent on mitigation projects per capita in the respective country. The data was first sorted from high to low. Then the list was divided into four nearly even sections and ranked accordingly. The value "1" was attributed to countries with a large amount of mitigation funds available per capita, whereas "4" was given to countries where the amount spent on mitigation projects per person was very low. Countries that did not appear in the World Bank list of mitigation projects were allocated "0" because the World Bank gave no money to these countries.

Global Urban Indicator

With an increasing population in urban areas, the impact of natural or man-made disasters on human settlements is becoming greater. These disasters require specific prevention, preparedness and mitigation instruments, which often do not exist in disaster-prone areas for economic and technical reasons. Major instruments are the existence and application of appropriate building codes, which prevent and mitigate the impacts of disasters, and hazard mapping, which informs the policy makers, population and professionals of disasters-prone areas.

Therefore we felt that the level of preparedness constituted a good proxy of a society with a low or high degree of coping capacity. The UN-Habitat 1998 GUI was used as an indicator to determine a country's preparedness (UN-Habitat, 2003). The GUI lists cities in more than 100 countries and determines on a yes/no basis whether they possess building codes based on hazard and vulnerability assessment, hazard mapping, and disaster insurance for public and private buildings.

The number of affirmative answers to the three criteria of disaster prevention was counted.

If 3 x yes ? high coping capacity, value = 1 (best, all of the three above-mentioned disaster prevention instruments exist)
If 2 x yes ? medium coping capacity, value = 2
If 1 x yes ? low coping capacity, value = 3
If 0 x yes ? very low coping capacity, value = 4 (worst case, none of the three above mentioned disaster prevention instruments exist)

If there was more than one dataset per country, the values for each individual city were added and divided by the number of cities included in the dataset.

A total of 87 countries were evaluated; the remaining countries on our list did not appear in the UN-Habitat database and were attributed an "x" for "not available". The reason why some countries did not answer the UN-Habitat questionnaire is not exactly known; some might not have any disaster preparedness instruments in place, others might have simply ignored the questionnaire.

Disaster Risk Index

This composite indicator has been chosen because the level of disaster risk in a given country can have a significant impact on the coping capacity of its population. Recurrent disasters, in particular, may drain a country's resources and thus erode its coping capacity. The introduction of this indicator to some extent also avoids unduly "penalising" countries that have not introduced disaster management measures – and thus would fare badly in the preparedness indicator – due to the simple fact that the disaster risk in those countries is low.

For the purpose of this chapter, we simplified and adapted the methodology of the UNDP DRI (UNDP, 2004; see also Chapter 8). We felt that such an adaptation was necessary for two main reasons: first, the UNDP DRI puts too much emphasis on the risk of loss of life. Based on Centre for Research on the Epidemiology of Disasters (CRED) data, our modified DRI takes into consideration the *severity* (both people killed and people *affected*), the *frequency* and the *diversity* of disasters. We also felt that the specific importance of *governance*, measured through the Corruption Perceptions Index (Transparency International, 2003), should be given more prominence in the concept of vulnerability. The lack of good governance, manifested by a high degree of corruption, contributes to problems of unequal access to resources and basic services. Under these conditions, although national and local authorities may have coping capacities, the most vulnerable could be neglected in the disaster reduction process.

For the purpose of our model, we decided to use a composite of four proxy indicators: *population density*, the *Human Development Index*, the

Human Poverty Index and governance (as measured with the *Corruption Perceptions Index*). For our analysis, countries in a very high disaster risk category were attributed the value "4", high risk countries received the value "3", medium risk the value "2" and countries classified as low disaster risk were given the value "1".

Classification of countries

For all countries, the rating (from 1 to 4) regarding the degree of preparedness was added to the ratings (from 1 to 4) for mitigation, IFRC volunteers and the DRI, giving a scale going from 4 to 16. Then the arithmetical average was calculated by dividing the total by the number of indicators available. Afterwards the countries were ranked in an ordinal scale and the list was finally divided into four almost even categories (~25 per cent each), which correspond to very low, low, medium and high levels of coping capacity.

Selected results and summary assessment of the results

The methodology developed above resulted in the following ranking of coping capacity (selection). Figure 21.1 shows the classification of coping capacity in descending order:

A *high* coping capacity was attributed to countries like Albania, Argentina, Cameroon, Croatia, Ghana, Kazakhstan, Lebanon, Malaysia, the Philippines, the Russian Federation, Senegal, Saint Lucia, Tunisia, Turkmenistan, Uzbekistan and others.

Countries with a *medium* coping capacity include Belarus, Brazil, Costa Rica, Dominica, Iran, Mexico, Morocco, Panama and Ukraine.

Among the countries with *low* coping capacities are: Cape Verde, Chad, the Dominican Republic, Ecuador, Egypt, El Salvador, Lesotho, Namibia, Nepal, Niger, Papua New Guinea, Syria, Uganda, Uruguay and Yemen.

At the bottom of the list are the countries with the *lowest* coping capacity such as: Bolivia, Cambodia, Ethiopia, Fiji, Guatemala, Guinea, Guinea-Bissau, Haiti, Honduras, Iraq, Mauritania, Nicaragua, Sao Tome and Principe and the Solomon Islands.

For obvious reasons, data availability and data quality impose constraints on the full validity of the results of the analysis. Nevertheless, they provide a first step towards a more systematic compilation of data on coping capacity. While it is difficult to identify clear patterns, there seems to be a certain prevalence of transition States (ex-Soviet Union or ex-Socialist countries) and of mid-income, relatively stable developing

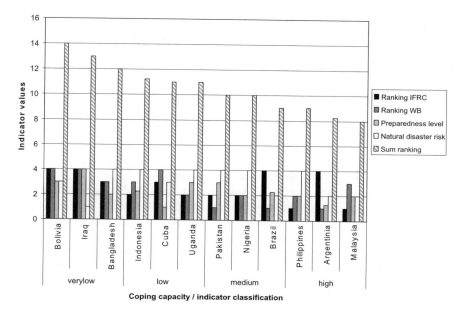

Figure 21.1 Coping capacity classification in descending order for selected countries.
Source: Author (based on the calculation explained above).

countries (e.g. Cameroon, Ghana, Malaysia, Philippines, Senegal and Tunisia) amongst the countries with better coping capacity.

As for the category of countries with lower coping capacity, this includes a significant number of States in Central America and the Caribbean (Guatemala, Haiti, Honduras, Nicaragua) but also many small island States (Fiji, Haiti, Mauritius, Sao Tomé and the Solomon Islands).

These results need to be taken with a grain of salt, though, since certain aspects could not be taken into account, for example the presence of armed conflict. This certainly points towards the need to improve the model. Furthermore, some results would not coincide with intuitive assessments, for example with respect to Cuba (category "low"), which an informed observer would have expected to see in a different category. This may also point towards the need to include further indicators into the model (e.g. civil protection resources of a country and number of international appeals for assistance). The problem, however, is that the data situation on this is patchy and incomplete.[2]

It is understood that a CCI index unavoidably provides a relatively simplified picture of reality. It can be used as a reference document when it comes to priority setting, but it should never be used as the sole instrument for decision-making. Moreover, the objective of this index is not to provide an exhaustive list of countries. Rather, it is to be used as a "mitigating" factor when assessing the global situation of a country. The index should be considered as a first approximation to what is admittedly a very complex subject.

Open questions and limitations

Clearly, the methodology presented here bears a number of inherent shortcomings. These can be summarised as follows:

- Indicators are only proxies. They cannot fully reflect a much more complex reality and cannot replace in-depth case studies.
- The data used are not always up to date (e.g. UN-Habitat data from 1998). The results, therefore, may not reflect possible recent changes in some countries' coping capacities.
- Data only reflect a situation aggregated at the *national* level. Pockets of low coping capacity inside a country (e.g. vulnerable populations living in remote areas) cannot be adequately reflected.
- Data only *quantitatively* measure coping capacity. No assessment is possible, for instance, regarding the *quality* of the Red Cross volunteers in terms of equipment or logistics. The presence of a high number of Red Cross volunteers in a country does not automatically mean that they would be deployable or effective. Neither does the existence of building codes necessarily imply that they are enforced.
- The World Bank mitigation project indicator may distort reality as some countries might not receive or do not request funding for *political* reasons (e.g. Cuba).
- The coping capacity can be different for *different disaster types*. Man-made disasters (conflict) in a country aggravate the situation but are not (yet) reflected in the model.
- Results very much depend on methodology and data availability. *Missing values* may distort the results.

These shortcomings notwithstanding, we nevertheless believe that the methodology and results shed some light into what otherwise still seems to be a dark tunnel. They can pave the way for more and more comprehensive efforts to address the complexity of the issue.

Conclusion and outlook

The CCI provides a rough overview of countries' levels of coping capacity in case of disasters. It does not replace an in-depth assessment of the situation on the ground. Hence it is only a supportive tool for assessing the global situation of a country and should be used – within the limits mentioned – as a complementary strategic planning tool only once the identified weaknesses have been remedied.

Further research needs have been identified in the course of preparing this analysis. Datasets should be as complete as possible and gaps must be filled. Furthermore, the data needs to be updated on a regular basis, especially the UN-Habitat Global Urban Indicators, which were due to be released in 2004, but are still not available.

There is increasing recognition of the need for continuous updating of data and related analytical tools, both within countries and regionally, in respect of trans-border or regional-scale risks. This requires improved availability and free exchange of data, coupled with retrospective studies of lessons learned and projections of future trends and scenarios. Common approaches to the maintenance of national datasets related to hazards and disaster consequences are widely recognised as inadequate (partial, outdated, sporadic and fragmented information). Therefore more standardised data collection and analysis methods are needed to enable countries to assess risks more systematically and to better evaluate risk management options.

A way forward would be to also include systematic assessment of subnational, family and individual coping capacity as well as indigenous knowledge. This, however, would be a very time-consuming, costly and lengthy endeavour. It would need to be sufficiently resourced and well prepared in advance in order to ensure comparability of different datasets. Although it will be very difficult to translate indigenous knowledge into measurable indicators, questionnaires could serve as a basis to identify coping capacity features.

Notes

1. See also introductory chapter from J. Birkmann in this volume
2. For example, the World Fire Statistics, published by the International Association of Fire and Rescue Services only covers 41 countries, most of which are developed countries.

This chapter reflects work in progress. It reflects the views of the authors and does not represent an official position of the European Commission.

The authors would like to thank Pascal Peduzzi and Hy Dao of the United Nations Environment Programme's Geneva office for contributing to the country ranking method.

REFERENCES

Bogardi, J. and J. Birkmann (2004) "Vulnerability Assessment: The First Step Towards Sustainable Risk Reduction", in D. Malzahn and T. Plapp, eds, *Disaster and Society: From Hazard Assessment to Risk Reduction*, Berlin: Logos Verlag Berlin, pp. 75–82.

Bohle, H.-G. (2001) "Vulnerability and Criticality: Perspectives from Social Geography", *IHDP Update* 2/2001, Newsletter of the International Human Dimensions Programme on Global Environmental Change, pp. 1–7.

Brushlinsky, N.N., J.R. Hall, S.V. Sokolov, and P. Wagner (2005) *World Fire Statistics. Report no. 10*, 2nd edn, available at http://www.ctif.org/index.php?page_id=2883/.

Center for Research on the Epidemiology of Disasters (CRED), *EM-DAT: The International Disaster Database*, available at http://www.em-dat.net/.

ECHO (Humanitarian Aid Department of the European Commission) (2005) *Global Needs Assessment and Forgotten Crisis 2005*, available at http://europa.eu.int/comm/echo/information/strategy/index_en.htm.

Intergovernmental Panel for Climate Change (IPCC) (2001) *Climate Change in 2001: Synthesis Report, Working Group II, Technical Summary*, available at http://www.ipcc.ch/pub/un/syreng/wg2ts.pdf.

International Federation of Red Cross and Red Crescent Societies (IFRC) (2003) *Partnerships in Profile 2002–2003: Profiles of National Red Cross and Red Crescent Societies 2002–2003*, available at http://www.ifrc.org/publicat/profile/index.asp/.

Schneiderbauer, S. and D. Ehrlich (2004) *Risk, Hazard and People's Vulnerability to Natural Hazards: A Review of Definitions, Concepts and Data*, Brussels: European Commission-Joint Research Centre (EC-JRC).

Transparency International (2003) *Corruption Perceptions Index*, available at http://www.transparency.org/policy_and_research/surveys_indices/cpi/2003/.

UNDP (United Nations Development Programme) (2004) *Reducing Disaster Risk: A Challenge for Development*, New York: UNDP, available at http://www.undp.org/bcpr/disred/documents/publications/rdr/english/rdr_english.pdf.

UN-Habitat (United Nations Human Settlements Programme) (2003) *Global Urban Indicators*, UN-Habitat Global Urban Observatory (GUO), available at http://www.unchs.org/programmes/guo/qualitative.asp/.

UN-ISDR (United Nations-International Strategy for Disaster Reduction) (2004) *Living with Risk: A Global Review of Disaster Reduction Initiatives*, Geneva: UN Publications.

World Bank (2005) *Hazard Risk Management: Mitigation Projects*, available at http://www.worldbank.org/hazards/projects/mitigation.htm.

22

A methodology for learning lessons: Experiences at the European level

Elisabeth Krausmann and Fesil Mushtaq

Abstract

In Europe, efforts to protect both the citizen and the environment face a continuing challenge from a wide range of risks that arise from both natural and technological hazards. The lessons learned from the systematic analysis of the evolution of past events and the circumstances that facilitated their occurrence are of paramount importance for future risk reduction and priority setting in terms of vulnerability management. This chapter proposes a methodology for learning lessons that addresses the following steps: accident and disaster investigation and reporting, data collection and analysis, generation and implementation of lessons learned. The methodology is demonstrated by introducing the Major Accident Reporting System (MARS) and Natural and Environmental Disaster Information Exchange System (NEDIES) knowledge bases, which are two of the European Community's systems for collecting existing and producing new lessons learned, and exchanging information on the management of technological accidents and natural disasters.

Background and rationale

Technological accidents and natural disasters can have a major impact on society, mostly due to the massive loss of lives and the disruption of community life they can cause, but also because of their potential adverse long-term effects on the environment and the economy. A thorough analysis of the causes, circumstances, evolution, consequences of and responses to these past accidents and disasters yields valuable lessons that can contribute towards future accident prevention and/or loss minimisation.

One of the reasons why accidents and disasters keep occurring is that the lessons from past events have either not been learned or publicised in a systematic way, or have not been translated into existing risk-management practices. This includes, for instance, the failure to integrate disaster and safety management into normal development planning in hazard-prone areas, which results in a misplaced emphasis on emergency relief instead of on prevention and vulnerability reduction, or the tendency to have blind confidence in structural risk-reduction measures while neglecting potentially more effective non-structural measures (Lebel et al., this volume, Chapter 19). Moreover, the adequacy of risk-management measures that were created by putting lessons learned into practice is not always properly monitored, even though that is fundamental to guaranteeing their maximum effectiveness. Therefore it is of paramount importance that all relevant information relating to accidents and disasters be collected and analysed meticulously, and that the lessons thus learned be mainstreamed into all stages of disaster or safety-management practices. Verification of the effectiveness of risk-management measures, their comparison with other measures implemented in places with different safety cultures and hence risk-management approaches and the widespread application and publicising of effective lessons learned is another vital step towards combating the occurrence of undesirable events and mitigating their consequences.

The Joint Research Centre (JRC) of the European Commission supports the learning goal by ensuring that the information available throughout Europe and beyond on the management of natural disasters and technological accidents is systematically exploited and the lessons learned disseminated. More specifically, it maintains the Major Accident Reporting System (JRC-MARS), which manages information on technological accidents in accordance with the provisions of the European Seveso directives (European Union, 1982 and 1997), and the Natural and Environmental Disaster Information Exchange System (JRC-NEDIES), which is a repository of lessons learned for the prevention, preparedness and response to natural disasters and non-Seveso technological accidents.

This chapter outlines a methodology for learning lessons and introduces the European Community's MARS and NEDIES knowledge bases to demonstrate the methodology in practice.

The methodology for learning lessons

Definitions and terminology

In the absence of a harmonised terminology (see also Birkmann, Chapter 1 and Thywissen, Chapter 24), we use the following definitions for the purpose of this article:

Disaster: a natural or man-made event resulting in widespread human, environmental, economic or material losses. The adverse consequences of a disaster may exceed the ability of the affected community or society to cope using its own resources.

Accident: an unintended and unforeseen event or series of events and circumstances that results in one or more specified undesirable consequences.

Hazard: a source of danger (Merriam-Webster Online Dictionary). A hazard does not necessarily lead to harm but represents only a potential for harm.

Risk: the combination of the frequency, or probability, of occurrence and the consequence of a specified hazardous event (ISO/IEC, 1999). Risk therefore includes the likelihood of a hazard actually causing injury, damage or harm.

Vulnerability: the degree to which a system is susceptible to, and unable to cope with, injury, damage or harm (EEA).

Lesson learned: knowledge gained from investigation, study or other activities in regard to the technical, behavioural, cultural, management or other factors, which led, could have led, or contributed to the occurrence of an accident (Rosenthal et al., 2004). Although originating from the chemical-process industry this definition is equally valid for natural disasters.

Description of the methodology

The methodology outlined below discusses the major steps in the lessons-learning procedure (Figure 22.1). Due to its generic nature it is applicable on any geographical scale (international, national, regional or local).

Step 1: Investigation of accidents and disasters

The investigation into the circumstances of a natural disaster or a technological accident should focus on identifying the underlying causes and assessing the consequences, thereby evaluating the effectiveness of existing systems for prevention, preparedness and mitigation. Consequently, the investigation should result in recommendations on how to prevent the re-

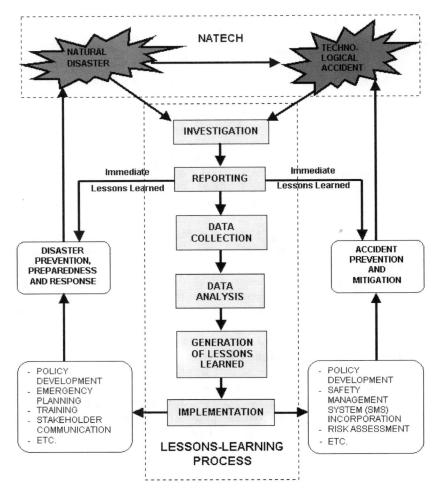

Figure 22.1 Schematic description of the lessons-learning methodology.
Source: Authors.

currence of the same or similar accidents or disasters, and to improve existing systems so as to reduce the consequences of future events.

Clearly, the outcome of the investigation in terms of lessons learned is determined by the protocol used to investigate and report the event, and in particular the scope of the investigation. Investigative objectives such as defence against litigation and regulatory actions or the identification of regulatory violations are usually not conducive to the generation of effective lessons learned, as the investigative findings are rarely openly disclosed or are often limited to establishing the party or agent responsible for an accident or a disaster instead of determining its root cause.

There are numerous approaches to accident or disaster investigation that all share the same underlying philosophy:

- *define the scope* to clarify ideas and set the boundaries within which the investigation is to be conducted
- *gather information* to understand the circumstances leading up to an undesirable event
- *analyse the information* to determine the root causes
- *make recommendations* to prevent/mitigate future accidents/disasters, and set priorities.

As a final step an investigation report addressing the steps listed above and containing a set of immediate lessons learned and accompanying recommendations specific to the investigated event should be produced.

Step 2: Reporting of accidents and disasters

Immediate lessons can be learned from individual accidents or disasters. These should normally be included in a distinct section of the final report of an investigation, and usually involve specific actions to be taken to prevent a recurrence. Complementary to this, more generic lessons can be learned from analysing the reports of similar accidents to find common elements or areas of particular concern. Essential to this is the quality, format and extent of such reports and their availability and accessibility.

Data collection, as discussed in the next section, is a very difficult and time-consuming activity. It is significantly assisted by the mandatory reporting of information that satisfies a pre-defined set of criteria, usually related to the extent of damage or harm, and that provides a minimum of useful information in an understandable and comparable format. Moreover, lesson-learning type studies benefit from systems that require the outcome of an investigation into an accident or a disaster to be reported to a centralised location where the data can be collected, analysed and shared. Rules are required for these systems to function effectively; this can be facilitated by targeted legislation.

Step 3: Data collection

The collection of relevant information on the evolution of a technological accident or natural disaster, as well as on the disaster- or safety-management measures implemented before and during the event, is another prerequisite for generating lessons learned. The reporting of information by designated bodies in fulfilment of regulatory requirements is a proactive and effective means by which to gather validated high-quality data. Data collection from other sources such as open literature or media sources can be extremely cumbersome and a significant effort has to be expended in corroborating the various data sources to guarantee the quality of the collected information.

Data collection through mandatory reporting is one approach to gathering information. It is based on the definition of data requirements considered indispensable for an in-depth analysis of an event and the learning of lessons by a group of experts. This approach typically yields a considerable amount of very detailed information (both quantitative and qualitative) in a specific field of application (e.g. the chemical industry).

In the absence of reporting obligations, information on past accidents or disasters can be collected by means of predefined templates that should be developed in close cooperation with authorities and experts in order to address their needs and to ensure the practical application of the lessons learned that have been generated. This approach results in the collection of data of a rather generic nature, which consequently enables the detection and appraisal of common elements, resulting in the development of lessons learned for specific hazards (e.g. landslides).

Irrespective of the approach chosen for the data collection, the compilation of accident and disaster information should be implemented by means of an interactive database that can be interrogated to support the data-analysis process.

Step 4: Data analysis

The purpose of data collection is to compile the information from the original investigation reports into a suitable format in order to perform a supplementary in-depth analysis. Additional findings pertaining to the identification of trends and areas of interest relating to the causes and consequences can be extracted from the examination of the compiled information on similar accidents or disasters.

The structuring of the collected data (e.g. in a database) is dependent on the type, quantity and range of information. There is a need to organise large quantities of data into distinct and well-defined sections to facilitate effective storage and retrieval. Fundamentally, the more extensive the information, the more sophisticated the structure required, and the greater the significance of the interrogation. Once the database is sufficiently populated with high-quality data, the user should have the possibility to perform statistical or trend analyses, preferably by means of a built-in query tool. In this way, patterns of accident or disaster causation can be detected, and particularly vulnerable risk receptors can be identified. If a database cannot be queried, then instead of being a valuable source it becomes an information sink. As a final step, the outcome of the data analysis needs to be evaluated by using either expert judgment or systematic tools.

It is essential for an effective data analysis that the collected information is of a high standard. Therefore, the reporting of the occurrence and the data collection should be subject to management procedures whereby the accuracy, consistency and completeness of the data are controlled.

Step 5: Generation of lessons learned

Lessons can be learned in all phases of disaster-risk-management: prevention, preparedness, response and recovery (both in the short and long term). The investigation of a single accident or disaster ideally results in the development of recommendations that contain immediate lessons learned specific to the event. The reporting and collection of supplementary information expands the data pool available for learning lessons, and thus permits the analysis of a number of similar occurrences, which results in the generation of lessons learned that are more widely applicable than the immediate lessons learned.

The *mono-hazard* type of analysis used to extract lessons after an undesirable event aims at providing input to risk-management practices with respect to one type of hazard (e.g. floods). For instance, this analysis makes it possible to identify commonly occurring causes of particularly serious accidents involving specific substances or industries, which may not be recognised within a single occurrence. In addition, this type of analysis lends itself to monitoring the progress made in disaster- or safety-management practices since the last event, and also to identifying technical and organisational measures that still need to be implemented or improved. The *cross-hazard* analysis compares and contrasts information across hazards and investigates the possibility of cross-fertilisation of lessons learned from one risk-management field to another. An important application of the latter is multi-hazard events, such as technological accidents triggered by a preceding natural disaster (so-called Natech disasters), or events with domino effects (e.g. a storm triggering floods and landslides or a technological accident affecting other installations in the vicinity).

Step 6: Implementation of lessons learned

The implementation of lessons learned into everyday disaster or safety management begins with the dissemination of the findings from the analyses of accidents or disasters. This can, for instance, be achieved by setting up platforms for the exchange of information between all the actors involved or by organising training courses or workshops in which expert knowledge is shared. The lessons learned can be targeted to responsible authorities, urban planners, the general public or any other stakeholder, but their implementation should be so directed as to have the greatest effectiveness on systems. Lessons learned should not only exist in the memory of people, because people can forget; rather the lessons must be incorporated into the memory of systems.

Mainstreaming lessons learned into disaster- or safety-management practices is frequently accompanied by policy development, either by encouraging the formulation of new legislation or by amending existing legislation (as with the amendment of the European Seveso II directive after the major accidents in Toulouse, Enschede and Baia Mare) (European Union, 2003). The results of lessons-learned type studies can filter directly into decision-making processes and land-use-planning policies in the vicinity of risk sites to support vulnerability management with respect to human targets and the natural or man-made environment. Another of the numerous areas that can benefit greatly from the integration of lessons learned is risk assessment, and in particular the hazard-identification and scenario-development stages.

The effectiveness of implemented lessons learned needs to be monitored continuously to verify the adequacy of the updated risk-management measures. This can, for instance, be achieved by comparing the success of risk-reduction practices that have been implemented in places with different safety cultures, and hence with different approaches to accident and disaster management.

Application of the methodology

The European Community's MARS and NEDIES knowledge bases serve the objective of reducing risk by storing and exploiting relevant data in a systematic way. They also contribute by feeding information back into prevention and/or mitigation practices in the form of lessons learned from past accidents or disasters that are then incorporated into guidelines and recommendations for facing future events. The methodology for learning lessons outlined in the previous section is the very foundation of the work carried out in MARS and NEDIES and will be described in more detail in the following examples.

Technological accidents falling under the provisions of the Seveso II Directive

The Major Accident Reporting System (MARS) was established to handle information on "major accidents" submitted by the Member States of the European Union to the European Commission in accordance with the provisions of the Seveso Directive, and later the Seveso II Directive, specifically Articles 14, 15, 19 and 20, and Annex VI (which provides the criteria for reporting in terms of the severity of consequences) (European Union, 1982 and 1997). The articles are principally concerned with responsibilities for the collection and submission of information relating to the circumstances of the accident, and with ensuring the distribution and analysis of the information in order to prevent major accidents from recurring. In many cases, the persons providing input to the MARS database are plant inspectors and accident investigators, working for the Member States' Competent Authorities, without English as their native language (English being the working language of MARS). Therefore, the Major Accident Hazards Bureau of the European Commission's Joint Research Centre that hosts and manages MARS carries out a quality-control check on all submitted reports before they are incorporated into the shared database.

The structure of MARS, the current format of which is shown in Table 22.1, was first established through a technical working group (TWG) of experts in the 1990s. Initially, the system consisted of paper reports kept in filing cabinets, but the database has been continuously improved since, and the current version is a stand-alone piece of software that the users have installed on their personal computers. Table 22.1 shows the information categories used for data collection, structuring and retrieval in MARS. The Short Report contains information submitted to MARS im-

Table 22.1 Description of the information categories used for data collection, structuring and retrieval in MARS

Report Profile	Short Report	Full Report A. Occurrence	Full Report B. Consequences	Full Report C. Response
Accident code	Accident type	Type of accident	Area concerned	Emergency measures
Date, time	Substances directly involved	Dangerous substances	Affected people	Seveso II duties
Reporting authority	Immediate sources	Source of accident	Ecological harm	Official action taken
Establishment	Suspected causes	Meteorological conditions	Natural heritage loss	Lessons learned
	Immediate effects	Causes	Material loss	Discussion about response
	Emergency measures taken	Discussion about occurrence	Disruption of community life	
	Immediate lessons learned		Discussion of consequences	

Source: Authors.

mediately after a major accident. The Full Report, which is submitted often after many years of accident investigation, contains more detailed information.

Recently, a new TWG of MARS users was set up to oversee the database's future development, but with an extended mandate to look further into related lessons-learning activities such as accident investigation and the implementation of lessons learned. Future MARS enhancements will be based on improving the user-friendliness of the system in terms of data entry, data analysis and extraction of meaningful information. At the same time it is necessary to maintain a high standard in the quality of the data submitted and to reduce delays in accident reporting; this is being addressed through continuous training and the active involvement of the users, and by making the database available online. The database in its current format is split into three sections:

- 1) the Report Profile identifying the date, time, location, etc. of an event
- 2) the Short Report containing free-text fields allowing the unrestrained input of information on the causes and consequences of an accident
- Full Report A, B and C, which expand the information in the Short Report using pre-defined selection lists that direct the input and allow for a more statistical-type of analysis to be carried out.

The non-confidential Short Reports are available for searching online (JRC, 2005b).

While the publishing of specific MARS data is restricted due to their confidentiality, some examples to illustrate the types of analyses possible in MARS are presented. Figure 22.2, for instance, shows a quantitative analysis of the human or organisational causes of the accidents contained in the Full MARS Reports. The numbers in the pie chart refer to the number of accidents as a function of accident cause.

This generic analysis indicates that in comparison to other management failures, inadequate procedures and equipment/system design are the largest contributors to the causes of accidents, and therefore greater attention needs to be paid to these aspects in the risk assessment. Analyses can also be directly based on the information contained in the lessons-learned sections of MARS. For instance, accidents that occurred during loading and unloading operations were examined with the following specific results in terms of lessons learned.

Improved equipment design

- Better isolation systems: installation of isolation keys on the rails; installation of quick-action isolation valves on loading/unloading parts of existing installations for toxic gases; installation of spark blocks; remote control devices to allow shut-off from a safe distance; nitrogen purging.

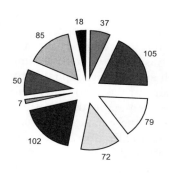

organisation: management organisation inadequate / attitude problem

organisation: organised procedures (none, inadequate, inappropriate, unclear)

organisation: training/instruction, supervision, staffing (none, inadequate, inappropriate)

organisation: process analysis (inadequate, incorrect)

organisation: design of plant / equipment/ system - user unfriendliness (inadequate, inappropriate)

organisation: manufacture / construction - installation (inadequate, inappropriate)

organisation: testing/inspecting/recording - maintenance/repair - isolation of equipment/system (none, inadequate, inappropriate)

person: operator error, health, wilful disobedience / failure to carry out duties, malicious intervention

- not identified - / - other -

Figure 22.2 Example of a quantitative analysis of the human and organisational causes of accidents contained in the MARS Full Reports.
Source: MARS Report.

- Detection systems: redundant safety systems including leakage probes and a permanent safeguarding system on a 24-hour basis, possibly extending over the whole site.
- General improvement of process control: redundant sensors, interlock systems, visual alarms etc; however, avoiding over-automation, which will result in over-confidence of the operators.
- Design and use of appropriate equipment: improve design of piping and piping components (lengths of hoses to allow redundancy, etc.).
- Improve plant layout: relocation of equipment.
- Improve mitigation systems: automatic fire-fighting system, containment, overpressure protection, appropriate clothing for operators.

Improved management

- Review and modification of operating procedures: mandatory permanent operator attendance at certain loading/unloading operations; supervisor authorisation required for some operations; restricted access to certain areas; verification of operations; handling conditions.
- Maintenance and inspection: better procedures required; increased frequency of rigorous testing; review of replacement strategies and frequencies; checks for integrity, correct functioning and fitting.
- Improved emergency planning: (internal and external) and public warning systems.
- Better chemical analysis: avoid concentration of impurities; use of alternative substances; discontinue manufacture of certain substances.

Natural disasters

The NEDIES knowledge base is currently the only repository of lessons learned of its kind. In contrast to MARS, NEDIES is not supported by any European legislation and therefore has to rely on voluntary reporting by authorities and practitioners or time-consuming data collection from open sources. Predefined templates, which are regularly updated to take account of feedback from the NEDIES users, are used to collect information on the circumstances of disasters, as well as on the disaster-management practices in place before and during the event, in a structured way (Table 22.2). This is similar to the data collection in MARS, with the difference being that the information gathered in NEDIES relies on the extensive use of free text-fields and is consequently much less specific. Therefore, the possibilities for searching the database are more generic in nature. Although the NEDIES knowledge base cannot be interrogated statistically for every type of accident or disaster it contains, its strength lies in the fact that it addresses many different types of risk. Consequently, it offers the opportunity to perform generic mono- or cross-hazard lessons-learned type studies that are widely applicable because of the broad horizontal information base they can draw from.

Occasionally, workshops are organised within the framework of the NEDIES project that bring together representatives of authorities and civil protection services and other experts from EU Member States and candidate countries for an exchange of lessons learned on the management of specific disasters that occurred in their respective countries. These workshops are a source of invaluable, validated information that is useful for mono-hazard analyses and they constitute a platform for the sharing of expert knowledge and opinion.

Subsequent to the expert workshops the NEDIES team summarises the discussions and produces reports on specific lessons learned, plus recommendations or guidelines, which are available for downloading from the NEDIES website (JRC-NEDIES). As an example, a selection of generic lessons learned for the management of landslide disasters is given in the following section (Hervás, 2003).

Lessons learned concerning landslide prevention measures

It is important to:
- identify precipitation thresholds that make it possible to define three alert levels (attention, warning and alarm)
- produce landslide hazard maps (and preferably risk maps) in residential and industrial areas, scaled in accordance with local land-use maps
- issue specific legislation that requires the elaboration of natural hazard/

Table 22.2 The NEDIES lessons-learned report format

GENERAL INFORMATION	
Starting date	
Duration	
Location(s) involved	
Administrative unit (NUTS3 Eurostat classification, if available) or region (province)	
Country	
Consequences to persons – Number of fatalities – Number of injured – Number of homeless	
Economic losses (in Euros) – Material losses – Response action costs	
Prediction made – Yes – No – Short comment (if any)	
INFORMATION ON DISASTER MANAGEMENT AND LESSONS LEARNED	
Short description of event *Prevention phase* – Prevention measures (specify if risk assessment, land use planning, building codes existed and were adopted) – Lessons learned *Preparedness phase* – Preparedness measures (specify if an emergency plan existed) – Lessons learned *Response phase* – Response actions (specify if an emergency plan was adopted) – Lessons learned *Information dissemination and related lessons learned* – Prior to event – During event – Following event	

Source: Authors.

risk maps, and couple land-use decisions to the hazard or risk level present

- provide sufficient financial resources for geological surveys in landslide-prone areas and for keeping them updated
- set up automatic warning and alarm systems connected to the head-quarters of emergency intervention bodies in the case of major active landslides.

Lessons learned concerning landslide preparedness measures

It is important to:

- develop a specific emergency plan that includes evacuation measures and the enlargement of the risk area when substantial slope movement has been ascertained, even before damage occurs
- establish systems for real-time weather information at a local scale in areas where landslides are most often related to intense rainfall
- identify diversion routes for key roads subject to landslide events, for use by emergency services.

Lessons learned concerning landslide response measures

It is important to:

- identify and organise the civil-protection emergency control centres, the persons in charge of the centres and their various functions
- ensure that rescue and protection services are staffed with sufficiently trained and adequately equipped personnel
- for landslides linked to rainstorms or floods, utilise houses or other permanent structures as reception centres for evacuees rather than tents or caravans.

Lessons learned concerning dissemination of information to the public

It is important to:

- provide regular and open information to residents of landslide hazard areas to raise awareness of the danger and to attain confidence in and promote collaboration with the rescue services
- issue general warnings through TV and radio, or directly warn the public by telephone using an automatic voice messaging system
- establish a notification centre and appoint an information manager to inform both the public and the media.

Limitations of the methodology

The methodology for learning lessons outlined in the previous sections is heavily dependent on the data reported or collected. Therefore anything that affects the quality, extent and accuracy of that information influences the final outcome in terms of lessons learned. The data can be affected in every step of the methodology from the investigation of an event to the data collection and analysis. This means that minimum standards are required for each step. The availability and accessibility of information is a further bottleneck in the methodology. External investigations, for example by judicial bodies, can be affected by a lack of free-flowing information as people may assume that the investigators' mandate is to assign blame, and so be reluctant to cooperate. In-house investigations may be more thorough, uncovering deeper underlying causes; however this information may then be restricted.

Additionally, there are limitations inherent in the development of databases. They are usually set up for a specific purpose or to answer particular questions. However, the field in which they are relevant is constantly evolving, and therefore information that is currently significant may at some point become irrelevant; conversely, information that is not being collected at present may actually be found to be indispensable in the future. Consequently, a great deal of effort needs to be invested in the design, application and evolution of databases if they are to be used for learning lessons.

Outlook and conclusions

The sharing of information and the application of lessons learned across different hazards is gaining importance in our societies, which are becoming ever more interdependent and vulnerable. This applies in particular to the management of multi-hazard events, which can benefit considerably from an integrated approach to risk reduction. This is illustrated by the example of Natech disasters that occurred during the floods in the Czech Republic in 2002. The flooding may not have been preventable but the ensuing technological accidents could have been averted, had the lessons learned from flood disaster management in terms of prevention, preparedness and response been directly integrated into the prevention and mitigation of major accidents. This could have resulted in different land-use-planning decisions or updated on-site/off-site emergency plans, which could have had a positive impact on the vulnerability of the affected installations. Another example for the cross-fertilisation between different fields is the exploitation of similarities in well-established emergency-response practices for some natural and technological hazards.

The proposed approach for learning lessons is straightforward but may be difficult to implement due to the subjectivity of the actors involved. For instance, accidents and disasters are investigated in different ways and reported in different styles, before they are finally collected within a dedicated database and analysed. Prescription and standardisation of practices would be detrimental to their functioning since they would lose their capacity to account for differences in value systems and safety cultures. However, the harmonisation of investigation and reporting approaches in terms of the outcome they produce would increase the comparability of results and is an example of how to improve the effective implementation of the methodology.

The lessons-learning methodology can be further developed by establishing a monitoring mechanism that feeds back on the effectiveness of the lessons learned integrated into risk-management practices. Unfortunately, this is very difficult to put into practice, since their success, or lack of success, can only be revealed by the occurrence of another accident or disaster under the same circumstances. The development of appropriate indicators is a requirement for measuring the performance of the lessons-learning mechanism. Consequently, indicators should be made a part of the core data upon which the accident or disaster analysis is based, and therefore should be subject to the same rigorous quality-control processes.

REFERENCES

European Environmental Agency (EEA) (2005) *Multilingual Environmental Glossary*, available at http://glossary.eea.eu.int/EEAGlossary.

European Union (1982) "Council Directive 82/501/EEC of 24 June 1982 on the Major-accident Hazards of Certain Industrial Activities", *Official Journal of the European Communities*, No. L230.

European Union (1997) "Council Directive 96/82/EC of 9 December 1996 on the Control of Major-accident Hazards Involving Dangerous Substances", *Official Journal of the European Communities*, No. L10.

European Union (2003) "Directive 2003/105/EC of the European Parliament and of the Council of 16 December 2003 amending Council Directive 96/82/EC", *Official Journal of the European Communities*, No. L345/97.

Hervás, J., ed. (2003) *Lessons Learned from Landslide Disasters in Europe*, EUR Report 20558 EN, European Communities.

International Organization for Standardization (ISO) and International Electrotechnical Commission (IEC) (1999) *Safety Aspects: Guidelines for their Inclusion in Standards*, ISO/IEC Guide 51, 2nd edn.

JRC (Joint Research Centre of the European Commission) (2005) *Major Accident Reporting System* (MARS), available at http://mahbsrv.jrc.it/mars/Default.html.

JRC (Joint Research Centre of the European Commission) (2005b) *MARS Database Search*, available at http://mahbsrv4.jrc.it/mars/servlet/GenQuery?servletaction=ShortReports.

JRC (Joint Research Centre of the European Commission) (2005) *Natural and Environmental Disaster Information Exchange System* (NEDIES), available at http://nedies.jrc.it.

Merriam-Webster Online Dictionary (2005), available at http://www.m-w.com/.

Rosenthal, I., P. Kleindorfer, H. Kunreuther, E. Michel-Kerjan and P. Schmeidler (2004) "Issues Affecting the Effective Use of Lessons Learned to Prevent and Mitigate Chemical-process Accidents", in *Proceedings of the OECD Workshop on Lessons Learned from Chemical Accidents and Incidents*, September 21–23, 2004, Karlskoga, Sweden.

23

Conclusion and recommendations

Jörn Birkmann

Introduction

In this book, more than 40 authors have presented approaches for measuring and assessing vulnerability and risk. The chapters have described different concepts, methodologies and procedures for measuring vulnerability and coping capacity, ranging from damage functions, retrospective loss and mortality assessment, macro trend analysis, identification of archetypes and patterns of vulnerability to self-assessment and participatory tools. This diversity shows it is not possible to draw a universal conclusion that fits all concepts and methodologies. Rather, the following reflections will focus on selected key aspects for future research and discuss some findings of the approaches presented in the book. In addition, recommendations for future research will be made.

Looking forward: key aspects for future research

A major conclusion from the review of the various approaches presented in this book is that more comparative assessments of existing methodologies and approaches in similar locations and situations are required; without them, it will be difficult to assess and judge the feasibility and range of the different approaches, including their potential limitations, overlaps and possible combinations. Thus there is an urgent need to strengthen collaborative research into vulnerability assessment, using different tools in similar case studies in order to explore and expand the collective expertise of different methodologies. A lesson one can learn already from the review of approaches presented in this book on "measuring vulnerability" is the fact that no single vulnerability assessment methodology is suitable and able to capture all the various features of vulnerability related to different social groups, economic sectors and environmental services. Therefore, different assessment methodologies have to be combined or used simultaneously to provide more comprehensive information for potentially affected households, urban and spatial planners, disaster managers, and community and political leaders in communities at risk. A challenge for future research is to respond to the need for integrated vulnerability and risk assessment using quantitative, qualitative, traditional and participatory methods at different scales.

Thus the major challenges for future research with regard to measuring vulnerability lie in combining different methodologies in order to provide a more comprehensive identification and understanding of vulnerability. In this context, the aspects to be addressed include the question of qualitative versus quantitative assessment methods, hazard-specific versus hazard-independent measurement, and how to merge these very different assessment types of vulnerability. Additionally, the complex matter of linking global and local approaches, the question of to what extent vulnerability assessment should be based on loss estimation versus context interpretation, as well as the crucial issue of complexity versus simplification has to be considered in more depth in future research.

Quantitative or qualitative?

A decision about whether to use qualitative or quantitative assessment tools depends both on the level of the approach (global, national, subnational or local) and on its focus (macro-economic, nation-state or individual actors and groups) and functions. The three global index projects presented by Pelling, Peduzzi, Dilley and Cardona (Chapters 7 to 10), as well as the approaches to measuring vulnerability in Tanzania (Chapter 12) and the CATSIM model (Chapter 20) show the capabilities and the coverage of most quantitative approaches for assessing and comparing vulnerabilities at global, national and sub-national scale. Additionally, the approach presented by Billing and Madengruber concentrates on the quantitative comparison of coping capacities between different countries (Chapter 21).

By contrast, the example presented by Wisner of "self-assessment" at local level (Chapter 17) illustrates the opportunities of participatory and more qualitative methodologies to assess vulnerability and coping capacity. Overall, the following remarks can be made:

Conclusions and recommendations

- While quantitative approaches based, for example, on global data have a high potential for measuring vulnerability with regard to experienced losses, such as mortality and economic loss (e.g. DRI, Hotspots, Human Security Index, see Chapters 7, 8, 9, 13 and 14), the abilities of these tools to measure context-dependent features and spatially specific characteristics of vulnerability such as coping capacity, institutional vulnerabilities and intangible assets are limited.
- Coping capacities, coping processes and adaptation strategies, as well as institutional aspects, are not sufficiently captured in the current global datasets and often reach beyond mere quantitative measures.
- Qualitative methodologies to capture vulnerability are particularly applicable and useful at the local and community level.
- Qualitative approaches are often limited in that they tend to lack continuous assessment; they are often used on a one-off basis. On the other hand, such approaches have a high potential, for example, to explore the role and function of social networks, which drive and determine important features of the vulnerability of different social groups and their coping and adaptation strategies.
- Research is needed for balancing qualitative and quantitative methods for measuring vulnerability and coping capacity. This encompasses the challenge of developing and testing more integrated assessment approaches, which capture vulnerability more comprehensively, combining quantitative, qualitative and participatory methods. Mayoux and Chambers (2005) underline, for example, that the potential of participatory approaches to generate accurate quantitative data is still underestimated. Furthermore, as Wisner (Chapter 17) recommends, it will be important for quantitative and qualitative as well as reflective and action-oriented methodologies to be transformed into continuous monitoring and correction measures.
- Future research should explore in more depth how to improve the application and implementation of vulnerability and risk assessment (e.g. indicators) into more traditional planning and decision-making processes, such as emergency and disaster mitigation plans, land use planning and community development strategies. This also means shifting the focus from assessing impacts to improving practice.
- However, we need also to acknowledge the limitations of measurability. Especially with regard to coping capacity, research into appropriate indicators and data will come up against profound limitations, such as the difficulties of measuring the robustness of social networks, institutions, trust and other intangible factors (see Chapters 11 and 18). This means that only some aspects of coping capacity can be quantified;

others, such as certain aspects of political circumstances, social cohesion and organisational structures, are still not adequate enough for a general quantification. Likewise, some methodologies, such as the assessment and measurement of institutional vulnerability, are in their initial phase and need to be examined further.

The hazard-specific versus hazard-independent focus

While some approaches presented in the book measure vulnerability with regard to a single or specific hazard, such as the one developed in Tanzania by Kiunsi and Meshack (Chapter 12) and Villagrán de León's sector approach (Chapter 16), others encompass a multi-hazard approach to assessing vulnerability and risk. Greiving, for example, shows an approach to multi-risk assessment encompassing vulnerability indicators that are relevant for different hazard types (Chapter 11). He argues that, especially for policy interventions and spatial-planning strategies, the identification of regions at high risk should be based on methodologies that consider vulnerability to multiple hazards in an aggregated manner. Kok et al. even argue that more comprehensive vulnerability assessment should focus on multiple stresses. Within the framework of the GEO-3 approach, they identified three critical areas (human health, food security and economic losses) as being closely related to vulnerability (Chapter 6). By contrast, Villagrán de León emphasises that vulnerabilities depend on the specific type of hazard in question. This means he clearly defines vulnerability as a characteristic of the specific hazard. Possibilities for combining hazard-specific and hazard-independent indicators are shown by Schneiderbauer and Ehrlich (Chapter 3).

Pelling (Chapter 7) stresses the problem of "hazard nesting", which was a major phenomenon with regard to Hurricane Katrina, which hit the US Gulf coast in September 2005. Hazard nesting describes what happens when an individual hazard phenomenon results in multiple hazard types. For example, Hurricane Katrina led to water pressure causing a break in a levee, which led in turn to the flooding of the city of New Orleans and ultimately also to chemical pollution of the water bodies. Regarding the question of hazard-specific versus hazard-independent indicators, the recommendations that can be made are given below.

Conclusions and recommendations

- Future research should explore more precisely how to combine hazard-dependent and hazard-independent indicators in order to cover both these aspects of vulnerability.
- Hazard-independent indicators of vulnerability tend to focus on gen-

eral and indirect features of vulnerability, such as income, poverty or education. In contrast, hazard-dependent indicators generally capture potential direct and hazard-specific impacts, such as the possibility of a building being flooded based on the assessment of the height of a building.

- Bollin and Hidajat (Chapter 14) point out that indicators have different meanings for specific hazards; as a consequence hazard-specific weighting factors should be used in order to combine the different figures into one index or final result. This underlines the need to investigate different methodologies of weighting with regard to the combination of different indicators and criteria. The approaches presented by Cardona, Bollin and Hidajat, Villagrán de León, and Greiving show examples of potential ways of using weighting factors to combine different indicators in a quantitative fashion (Chapters 10, 14, 16 and 11).
- Future research should also take into account the phenomenon of hazard nesting. This means it will be necessary to develop methodologies to assess primary and secondary effects of hazards of natural origin and eventually also the link to natural technological hazards, without simply measuring the same effect twice. Current data sources mainly cover the primary effects of hazards.
- Finally, one has to assess and measure the potential vulnerability of societies, economies and environmental services with regard to a cascade of hazards. The methodology of lessons learned introduced by Krausmann and Mushtaq might be a useful entry point to explore this interface of hazard nesting as well as the link between "Natural and Technological" (NaTech) hazards (Chapter 22).

Linking global and local

While global index projects allow a first visualisation and comparison of vulnerability and risk worldwide, the situational context-specific approaches (place and time specific), such as self-assessment, explore place or spatially specific features and patterns of vulnerability of selected communities. The question of how to integrate these "spatial" and "time" dependencies of vulnerability is yet to be answered adequately.

As shown in this book, divergent approaches to measuring vulnerability deal with the "time" dimension in different ways. The DRI, for example, compares different nations with regard to their relative vulnerability on the basis of mortality data going back to 21 years for flooding and cyclones and 36 years for earthquakes (Chapters 7 and 8). This index identifies Venezuela as a country highly vulnerable to flood risk since, during the period between 1980 and 2000, it faced a major flood event. The time span of 21 or 36 years may seem long with regard to a human life; however, an event like the Northern Sumatra earthquake that caused the devastating tsunami in the Indian Ocean with a magnitude of 8.7 (Richter scale) only occurs approximately every 230 years (Carpenter, 2005). That said it is difficult to incorporate low-frequency, potential future hazards when basing the assessment of vulnerability solely on past experience, especially mortality data. The long-tailed high-impact events may occur only once in a 100 or 1,000 years, which makes a 20-year average difficult.

Moreover, global indexing programmes such as the DRI and the Hotspots project have revealed that, on a global scale, one can observe an inverse relationship between those countries with the highest number of people killed and those countries with the highest absolute economic losses (Chapters 7 and 9). Although it is evident that countries with heavy human losses have to be viewed as priority countries for humanitarian aid, the high economic losses in developed countries may point out the need to analyse vulnerability differently in different countries or with regard to the different degree of development of a country. That means the focus on mortality as the main characteristic of vulnerability is problematic, particularly since in many regions floods for example occur regularly and catastrophically without significant loss of life, but with very significant loss of property and livelihoods. Local assessment methodologies are able to capture vulnerability more place-specifically; however, they also face the problem of how to integrate an appropriate "time" and "spatial" dimension in assessing vulnerability.

In this context, Queste and Lauwe (Chapter 4) draw attention to the question of whether different perspectives, responsibilities and requirements on vulnerability assessment can play a major role if, for example, in a federally structured country like Germany different agencies at varying levels have to deal with risk management and vulnerability reduction. Here effective cooperation and the use of similar assessment methods would require that the indicators used allow for up- and down-scaling to a certain degree. Queste and Lauwe argue that there is still a lack of vulnerability indicators which can be used for concrete disaster risk management on different geographical scales.

Conclusions and recommendations

- We still have only a limited understanding of how changing place-based socio-economic and environmental conditions affect vulnerability. Global indexing projects and national vulnerability and risk profiling are often too general to permit useful exploration of these issues. On the other hand one should also acknowledge that the global index projects – DRI, Hotspots and the system of indicators for disaster risk management in the Americas – are not intended to be verifiable against specific disaster-related outcomes, which might even be very difficult for place- and time-specific approaches. However, research is needed to give a more precise idea of how to integrate the "time" and the "spatial" dependency of vulnerability into measurement tools, especially with regard to coping capacity and adaptation.

- Global and international indexing projects (e.g. Hotspots, DRI and Americas Programme), which do not account for spatially specific features of vulnerability, are very useful for identifying regions and countries with high levels of risk and vulnerability. Although these approaches do not lead to an identical list of highly vulnerable countries and countries at risk, they could serve as a first screening for hotspots or priority countries, while local and sub-national approaches might also consider spatially specific aspects of vulnerability. These spatially specific measures should reflect the socio-economic, environmental and spatial contexts. This requires further analysis on how to combine, contextualise and link global indexing methodologies (first lens) with local and sub-national assessment methodologies (second lens).
- Particular emphasis should also be given to the question of how these approaches can be used to stimulate future actions that will be undertaken to reduce vulnerability and risk. This means research has to also address the question of how these approaches can be applied in decision-making processes, and by whom at different levels.
- Both global indexing and locally specific assessment tools, as well as combined approaches, need to be strengthened in terms of their applicability in traditional planning and decision-making processes (e.g. urban planning, disaster emergency plans) as well as non-traditional or new planning tools, such as innovative education and awareness raising programmes.
- Furthermore, more emphasis should be given to the links between vulnerabilities of various elements at different scales. Schneiderbauer and Ehrlich describe these links within the framework of levels of vulnerability (Chapter 3). Lebel et al. (Chapter 19) emphasise that institutions operating at the scale of basins or regions might influence the vulnerabilities of individuals and households. Therefore future research should address more precisely these links of vulnerability between different scales as well as the issue of the up- and down-scaling of different indicators and tools to measure vulnerability.

Reliable loss estimation versus fuzzy context interpretation

The review of current approaches shows the divergence between reliable loss data (implying a retrospective focus) and forward-looking assessment based on broader and general development and context indicators such as population growth, poverty level and literacy rate. These differences can be illustrated, for example, by comparing the Hotspot and DRI approaches (Chapters 7, 8 and 9) with the community-based disaster-risk indicators of Bollin and Hidajat (Chapter 14). In the community-based disaster-risk indicators project, vulnerability is measured with context variables such as population density, demographic pressure (population growth), poverty level, literacy rate, decentralisation, community participation and economic diversification. These indicators represent important factors that can determine and drive the vulnerability of communities; however they are not necessarily able to explain the vulnerabilities revealed in the past. In contrast, assessments based on experienced events, known damage and revealed vulnerabilities often appeal to decision makers and the general public; they seem to have statistical rigour since they include actual losses and fatalities experienced due to a hazard event.

However, the losses experienced in the past are not necessarily a reliable indicator for estimating present and future vulnerabilities (Birkmann, Chapter 2). On the other hand the vulnerability patterns revealed can serve as an important basis for investigating and measuring the pre-existing and emergent vulnerability and to provide essential information to promote disaster resilience within the reconstruction process (Chapter 18). However, the calculation of future vulnerability based on previously experienced losses is particularly difficult for low-frequency hazards, such as tsunamis. Nevertheless, estimating vulnerability in areas where a hazardous event took place recently is often important in order to understand the various vulnerability profiles and to estimate the vulnerability level and unusual difficulties different groups experience in the recovery process, as Birkmann, Fernando and Hettige illustrate through some selected findings of the local vulnerability assessment surveys in Sri Lanka (Chapter 18).

Another option for measuring vulnerability and coping capacity are self-check approaches, such as that of the Asian Disaster Reduction Center (ADRC) for earthquakes, presented by Arakida (Chapter 15). These approaches can give useful insights into the disaster-related knowledge and preparedness of different households and different governmental agencies.

Conclusions and recommendations

- Vulnerability assessment must go beyond retrospective loss estimation and mortality assessment.
- On the other hand, the analysis of specific cases of severe hazardous events that have taken place recently (Indian Ocean tsunami, Hurricane Katrina) is often crucial for understanding the divergence between general contexts, such as poverty or theoretical rules and capaci-

ties on "paper", and the reality of the actual vulnerabilities revealed as well as the actions undertaken during an extreme event. This is particularly important in terms of assessing institutional capacities to reduce risk and vulnerability, clearly revealed by the lack of effective and reliable institutional capacities during the Hurricane Katrina catastrophe.

- Additional research is needed to examine those contextual features and characteristics contributing to vulnerability that can be measured and those that can only be captured through qualitative assessment tools, such as the various influences of armed conflicts on the vulnerability of the people exposed. This is relevant, for example, not only for countries such as Sri Lanka and Indonesia that were hit by the Indian Ocean tsunami of December 2004, but also for other regions affected by civil unrest such as Kashmir, which was struck by an earthquake in October 2005. The underlying conflict situations could hamper both rescue and rehabilitation actions.
- Furthermore, approaches that combine forward-looking context analysis with retrospective loss estimation should be tested. An example is the Human Security Index proposed by Plate, which assesses the vulnerability of individuals or households on the basis of their income above the minimum subsistence level compared with the economic losses experienced by the respective entity (Chapter 13).
- Finally, more research is needed with regard to sub-national and local approaches for addressing spatially specific root causes of vulnerability and exploring existing coping capacities and potential intervention tools to reduce vulnerability and promote the disaster resilience of communities. This is especially important for the development of policy recommendations, as is shown implicitly in the Tanzania case study (Chapter 12) and the research on vulnerabilities of different groups in coastal communities in Sri Lanka (Chapter 18). The identification and assessment of mitigation strategies and disaster risk management performance is especially important (even though often controversial) which, however, has only been captured by just a few approaches (see Chapters 10 and 18).

Complexity versus simplification

All the approaches presented deal explicitly or implicitly with the question of how to simplify the complex concept of vulnerability in order to be able to measure it. Once a quantitative assessment is required, it is necessary to simplify the notion of vulnerability in terms of measurable components. Queste and Lauwe (Chapter 4), for example, point out that from a practitioner's point of view the collection of data should be kept as simple as possible in order to avoid mistakes during the subsequent processes of comparison and analysis. Quantitative as well as qualitative approaches to measuring vulnerability at different levels are based on a selection of specific features and characteristics in order to estimate the status or the development level of a nation, community, group or economic sector with regard to vulnerability (Chapter 2). However, complexity and simplification depend on the scale of the approach selected; normally, global measurement tools need to be limited to a small set of data that is available for all analytical units around the globe, as the DRI and Hotspots show. Local approaches, in contrast, are generally more open to a large number of input variables available for the specific and relatively small-size locations. Another option to simplify complex processes is the archetype approach presented by Kok et al. (Chapter 6), which shows new ways of generalising processes and trends through the development of blueprint scenarios that help to provide a better understanding of basic processes leading to vulnerability, with a special focus on the human–environmental interaction. Regarding this interaction, Renaud (Chapter 5) also calls for a stronger focus on the impact that environmental degradation has on societies and their vulnerability.

On the other hand, the degree of complexity or simplification is linked to the thematic scope of the approach: that is, whether the approach accounts for susceptibility alone, or whether it also encompasses coping capacity, exposure and adaptive capacities (e.g. see Turner et al., 2003). According to Villagrán de León, if many elements are included within vulnerability – such as coping capacity, resilience and exposure – major complications arise, firstly regarding decisions about how to assess each of these components (identification and quantification), and secondly, with respect to how to combine the different figures in order to arrive at a final result (Chapter 16).

In this context a careful balance is required between broad measurement approaches encompassing various characteristics of vulnerability – implying a large number of input variables or characteristics – and the alternative of focusing only on a very few indicators to describe the complex processes behind vulnerability that might produce greater transparency. This balancing act can be seen in various approaches presented in the book, such as in the overview given by Pelling regarding the global index projects (Chapter 7), the approach proposed by Cardona (Chapter 10), the research in Sri Lanka by Birkmann, Fernando and Hettige (Chapter 18), the vulnerability assessment conducted by Kiunsi and Meshack in Tanzania (Chapter 12), the quantitative model to compare coping capacities at national level by Billing and Madengruber (Chapter

21) and the discussion on how to measure coping capacity with participatory self-assessment methodologies, presented by Wisner (Chapter 17).

Conclusions and recommendations

- Although vulnerability assessment can be either simple or complex, it invariably deals with complex phenomena and multidimensional problems. In this regard, it is important to explore the added value of very precise and specific assessment methodologies compared to general overviews that can be provided by highly aggregated approaches or those that use only a very limited number of quantitative key indicators or qualitative criteria.
- Simplification of the complex interactions that determine and drive vulnerability is necessary in any approach to measure or describe vulnerability. In this context, it is useful to promote a more harmonised and comprehensive use of the terminology within vulnerability research. Differing interpretations of the same term (see Thywissen, Chapter 24) hamper efforts to derive appropriate indicators to measure the different facets of vulnerability. The promotion of a common language to describe key components of vulnerability is therefore an important task, although it is necessary to acknowledge the different schools of thinking and their justifications.
- Although there is no intention to promote a single and universal catalogue for measuring vulnerability, it would be a step forward if the different interpretations of terms by different disciplines could be harmonised to create a common language for describing major components of vulnerability.
- On the other hand, as this publication itself shows, it seems futile to try to develop a universal set of indicators. Therefore, a second option and task for future research should be the formulation of procedural requirements for the development of appropriate tools to measure vulnerability. Currently, many approaches do not provide a transparent procedure or sufficient information regarding their selection choices for the indicators and criteria used. This reflective element is underdeveloped in current approaches.
- Often the lack of data is a main argument for the selection of specific parameters and criteria used to estimate vulnerability. An alternative could be the development of surrogate indicators.
- Furthermore, the appropriate degree of simplification is also determined by the function and nature of the target group that the approach focuses on. Although one can agree with Bollin and Hidajat (Chapter 14) that indices are often appealing because of their ability to summarise a great deal of information in a way that is easy for non-experts to

visualise and understand, one has to acknowledge that a single number is often not sufficient for making policy recommendations. That said, Downing et al. (2006) have stressed that, instead of a single, composite number, vulnerability can also be viewed as a profile, using diagrams to visualise the different vulnerability profiles.

- There is often too little understanding of the close link between the definition or assigned function of the approach and the target group, on the one hand, and the level of complexity and simplification, on the other. Research is needed to explore how different target groups deal with, understand and respond to the differing levels of complexity of the information reaching them with regard to vulnerability and risk.

- Arguments for or against a specific level of simplification and aggregation should also be based on considerations of the target group and functions that the approach has. However, this is not an easy task either, since some approaches focus on a range of target groups; examples include the Americas Indicator Programme presented by Cardona (Chapter 10) and the approach to measuring vulnerability of coastal communities in Sri Lanka described by Birkmann, Fernando and Hettige (Chapter 18).

- Furthermore, it seems important to explore how a modular system of indices, indicators, criteria and qualitative descriptions could allow for a more flexible handling of complexity and aggregation. Comparative studies are needed to show how different assessment methodologies are being used in similar situations and locations; such studies should focus on and identify the various potentials of these tools. In addition to these comparative approaches, future research should also examine how to combine very quantitative assessment methodologies (e.g. damage functions, highly aggregated indicators) with qualitative assessment tools (e.g. self-assessment), for example at the local and household level.

- Not enough has been done to exploit the opportunities of a combined and modular approach to measuring vulnerability that takes into account different levels of aggregation, different datasets and assessment methodologies. A prerequisite for a more effective and in-depth combination also requires stronger interdisciplinary cooperation between very different disciplines and schools of thought such as sociology, disaster management and space technology. Often there is a tendency to still stick to a traditional disciplinary focus, although interdisciplinary and multidisciplinary approaches are imperative for more comprehensive and effective vulnerability and risk reduction strategies.

Measuring without goals?

The majority of current approaches are functioning without precise goals; a review of the approaches presented shows indirectly that the measuring of vulnerability, coping capacity and performance in risk reduction would benefit from having more clearly defined targets and standards of vulnerability reduction. Although some approaches were able to establish indicators and evaluation tools without setting specific goals (examples include the approach of the system of indicators for disaster risk management in the Americas (Chapter 10) or the assessment of institutionalised capacities (Chapter 19)), it is increasingly evident that a benchmarking and assessment of vulnerability and vulnerability reduction strategies requires precise goals and standards of disaster and vulnerability reduction at various scales. Thus the following recommendations can be made.

Conclusions and recommendations

- The systematised and logical development of the measurement of vulnerability should be based on goals. These goals are still either missing or, at best, available solely for a few regions and communities. Therefore, vulnerability assessment and measurement should also promote the formulation of specific goals for vulnerability reduction, which could themselves serve as a basis for measurement. Especially if the aim is to promote more proactive efforts towards vulnerability reduction, we need to define the goals of vulnerability reduction beforehand in order to get away from the reactive focus on disaster relief and rescue operations.
- Finally, one of the most important goals in developing indicators to measure vulnerability and coping capacity is to help bridge the gaps between the theoretical concepts and day-to-day decision-making. For example, White et al. (2001) argue that in the past, improved knowledge was not by itself sufficient to reverse the upward trend in disaster statistics. Weichselgartner and Obersteiner (2002) came to a similar conclusion, stating that sufficient progress has not been made in converting theoretical research findings into concrete actions in practical disaster management. Therefore, the crucial feature of any vulnerability indicators is their relevance to policy and decision-making processes. This publication offers various examples of methodologies and approaches to measuring vulnerability, and of ways to derive practical recommendations from these approaches that are useful for political decision-making processes. Those approaches that also incorporated coping capacity, intervention tools and manageability, such as the Tanzanian case study (Kiunsi and Meshack, Chapter 12), the Americas

Indicator Programme (Cardona, Chapter 10), and the results of measuring coastal vulnerability in Sri Lanka (Birkmann et al., Chapter 18), showed among other aspects the opportunities and difficulties inherent in assessing and evaluating the current preparedness status for dealing with the negative impacts of hazardous events. These capacities and intervention tools should be the primary target of policy interventions; but how can they be addressed in regions where governance is weak (e.g. civil war regions in Sri Lanka) and where no precise and spatially differentiated vulnerability reduction standards exist?

- Thus, more research and development is required for the improvement of data regarding the various vulnerabilities of the social, economic and environmental systems, but we also need to address the current lack of clearly defined goals for vulnerability and risk reduction.
- We should promote the development of proposals for setting risk and vulnerability reduction goals for various scales and regions, based on scientific expertise. This would allow us to overcome the descriptive analytic focus of current concepts and move towards the use of indicators as an effective benchmarking and evaluation tool in political decision-making.
- The question of whether or not tools for measuring vulnerability will have an impact on vulnerability reduction depends not only on their structure, data and thematic focus, but also on the willingness of political decision makers to define precise targets and goals for vulnerability reduction in the future. The methodologies shown in this book are an important prerequisite for such a standard and goal setting.
- Furthermore, the debate about environmental pollution and global environmental policies has shown that improved measurement approaches and tools to estimate the potential future consequences of these hazards and countermeasures have had a significant impact on standards for sustainable environmental management. Thus, both the political and the research communities need to strengthen their efforts towards the establishment and promotion of concepts, methodologies and goals to identify, measure and assess, as well as reduce, the vulnerability of societies to hazards of natural origin before – and also after – disasters occur. This need became evident for example in the context of the Indian Ocean tsunami in 2004.

This chapter, but also the whole book, illustrates that many fundamental questions concerning the identification, measurement and assessment of vulnerability are still subject to discussions within the scientific and professional community from all over the world. We have to acknowledge that a range of approaches are needed to capture the multifaceted nature of vulnerability and to serve the specific needs of different end-user groups. However, the various approaches presented in the book show some clear trends pointing at the priorities set within the discussion. An example is the intention to focus more precisely on the governance and institutional dimensions of vulnerability as well as on intervention tools and measures undertaken to reduce vulnerability and risk.

It is noteworthy that the second meeting of the UNU-EHS Expert Working Group on Measuring Vulnerability to Hazards of Natural Origin revealed a set of differing views. For instance, the debate between qualitative and quantitative approaches was still a major issue, often linked to the question of the level of the approach. Furthermore, the issue of complexity versus simplification is crucial, particularly in connection with the question concerning the goals and functions the vulnerability measurement tools and methodologies should fulfil at different levels. In this context, it was interesting to notice the different standpoints put forward by social scientists and experts with a natural science or engineering background. While the social scientists argued in favour of incorporating a broad variety of aspects and issues in the measurement of vulnerability, engineers and natural scientists preferred to start with a narrower focus using quantitative measures, expecting to improve and diversify them over the course of the development.

Finally, a kind of consensus was reached that future research into vulnerability should address quality as well as quantity, aiming at learning more about both (see in detail, Birkmann and Wisner, 2006).

Overall, it became evident that a stronger exchange between ideas from the various disciplines as well as from very different regions is useful not only in order to achieve a better overview of the various approaches, but also to promote a stronger interdisciplinary cooperation and combination of different techniques to measure vulnerability. With the establishment of the Expert Working Group, as a scientific platform, UNU-EHS intends to contribute to the above-mentioned goals and to move the scientific debate towards policy-relevant goals and impacts. Lastly, the UNU-EHS Expert Working Group on Measuring Vulnerability is expected to play an important role within the follow-up mechanism of the Hyogo Framework for Action, which, as mentioned at the beginning of the book in Chapter 1, defined the development of indicators and the identification of vulnerability as an important prerequisite for effective disaster risk reduction.

REFERENCES

Birkmann, J. and B. Wisner (2006) "Measuring the Un-Measurable, the Challenge of Vulnerability", in Source No. 5/2006, Publication Series of UNU-EHS

(United Nations University – Institute for Environment and Human Security), available at http://www.ehs.unu.edu.

Carpenter, G. (2005) "Indian Ocean Tsunami", in *Speciality Practice Briefing: An Update from the Property Specialty*, Issue No. 1, available at http://www.guycarp.com/portal/extranet/pdf/GCBriefings/Indian%20Ocean%20Tsunami%2001_07_05.pdf;jsessionid=%40168daa%3a1075050c7fd?vid=1/.

Downing, T., J. Aerts, J. Soussan, S. Bharwani, C. Ionescu, J. Hinkel, R. Klein, L. Mata, N. Matin, S. Moss, D. Purkey and G. Ziervogel (2006) "Integrating Social Vulnerability into Water Management", *Climate Change* (in preparation).

Mayoux, L. and R. Chambers (2005) "Reversing the Paradigm: Quantification, Participatory Methods and Pro-poor Impact Assessment", *Journal of International Development* 17: 271–298.

Turner, B.L., R. Kasperson, P. Matson, J.J. McCarthy, R. Corell, L. Christensehn, N. Eckley, J. Kasperson, A. Luers, M. Martello, C. Polsky, A. Pulsipher and A. Schiller (2003) "A Framework for Vulnerability Analysis in Sustainability Science", *PNAS*, 100(14): 8074–8079.

Weichselgartner, J. and M. Obersteiner (2002) "Knowing Sufficient and Applying More: Challenges in Hazard Management", *Environmental Hazards* 4: 73–77.

White, G.F., R.W. Kates and I. Burton (2001) "Knowing Better and Losing More: The Use of Knowledge in Hazard Management", *Environmental Hazards* 3(3/4): 81–92.

24

Core terminology of disaster reduction: A comparative glossary

Katharina Thywissen

Introduction

The extent of disasters and their foreboding trend to increase in frequency and severity imply that the problem of disasters will have to be addressed by the world community in the coming years. In the course of the International Decade for Natural Disaster Reduction (IDNDR), 1990–1999, and of many other initiatives spawned over the last few years, disaster reduction has gained a lot of momentum and attention. In addition, the World Conference on Disaster Reduction in Kobe, Japan (17–20 January 2005) and the Boxing Day Tsunami (26 December 2004) in the Indian Ocean exposed the need for action to a global audience. Disasters take a devastating toll on countries' development, economies and environment in all regions of the world and thereby severely compromise human security and livelihoods.

The paradigm shift

There has been a paradigm shift in some vital concepts evolving around the understanding of human livelihood. The human being is moving more and more into the centre of attention. The general understanding of security has shifted from the nationalistic and militaristic perspective to a more individualistic and humanitarian one: human security (Commission on Human Security (CHS), 2003). Another paradigm shift has taken place in the shift from income poverty (lack of financial affluence) to human poverty (lack of well-being). This shift has been paralleled in disaster management by a shift from seeing disasters as extreme events created by natural forces, to viewing them as manifestations of unresolved development problems (Yodmani, 2001).

Approaches in disaster reduction have become much more complex and emphasis has shifted from relief to mitigation. Consequently, vulnerability, resilience and coping capacities have gained a more prominent role and more light is being shed on social, economic, political and cultural factors.

Integrated disaster reduction depends on collaboration and exchanges between experts from a multitude of disciplines and competencies. Those range from science, through policy building and civil society, to disaster relief and rehabilitation. Approaches can be quantitative in nature as well as qualitative or descriptive, and many fields have cultivated their own understanding, and hence their own definitions, of disaster-related terms. As a consequence, communication within the disaster reduction community is often encumbered and misunderstandings are common.

"Babelonian confusion"

A shared language and shared concepts are crucial stepping-stones in widening the understanding and effectiveness of disaster reduction. A term is defined in order to explain its content and context in a logically consistent way while ensuring the widespread acceptance of peers. Definitions of the same terms may have developed simultaneously and separately in different disciplines. As a result multidisciplinarity often results in the same term being defined in different ways. The resulting situation is often perceived as the proverbial *Babelonian confusion*. Most of these sometimes colliding definitions are valid in their respective contexts and cannot be discarded. Therefore, in order to enable collaboration and communication free of misunderstanding, it is crucial to disseminate the different definitions across the disciplines, with the goal that eventually a common vocabulary of unique, well-formulated definitions and concepts will emerge.

Terms and concepts are not just an academic exercise but have real importance in the practical world. The language used by workers in the disaster field frames, focuses and limits the kinds of questions they ask (Handmer and Wisner, 1998). Before work on disaster risk reduction can be carried through, differing perceptions, interests and methodologies have to be recognised and a broad consensus on targets, strategies and methodologies has to be reached (Yodmani, 2001). Clearly, definitions and concepts are needed at every level of disaster reduction.

Common, coordinated and consistent approaches to risk reduction can only be achieved if there is a common agreement as to the structure of the problem and the basic notions, concepts and terms used in its definition (Lavell, 2003).

The moral aspect of disaster reduction

To an unknown extent environmental deterioration and climate change have been exacerbated by today's developed countries, and the developing countries are repeating the same processes and harmful activities, but exponentially, due to the sheer size of their populations. The resulting increase of disaster frequencies should alarm all countries equally, but the developed countries are facing this situation with a heightened responsibility for the poor countries because it is the people in developing countries who suffer most from disasters. As the World Bank (2005) comments:

developing countries suffer the greatest costs when disaster hits: more than 95 percent of all deaths caused by disasters occur in developing countries; and losses due to natural disasters are 20 times greater (as a percent of GDP) in developing countries than in industrial countries.

But it seems that even developed countries are spurred to act upon their contribution to climate change irrespective of their signing the Kyoto Protocol. For Allen and Lord (2004) report that in July 2004 eight US States and New York City have filed charges against five US power companies for their contribution to climate change.

If the UN Millennium Development Goals (MDGs) carry any clout, the direct link between poverty and disaster impact implies a moral obligation for the international community to address both these concerns in a concerted way. Cannon (1994) points out that:

it may be true that most of the suffering in disasters is experienced by poor people, it may not be the case that all poor suffer. Nor is it only the poor who suffer, but the impact of hazards may well be a factor in creating newly impoverished people.

Risk usually involves a decision by the person at risk (to take a certain risk or not), always presuming the individual knows about the risk. According to Cardona (2003) and Lavell (2003), risk must be associated with decision if it is to have any relevance as a notion and concept. Thus, one objective of disaster reduction is to raise awareness and make sure that people know of the risks. Another objective is to see to it that people are in a situation to make choices, which directly leads to poverty reduction because poverty, by definition, reduces people's choices.

With risk also comes responsibility, and the question of morality arises. However, there can be no direct moral valuation of risk because the level of acceptable risk is highly subjective and highly variable. What complicates the matter further is the fact that the perception of probability connected with the risk varies from individual to individual and group to group (Luhmann, 1993).

The UNU-EHS stance

UNU-EHS (United Nations University – Institute for Environment and Human Security) as a member institute of the UNU, forms a bridge between the UN and the academic world, acts as a think tank for the UN and provides a platform for dialogue and exchange of ideas. UNU-EHS aims to improve the in-depth understanding of the cause–effect relationships that lead up to disasters in order to find possible ways to increase human security. As an academic institution, UNU-EHS aims to strengthen the capabilities of individuals and institutions to address the potential impacts of hazards and their associated risks and vulnerabilities, turning research results into practical knowledge through training and other forms of human capacity building. Therefore common terminology and definitions are essential prerequisites for a focused scientific debate, interdisciplinary approaches and ultimately for improved disaster reduction.

The comparative glossary

In this glossary core terms from the cause-and-effect chain of disasters have been selected and their definitions put up for discussion among peers. There are already a number of listings of terms published (e.g. ISDR, UNDP-BCPR, UNEP, IPCC, DKKV, BBK, CEDIM). However, the lists generally do not juxtapose the different definitions of various disciplines; rather, they lay out their own definitions in an attempt to put an end to the *Babelonian confusion*. The comparative glossary given in this chapter, in contrast, aims to inform experts from different disciplines about the various – sometimes contradictory – definitions currently used or referred to in the field of disaster mitigation. Even if some terms are defined differently by different disciplines, it is vitally important to make those differences in terminology known across the disaster reduction community in order to avoid misunderstandings and to enhance knowledge, mutual understanding and efficiency in disaster reduction.

This comparative glossary does not claim to be exhaustive; rather, it focuses on a selection of terms that typically are used across multiple disciplines and that are central to the cause-and-effect chain of disaster reduction. The listing of definitions concludes this chapter and shows the relationships between the main terms while being as concise as possible and as diverse and elaborate as necessary.

These terms and definitions have been collected from the literature, including several reports that already offer glossaries of disaster reduction terms. Disciplines and sectors represented so far include: the insurance industry, United Nations, natural, social and multidisciplinary sciences, economics, engineering, governance/policy, civil society and disaster relief.

Core terminology of disaster reduction

Term	Definition	Source	Discipline
Capacity	"The maximum amount of risk that can be accepted in insurance. One factor in determining capacity is government regulations that define minimum solvency requirements. Capacity also refers to the amount of insurance coverage allocated to a particular policyholder or in the marketplace in general."	Swiss Re (2005)	Insurance industry
Capacity	"A combination of all the strengths and resources available within a community, society or organization that can reduce the level of risk, or the effects of a disaster. Capacity may include physical, institutional, social or economic means as well as skilled personal or collective attributes such as leadership and management. Capacity may also be described as capability."	UN/ISDR (2004)	United Nations
Capacity building	"Efforts aimed to develop human skills or societal infrastructures within a community or organization needed to reduce the level of risk. In extended understanding, capacity building also includes development of institutional, financial, political and other resources, such as technology at different levels and sectors of the society."	UN/ISDR (2004)	United Nations

Term	Definition	Source	Discipline
Capacity, adaptive	"The potential or ability of a system, region, or community to adapt to the effects or impacts of climate change. Enhancement of adaptive capacity represents a practical means of coping with changes and uncertainties in climate, including variability and extremes. In this way, enhancement of adaptive capacity reduces vulnerabilities and promotes sustainable development." (Goklany, 1995; Burton, 1997; Cohen et al., 1998; Klein, 1998; Rayner and Malone, 1998; Munasinghe, 2000; Smit et al., 2000) quoted in IPCC (2001)	IPCC (2001) p. 881	Science (multidisciplinary)
Capacity, adaptive	"The degree to which adjustments in practices, processes, or structures can moderate or offset the potential for damage or take advantage of opportunities created by a given change in climate."	IPCC (2001)	United Nations
Catastrophe	"In the English speaking world a differentiation is sometimes made between disaster and catastrophes. In the latter, most or all people living in a community are affected, as are the basic supply centers, so that help from neighbours is largely impossible (the affected people helping each other is a general phenomenon in disasters with a lower degree of severity)."	Quarantelli (1998)	Science (multidisciplinary)

454

Climate change	IPCC (2001)	United Nations
	"Climate change refers to a statistically significant variation in either the mean state of the climate or in its variability, persisting for an extended period (typically decades or longer). Climate change may be due to natural internal processes or external forcings, or to persistent anthropogenic changes in the composition of the atmosphere or in land use. Note that the → Framework Convention on Climate Change (UNFCCC), in its Article 1, defines 'climate change' as: 'a change of climate which is attributed directly or indirectly to human activity that alters the composition of the global atmosphere and which is in addition to natural climate variability observed over comparable time periods'. The UNFCCC thus makes a distinction between 'climate change' attributable to human activities altering the atmospheric composition, and 'climate variability' attributable to natural causes."	
Climate change	UN/ISDR (2004)	United Nations
	"The climate of a place or region is changed if over an extended period (typically decades or longer) there is a statistically significant change in measurements of either the mean state or variability of the climate for that place or region." "Changes in climate may be due to natural processes or to persistent anthropogenic changes in atmosphere or in land use. Note that the definition of climate change used in the United Nations Framework Convention on Climate Change is more restricted, as it includes only those changes which are attributable directly or indirectly to human activity."	

Term	Definition	Source	Discipline
Coping capacity	"The ability to cope with threats includes the ability to absorb impacts by guarding against or adapting to them. It also includes provisions made in advance to pay for potential damages, for instance by mobilizing insurance repayments, savings or contingency reserves."		United Nations
Coping capacity	"Refers to the manner in which people and organisations use existing resources to achieve various beneficial ends during unusual, abnormal, and adverse conditions of a disaster event or process. The strengthening of coping capacities usually builds resilience to withstand the effects of natural and other hazards."	Europ. Spatial Planning Observ. Netw. (2003)	Science (multidisciplinary)
Coping capacity	"Is a function of: perception (of risk and potential avenues of action – the ability to cope is information contingent); possibilities (options ranging from avoidance and insurance, prevention, mitigation, coping); private action (degree to which special capital can be invoked); and public action." (e.g. Webb and Harinarayan 1999, Sharma et al. 2000) quoted in IPCC (2001)	IPCC (2001)	Science (multidisciplinary)
Coping capacity	"The means by which people or organizations use available resources and abilities to face adverse consequences that could lead to a disaster. In general, this involves managing resources, both in normal times as well as during crises or adverse conditions. The strengthening of coping capacities usually builds resilience to withstand the effects of natural and human-induced hazards."	UN/ISDR (2004)	United Nations

Term	Definition	Source	Field
Coping capacity	"The manner in which people and organisations use existing resources to achieve various beneficial ends during unusual, abnormal and adverse conditions of a disaster phenomenon or process."	UNDP-BCPR (2004)	United Nations
Cost	"Means the measurable economic losses due to failures, such as the loss of crops that not irrigated on time or the production in a factory, and any other loss incurred by failure to supply an adequate quantity of good quality water at the time it is required. Losses are very difficult to measure and quantify, especially those associated with the quality of life of urban consumers."	Shamir (2002)	Engineering
Disaster	"A disaster is an unusually severe and/or extensive event that usually occurs unexpectedly and has such a severe impact on life and health of many people and/or causes considerable material damage and/or impairs or endangers the life of a large number of people for a long period of time to such an extent that resources and funding available at local or regional level cannot cope without outside help. The disaster qualifies as such when it becomes apparent that the available resources and funding are inadequate for the necessary and prompt relief. Relief provision systems that are capable of evolving from every day use and which integrate all the necessary components are required for effectively managing disasters." 30.11.98 Report of the working group of the Permanent Conference on Disaster Reduction and Disaster Protection.	DKKV (2002) p. 2	Science (multidisciplinary)

457

Term	Definition	Source	Discipline
Disaster	"External danger, the loss of development potential and the helplessness of the affected population; a serious disruption of the functioning of a society causing widespread human, material or environmental losses which exceed the ability of the affected society to cope using only its own resources."	DKKV (2002)	United Nations/DKKV
Disaster	"A hazard might lead to a disaster. A disaster by itself is an impact of a hazard on a community or area – usually defined as an event that overwhelms the capacity to cope with it."	Europ. Spatial Planning Observ. Netw. (2003)	Science (multidisciplinary)
Disaster	"Disasters combine two elements: events and vulnerable people. A disaster occurs when a disaster agent (the event) exposes the vulnerability of individuals and communities in such a way that their lives are directly threatened or sufficient harm has been done to their community's economic and social structures to undermine their ability to survive. A disaster is fundamentally a socio-economic phenomenon. It is an extreme but not necessarily abnormal state of everyday life in which the continuity of community structures and processes temporarily fails. Social disruption may typify a disaster but not social disintegration."	IFRC (1993) pp. 12–13	Disaster relief
Disaster	"The result of a vast ecological breakdown in the relations between man and his environment, a serious and sudden event (or slow, as in drought) on such a scale that the stricken community need extraordinary efforts to cope with it, often with outside help or international aid."	*Journal of Prehospital and Disaster Medicine* (2004)	Disaster relief

| Disaster (risk) reduction | "The conceptual framework of elements considered with the possibilities to minimize vulnerabilities and disaster risks throughout a society, to avoid (prevention) or to limit (mitigation and preparedness) the adverse impacts of hazards, within the broad context of sustainable development. The disaster risk reduction framework is composed of the following fields of action, as described in ISDR's publication 2002 *Living with Risk: a global review of disaster reduction initiatives*, page 23:
• Risk awareness and assessment including hazard analysis and vulnerability/capacity analysis;
• Knowledge development including education, training, research and information;
• Public commitment and institutional frameworks, including organisational, policy, legislation and community action;
• Application of measures including environmental management, land-use and urban planning, protection of critical facilities, application of science and technology, partnership and networking, and financial instruments;
• Early warning systems including forecasting, dissemination of warnings, preparedness measures and reaction capacities." | UN/ISDR (2004) | United Nations |

459

Term	Definition	Source	Discipline
Disaster mitigation	"A collective term used to encompass all activities undertaken in anticipation of the occurrence of a potentially disastrous event, including preparedness and long-term risk reduction measures. The process of planning and implementing measures to reduce the risks associated with known natural and man-made hazards and to deal with disasters which do occur. Strategies and specific measures are designed on the basis of risk assessments and political decisions concerning the levels of risk which are considered to be acceptable and the resources to be allocated (by the national and sub-national authorities and external donors). Mitigation has been used by some institutions/ authors in a narrower sense, excluding preparedness. It has occasionally been defined to include post-disaster response, then being equivalent to disaster management, as defined in this glossary."	Dept. of Humanitarian Affairs (1994)	United Nations
Disaster risk management	"The systematic process of using administrative decisions, organization, operational skills and capacities to implement policies, strategies and coping capacities of the society and communities to lessen the impacts of natural hazards and related environmental and technological disasters. This comprises all forms of activities, including structural and non-structural measures to a void (prevention) or to limit (mitigation and preparedness) adverse effects of hazards."	UN/ISDR (2004)	United Nations

460

Term	Definition	Source	Discipline
Disaster, remarks on	"In summary, it can be determined that there is a problem of definition which affects the interpretation of vulnerability to disasters. Therefore, a list of important questions often cannot be answered clearly: When does a disaster begin? Who decides about shortcomings in the coping capacity of a society? When does the disaster end? What are the appropriate indicators for disasters? In addition, many definitions do not take differing vulnerabilities of population groups into account."	Feldbrügge and von Braun (2002)	Science (multidisciplinary)
Early warning	"The provision of timely and effective information, through identified institutions, that allows individuals exposed to a hazard to take action to avoid or reduce their risk and prepare for effective response." "Early warning systems include a chain of concerns, namely: understanding and mapping the hazard; monitoring and forecasting impending events; processing and disseminating understandable warnings to political authorities and the population, and undertaking appropriate and timely actions in response to the warnings."	UN/ISDR (2004)	United Nations
Exposure	"The economic value or the set of units related to each of the hazards for a given area. The exposed value is a function of the type of hazard."	Europ. Spatial Planning Observ. Netw. (2003)	Science (multidisciplinary)
Exposure	"The degree to which a risk or portfolio of risks is subject to the possibility of loss; basis for calculating premiums in (re)insurance."	MunichRe (2002) p. 259	Insurance industry

461

Term	Definition	Source	Discipline
Exposure	"Elements at risk, an inventory of those people or artefacts that are exposed to a hazard."	UNDP-BCPR (2004)	United Nations
Forecasting	"Definite statement or statistical estimate of the occurrence of a future event (UNESCO, WMO). This term is used with different meanings in different Disciplines."	UN/ISDR (2004)	United Nations
Hazard	"Natural hazard: the probability of occurrence, within a specific period of time in a given area, of a potentially damaging natural phenomenon." "In general, the concept of hazard is now used to refer to latent danger or an external risk factor of a system or exposed subject. Hazard can be expressed mathematically as the probability of occurrence of an event of certain intensity, in a specific site and during a determined period of exposure time."	Cardona (2003)	Science (multidisciplinary)
Hazard	"A property or situation that under particular circumstances could lead to harm. More specific, a hazard is a potentially damaging physical event, phenomenon or human activity, which may cause the loss of life or injury, property damage, social and economic disruption or environmental degradation. Hazards can be single, sequential or combined in their origin and effects. Each hazard is characterised by its location, intensity and probability."	Europ. Spatial Planning Observ. Netw. (2003)	Science (multidisciplinary)

Term	Definition	Source	Organization
Hazard	"The probability of the occurrence of a disaster caused by a natural phenomenon (earthquake, cyclone), or by failure of man-made sources of energy (nuclear reactor, industrial explosion) or by uncontrolled human activity (overgrazing, heavy traffic, conflicts). Some authors use the term in a broader sense, including vulnerability, elements at risk and the consequences of risk."	*Journal of Prehospital and Disaster Medicine* (2004)	United Nations/science (multidisciplinary)
Hazard	"The probability of occurrence associated with an extreme event that can cause a failure." (UNDRO quoted in Plate, 2002)	Plate (2002)	United Nations
Hazard	"... there is a distinction between an event, a hazard, and a disaster. A natural event, whether geological, climatological, etc., is simply a natural occurrence, whereas a hazard, geological or otherwise, is the potential danger to human life or property."	Rahn (1996) p. 489	Science
Hazard	"A potentially damaging physical event, phenomenon or human activity that may cause the loss of life or injury, property damage, social and economic disruption or environmental degradation. Hazards can include latent conditions that may represent future threats and can have different origins: natural (geological, hydrometeorological and biological) or induced by human processes (environmental degradation and technological hazards). Hazards can be single, sequential or combined in their origin and effects. Each hazard is characterised by its location, intensity, frequency and probability."	UN/ISDR (2004)	United Nations

Term	Definition	Source	Discipline
Hazard, natural	"Natural hazards are dynamic phenomena that involve people not only as victims but also as contributors and modifiers." (Kates 1996) quoted in Rashed and Weeks (2003).	Rashed and Weeks (2003)	Science (multidisciplinary)
Hazard, natural	"Natural processes or phenomena occurring in the biosphere that may constitute a damaging event."	UNDP-BCPR (2004)	United Nations
Human security	"Human Security can no longer be understood in purely military terms. Rather, it must encompass economic development, social justice, environmental protection, democratization, disarmament, and respect for human rights and the rule of law."	Annan (2005)	United Nations
Human security	"The Commission on Human Security's definition of human security: to protect the vital core of all human lives in ways that enhance human freedoms and human fulfilment. Human security means protecting fundamental freedoms – freedoms that are the essence of life. It means protecting people from critical (severe) and pervasive (widespread) threats and situations. It means using processes that build on people's strengths and aspirations. It means creating political, social, environmental, economic, military and cultural systems that together give people the building blocks of survival, livelihood and dignity."	Commission on Human Security (2003)	United Nations
Human security	"In policy terms, human security is an integrated, sustainable, comprehensive security from fear, conflict, ignorance, poverty, social and cultural deprivation and hunger, resting upon positive and negative freedoms."	Van Ginkel and Newman (2000)	United Nations

464

Human security	"Human Security is about attaining the social, political, environmental and economic conditions conducive to a life in freedom and dignity for the individual."	Hammerstad (2000)	United Nations
Human security	[To achieve] "human security, recognizing the inter linkages of environment and society, and acknowledging that that our perceptions of our environment and the way we interact with our environment are historically, socially, and politically constructed. In this context human security is achieved when and where individuals and communities: ● have the options necessary to end, mitigate, or adapt to threats to their human, environmental, and social rights ● have the capacity and the freedom to exercise these options; and ● actively participate in attaining these options. Human security embodies the notion that problems must always be addressed from a broader perspective that encompasses both poverty and issues of equity (social, economic, environmental, or institutional) as it is these issues that often lead to insecurity and conflict."	Lonergan, Gustavson and Carter (2000)	United Nations
Indicator	"An indicator provides evidence that a certain condition exists or certain results have or have not been achieved (Brizius and Campbell: A-15). Indicators enable decision-makers to assess progress towards the achievement of intended outputs, outcomes, goals, and objectives. As such, indicators are an integral part of a results-based accountability system."	Horsch (2004)	Policy making

465

Term	Definition	Source	Discipline
Indicator	"Indicators are statistical measurements, rates, and indices of financial and social trends, used to help economists and financial analysts determine the business growth patterns and the overall direction of the economy."	Investor Dictionary (2004)	Economics
Indicator	"An indicator is the representation of a trend. It trends measurable change in some social, economic, or environmental system over time. Generally an indicator focuses on a small, manageable, and telling piece of a system to give people a sense of the bigger picture."	King County Indicators Initiative Partners (2004)	Civil society
Indicator	"Indicators, in a simple way, can be defined as surrogates or proxy measures of some abstract, multi-dimensional concepts."	Wong (2001)	Science (multidisciplinary)
Livelihood	"The means by which an individual or household obtains assets for survival and self-development. Livelihood assets are the tools (skills, objects, rights, knowledge, social capital) applied to enacting the livelihood."	UNDP-BCPR (2004)	United Nations
Mitigation	"Intervention in the measurement system in order to achieve the desired level of resilience can be defined as mitigation."	Petak (2002)	Geoscience
Mitigation	"Structural and non-structural measures undertaken to limit the adverse impact of natural hazards, environmental degradation and technological hazards."	UN/ISDR (2004)	United Nations

Preparedness	"Activities and measures taken in advance to ensure effective response to the impact of hazards, including the issuance of timely and effective early warnings and the temporary evacuation of people and property from threatened locations."	UN/ISDR (2004)	United Nations
Prevention	"Activities to provide outright avoidance of the adverse impact of hazards and means to minimize related environmental, technological and biological disasters. Depending on social and technical feasibility and cost/benefit considerations, investing in preventive measures is justified in areas frequently affected by disasters. In the context of public awareness and education, related to disaster risk reduction changing attitudes and behaviour contribute to promoting a 'culture of prevention'."	UN/ISDR (2004)	United Nations
Recovery	"Decisions and actions taken after a disaster with a view to restoring or improving the pre-disaster living conditions of the stricken community, while encouraging and facilitating necessary adjustments to reduce disaster risk. Recovery (rehabilitation and reconstruction) affords an opportunity to develop and apply disaster risk reduction measures."	UN/ISDR (2004)	United Nations
Relief/response	"The provision of assistance or intervention during or immediately after a disaster to meet the life preservation and basic subsistence needs of those people affected. It can be of an immediate, short-term, or protracted duration."	UN/ISDR (2004)	United Nations

467

Term	Definition	Source	Discipline
Resilience	"The ability to resist downward pressures and to recover from a shock. From the ecology literature: property that allows a system to absorb and use (even benefit from) change. Where resilience is high, it requires a major disturbance to overcome the limits to qualitative change in a system and allow it to be transformed rapidly into another condition. From the sociology literature: ability to exploit opportunities, and resist and recover from negative shocks."	Alwang, Siegel and Jorgensen (2001)	Social sciences/science (multidisciplinary)
Resilience	"The capacity that people or groups may possess to withstand or recover from emergencies and which can stand as a counterbalance to vulnerability."	Buckle (1998)	Disaster relief
Resilience	"Qualities of people, communities, agencies, infrastructure that reduce vulnerability. Not just the absence of vulnerability rather the capacity to 1) prevent, mitigate losses and then if damage occurs 2) to maintain normal living conditions and to 3) manage recovery from the impact."	Buckle, Marsh and Smale (2000)	Disaster relief/social science
Resilience	"A measure of how quickly a system recovers from failures." (Emergency Mngm. Australia, 1998, quoted in Buckle et al., 2000)	Buckle, Marsh and Smale (2000)	Disaster relief
Resilience	"Not just the absence of vulnerability. Rather it is the capacity, in the first place, to prevent or mitigate losses and then, secondly, if damage does occur to maintain normal living conditions as far as possible, and thirdly, to manage recovery from the impact."	Buckle, Marsh and Smale (2000)	Disaster relief

Term	Definition	Source	Field
Resilience	"Resilience is a measure of the recovery time of a system."	Correia, Santos and Rodrigues (1987)	Engineering
Resilience	"The capacity of a group or organization to withstand loss or damage or to recover from the impact of an emergency or disaster. The higher the resilience, the less likely damage may be, and the faster and more effective recovery is likely to be."	Department of Human Services (2000)	Disaster relief
Resilience	Details of Resilience might be inherently unknowable, especially in the case of complex communities undergoing constant change.	Handmer (2002)	Disaster relief
Resilience	"Resilience is the flip side of vulnerability – a resilient system or population is not sensitive to climate variability and change and has the capacity to adapt."	IPCC (2001) p. 89	United Nations
Resilience	"The capacity of a system, community or society potentially exposed to hazards to adapt by resisting or changing in order to reach and maintain an acceptable level of functioning and structure. This is determined by the degree to which the social system is capable of organizing itself to increase its capacity for learning from past disasters for better future protection and to improve risk reduction measures."	UN/ISDR (2004)	United Nations
Resilience	"The capacity of a system, community or society to resist or to change in order that it may obtain an acceptable level in functioning and structure. This is determined by the degree to which the social system is capable of organising itself, and the ability to increase its capacity for learning and adaptation, including the capacity to recover from a disaster."	UNDP-BCPR (2004)	United Nations

469

Term	Definition	Source	Discipline
Resiliency	"Pliability, flexibility, or elasticity to absorb the event. Resiliency is offered by types of construction, barriers, composition of the land (geological base), geography, bomb shelters, location of dwelling, etc. As resiliency increases, so does the absorbing capacity of the society and/or the environment. Resiliency is the inverse of vulnerability."	*Journal of Prehospital and Disaster Medicine* (2004)	Science (multidisciplinary)
Resiliency	"Resiliency to disasters means a locale can withstand an extreme natural event with a tolerable level of losses. It takes mitigation actions consistent with achieving that level of protection."	Mileti (1999)	Geosciences
Risk	(In this definition risk and hazard are used as synonyms) "Risk is characterized by a known or unknown probability distribution of events. These events are themselves characterized by their magnitude (including size and spread), their frequency and duration, and their history."	Alwang, Siegel and Jorgensen (2001)	Social sciences
Risk	"Risk: the expected number of lives lost, persons injured, damage to property and disruption of economic activity due to a particular natural phenomenon, and consequently the product of specific risk and elements at risk." "Thus, risk is the potential loss to the exposed subject or system, resulting from the 'convolution' of hazard and and vulnerability. In this sense, risk may be expressed in mathematical form as the probability of surpassing a determined level of economic, social or environmental consequences at a certain place and during a certain period of time."	Cardona (2003)	Science (multidisciplinary)

Risk	"Risk is a function of the probability of the specified natural hazard event and vulnerability of cultural entities."	Chapmann (1994)	Natural sciences
Risk	"Risk can be defined as the probability that a system is not in a satisfactory state."	Correia, Santos and Rodrigues (1987)	Engineering
Risk	"'Risk' is the probability of a loss, and this depends on three elements, hazard, vulnerability, and exposure. If any of these three elements in risk increases or decreases, then the risk increases or decreases respectively."	Crichton (1999)	Natural sciences/ insurance industry
Risk	Risk is "the probability of an event multiplied by the consequences if the event occurs."	Einstein (1988)	Natural sciences
Risk	"A combination of the probability or frequency of occurrence of a defined hazard and the magnitude of the consequences of the occurrence. More specific, a risk is defined as the probability of harmful consequences, or expected loss (of lives, people, injured, property, livelihoods, economic activity disrupted or environment damaged) resulting from interactions between natural or human induced hazards."	European Spatial Planning Observ. Netw. (2003)	Science (multidisciplinary)
Risk	"The risk associated with flood disaster for any region is a product of both the region's exposure to the hazard (natural event) and the vulnerability of objects (society) to the hazard. It suggests that three main factors contribute to a region's flood disaster risk: hazard, exposure, and vulnerability."	Hori et al. (2002)	Geosciences

Term	Definition	Source	Discipline
Risk	"The objective (mathematical) or subjective (inductive) probability that the hazard will become an event. Factors (risk factors) can be identified that modify this probability. Such risk factors are constituted by personal behaviours, life-styles, cultures, environmental factors, and inherited characteristics that are known to be associated with health-related questions. Risk is the probability of loss to the elements at risk as the result of the occurrence, physical and societal consequences of a natural or technological hazard, and the mitigation and preparedness measures in place in the community. Risk is the expected number of lives lost, persons injured, damage to property and disruption of economic activity due to a particular natural phenomenon, and consequently the product of specific risk and elements at risk." UNDRO	*Journal of Prehospital and Disaster Medicine* (2004)	Science (multidisciplinary)
Risk	"Risk indicates the degree of potential losses in urban places due to their exposure to hazards and can be thought of as a product of the probability of hazards occurrence and the degree of vulnerability."	Rashed and Weeks (2003)	Geosciences
Risk	"Risk of a system may be defined simply as the possibility of an adverse and unwanted event. Risk may be due solely to physical phenomenon such as health hazards or to the interaction between manmade systems and natural events, e.g. a flood loss due to an overtopped levee. Engineering risk for water resources systems in general has also been described in terms of a figure of merit which is a function of performance indices, say for example, reliability, incident period, and reparability …"	Shrestha (2002)	Engineering

472

Term	Definition	Source	Sector
Risk	"Used in an abstract sense to indicate a condition of the real world in which there is a possibility of loss; also used by insurance practitioners to indicate the property insured or the peril insured against."	Swiss Re (2005)	Insurance industry
Risk	"The expected number of lives lost, persons injured, damage to property and disruption of economic activities due to a particular natural phenomenon, and consequently the product of specific risk and element at risk. Specific risk: The expected degree of loss due to a particular natural phenomenon and as a function of both, natural hazard and vulnerability."	Tiedemann (1992)	Insurance industry
Risk	"The probability of harmful consequences, or expected loss of lives, people injured, property, livelihoods, economic activity disrupted (or environment damaged) resulting from interactions between natural or human induced hazards and vulnerable conditions. Risk is conventionally expressed by the equation: Risk = Hazard × Vulnerability."	UNDP-BCPR (2004)	United Nations
Risk	"The probability of exposure to an event, which can occur with varying severity at different geographical scales, suddenly and expectedly or gradually and predictably, and to the degree of exposure."	UNEP (2002)	United Nations

473

Term	Definition	Source	Discipline
Risk assessment	"Risk assessment has 2 parts: 'Objective' risk assessment of experts and 'Subjective' risk assessment of lay people, called risk perception. Experts base their risk assessment on objective and quantifiable data (probabilities, severity of consequences). Lay people base their risk perception on subjective characteristics: voluntarism, potential consequences on future generations, catastrophicity, dread, number of people exposed, known by science and/or people exposed. In addition, the subjective risk assessment depends on the risk target (children, women, personal, family, world, etc.)."	Chauvin and Hermand (2002) p. 2	Psychology
Risk assessment	"Risk assessment of natural disasters is defined as the assessment on both the probability of natural disaster occurrence and the degree of danger caused by natural disasters."	Hori et al. (2002)	Geosciences
Risk management	"The culture, processes and structures that are directed towards the effective management of potential opportunities and adverse effects.	Britton (2002)	Science (multidisciplinary)
Risk management	"Risk management is a methodology for giving rational considerations to all factors affecting the safety or the operation of large structures or systems of structures. It identifies, evaluates, and executes, in conformity with other social sectors, all aspects of the management of the system, from identification of loads to the planning of emergency scenarios for the case of operational failure, and of relief and rehabilitation for the case of structural failure."	Plate (2002) p. 211	Engineering

Term	Definition	Reference	Discipline
Risk management process	*The Standard* defines the process of risk management as "the systematic application of management policies, procedures and practices to the tasks of establishing the context, identifying, analysing, evaluating, treating, monitoring and communicating risk".	Britton (2002)	Science (multidisciplinary)
Risk, acceptable	"The probability of occurrences of physical, social, or economic consequences of an earthquake that is considered by authorities to be sufficiently low in comparison with the risks from other natural or technological hazards that these occurrences are accepted as realistic reference points for determining design requirements for structures, or for taking social, political, legal, and economic actions in the community to protect people and property."	*Journal of Prehospital and Disaster Medicine* (2004)	Science (multidisciplinary)
Risk, acceptable	"The level of loss a society or community considers acceptable given existing social, economic, political, cultural, technical and environmental conditions." "In engineering terms, acceptable risk is also used to assess structural and non-structural measures undertaken to reduce possible damage at a level which does not harm people and property, according to codes or 'accepted practice' based, among other issues, on a known probability of hazard."	UN/ISDR (2004)	United Nations
Risk, acceptable	"The acceptable probability of losing one's life from an action or an event based on equation: $P_{Epj}(x_{d}) \leq P_{PAcc} = \frac{\beta_i \cdot 10^{-4}/year}{v_{ij}}$	Vrijling, van Hengel and Houben (1995) p. 218	Engineering
Risk, seismic	"Seismic risk consists of the components seismic hazard, seismic vulnerability, and value of elements at risk (both, in human and economic terms)."	Wahlström et al. (2004) and Sinha and Goyal (2004)	Science (multidisciplinary)

Term	Definition	Source	Discipline
Sustainability	"Sustainability is the ability of a locality to tolerate – and overcome – damage, diminished productivity, and reduced quality of life from an extreme event without significant outside assistance."	Mileti (1999)	Geosciences
Sustainability	"In general, there is a consensus that sustainability should encompass social equity, economic growth and environmental protection. Some of the most widely quoted definitions include: • 'Meeting the needs of the present without compromising the ability of future generations to meet their own needs.' (UNCED, 1987) • 'Improving the quality of human life while living within the carrying capacity of supporting ecosystems.' (IUCN/UNEP/WWF, 1991) • 'To equitably meet developmental and environmental needs of present and future generations.' (UNCED, 1992)"	Wong (2001)	United Nations
Sustainable development	"Development that meets the needs of the present without compromising the ability of future generations to meet their own needs. It contains within it two key concepts: the concept of 'needs', in particular the essential needs of the world's poor, to which overriding priority should be given; and the idea of limitations imposed by the state of technology and social organization on the environment's ability to meet present and the future needs (Brundtland Commission, 1987). Sustainable development is based on socio-cultural development, political stability and decorum, economic growth and ecosystem protection, which all relate to disaster risk reduction."	UN/ISDR (2004)	United Nations

Term	Definition	Source	Discipline
Sustainable development	"Development that meets the needs of the present without compromising the ability of future generations to meet their own needs. It contains within it two key concepts: the concept of 'needs', in particular the essential needs of the world's poor, to which overriding priority should be given; and the idea of limitations imposed by the state of technology and social organisation on the environment's ability to meet present and future needs."	UNDP-BCPR (2004)	United Nations
Vulnerability	"Vulnerability should be recognized as a key indicator of the seriousness of environmental problems such as global warming."	Adger, Kelly and Bentham 2001	Science (multidisciplinary)
Vulnerability	"The insecurity of the well-being of individuals, households or communities in the face of a changing environment." (Moser and Holland, 1989, quoted in Alwang, Siegel and Jorgensen, 2001).	Alwang, Siegel and Jorgensen (2001)	Social sciences
Vulnerability	"Is the characteristic of a person or a group in terms of their capacity to anticipate, cope with, resist, and recover from the impact of a natural disaster" (Blakie et al. 1994 p. 9 quoted in Alwang, Siegel and Jorgensen, 2001). "The Extent of a disaster cannot be measured without knowledge of the resilience of the affected groups; this resilience plays out over time."	Alwang, Siegel and Jorgensen (2001)	Disaster management
Vulnerability	"Summarizing livelihood and environmental literature: vulnerability is the exposure of individuals or groups to livelihood stress as a result of environmental change."	Alwang, Siegel and Jorgensen (2001)	Science (multidisciplinary)

477

Term	Definition	Source	Discipline
Vulnerability	"Vulnerability concept consists of two opposing forces: On one hand, the processes that cause vulnerability that can be observed; on the other hand, the physical exposure to hazards (earthquakes, storms, floods, etc.). Vulnerability develops then from underlying reasons in the economic, demographic and political spheres into insecure conditions (fragile physical environment, instable local economy, vulnerable groups, lack of state or private precautions) through the so-called dynamic processes (e.g., lack of local institutions, under-developed markets, population growth, and urbanization)."	Blaikie et al. (1994) pp. 21–26	Social sciences
Vulnerability	"Vulnerability (in contrast to poverty which is a measure of current status) should involve a predictive quality: it is supposedly a way of conceptualizing what may happen to an identifiable population under conditions of particular risk and hazards. Is the complex set of characteristics that include a person's • initial well-being (health, morale, etc.) • self-protection (asset pattern, income, qualifications, etc.) • social protection (hazard preparedness by society, building codes, shelters, etc.) • social and political networks and institutions (social capital, institutional environment, etc.)."	Cannon, Twigg and Rowell (2003)	Social sciences

Term	Definition	Source	Discipline
Vulnerability	"Vulnerability: the degree of loss to a given element at risk or set of such elements resulting from the occurrence of a natural phenomenon of a given magnitude and expressed on a scale from 0 (no damage) to 1 (total loss). On the other hand, vulnerability may be understood, in general terms, as an internal risk factor, mathematically expressed in terms of the feasibility that the exposed subject or system will be affected by the phenomenon that characterizes the hazard."	Cardona (2003)	Science (multidisciplinary)
Vulnerability	"Vulnerability … is not the same as poverty. In means not lack or want, but defencelessness, insecurity, and exposure to risk, shocks and stress…. Vulnerability here refers to exposure to contingencies and stress, and difficulty in coping with them."	Chambers (1989)	Social sciences
Vulnerability	"The vulnerability of a given entity (system, sector, region, etc.) with respect to Global Change may tentatively be defined as the expected damage as resulting from the expected environmental perturbations in view of the expected transformation and adaptation processes."	Corell, Cramer and Schellnhuber (2001)	Natural sciences
Vulnerability	"Vulnerability expresses the severity of failure in terms of its consequences. The concern is not how long the failure lasts but how costly it is."	Correia, Santos and Rodrigues (1987)	Engineering
Vulnerability	"Is a broad measure of the susceptibility to suffer loss or damage. The higher the vulnerability, the more exposure there is to loss and damage."	Department of Human Services (2000)	Social sciences

479

Term	Definition	Source	Discipline
Vulnerability	"The degree of loss to a given element at risk (or set of elements) resulting from a given hazard at a given severity level" In contrast to the concept of risk, here the probability of the occurrence of a hazard is not considered." (UNDP-BCPR/UNDHA, 1994: 38–39; see also UNDHA, 1992). "Vulnerability has process character and is not static."	Feldbrügge and von Braun (2002)	United Nations
Vulnerability	"Vulnerability (V) = Hazard – Coping with: Hazard = H (Probability of the hazard or process; shock value; predictability; prevalence; intensity/strength); and Coping = C (Perception of risk and potential of an activity; possibilities for trade; private trade, open trade)."	Feldbrügge and von Braun (2002) p. 11	Science (multidisciplinary)
Vulnerability	"Determinants of disaster vulnerability: • demographic factors: population growth, urbanization, settlements near coastal areas, etc. • the state of economic development: poverty, modernization processes • environmental changes: climate changes, degradation and depletion of resources (straightening the courses of rivers, deforestation, etc.) • political factors, an increase in tangible assets, which leads to an increase in damages, effects of disaster protection structures and research, and the interactions of the causes of disasters."	Feldbrügge and von Braun (2002) p. 14	Science (multidisciplinary)

480

Vulnerability	"The likelihood that some socially defined group in society will suffer disproportionate death, injury, loss or disruption of livelihood in an extreme event, or face greater than normal difficulties in recovering from a disaster."	Handmer and Wisner (1998)	Science (multidisciplinary)
Vulnerability	"The characteristics of a person or group in terms of their capacity to anticipate, cope with, resist and recover from the impact of a natural or man-made hazard."	IFRC (1999)	Disaster management
Vulnerability	"Vulnerability is defined as the extent to which a natural or social system is susceptible to sustaining damage from climate change. Vulnerability is a function of the sensitivity of a system to changes in climate (the degree to which a system will respond to a given change in climate, including beneficial and harmful effects), adaptive capacity (the degree to which adjustments in practices, processes, or structures can moderate or offset the potential for damage or take advantage of opportunities created by a given change in climate), and the degree of exposure of the system to climatic hazards."	IPCC (2001) p. 89	United Nations
Vulnerability	"The potential loss in value of an element at risk from the occurrence and consequences of natural and technological hazards. The factors that influence vulnerability include: demographics, the age and resilience of the built environment, technology, social differentiation and diversity, regional and global economies, and political arrangements. Vulnerability is a result of flaws in planning, siting, design, and construction. Vulnerability is the degree of loss to a given element at risk, or set of such elements, resulting from the occurrence of a natural phenomenon of a given magnitude and expressed on a scale from 0 (= no damage) to 1 (= total loss)." UNDRO	Journal of Prehospital and Disaster Medicine (2004)	Science (multidisciplinary)

481

Term	Definition	Source	Discipline
Vulnerability	"Vulnerability is provisionally defined as the degree to which a system is sensitive to and unable to cope with adverse impacts of global change stimuli. Vulnerability is therefore a function of a system's exposure to global change stimuli and its adaptive capacity, that is, its ability to cope with these stimuli."	Klein (2003)	Science (multidisciplinary)
Vulnerability	"Vulnerability defines the inherent weakness in certain aspects of the urban environment with are susceptible to harm due to social, biophysical, or design characteristics."	Rashed and Weeks (2003)	Science (multidisciplinary)
Vulnerability	Is the predisposition of being susceptible to injuries, attacks or to have difficulties to reconstitute a compromised state of health. All depends on the vulnerable components placed at the centre of our system: 1.) vulnerability of human beings to natural hazards of the planet, depending on their systems, behaviours and reactions of individuals. 2.) formally more or less fragile natural environments that have been settled, often in excess, and that have become vulnerable due the increase human activity. 3.) Nature itself. 4.) vulnerabilities: man, goods, activities, and the environment. Translated from Reveau (2004).	Reveau (2004)	Science (multidisciplinary)
Vulnerability	"Vulnerability is usually defined as the capacity of a system to be wounded from a stress or perturbation. It is a function of the probability of occurrence of the perturbation and its magnitude, as well as of the ability of the system to absorb and recover from such perturbation."	Suarez (2002)	Science (multidisciplinary)

Term	Definition	Source	Organization
Vulnerability	"The degree of loss to a given element at risk or set of such elements resulting from the occurrence of a natural phenomenon of a given magnitude and expressed on a scale from 0 (no damage) to 1 (total loss) or in percent of the new replacement value in the case of damage to property."	Tiedemann (1992) and Buckle, Marsh and Smale (2000)	
Vulnerability	"The conditions determined by physical, social, economic, and environmental factors or processes, which increase the susceptibility of a community to the impact of hazards. For positive factors, which increase the ability of people to cope with hazards, see definition of capacity."	UN/ISDR (2004)	United Nations
Vulnerability	"A human condition or process resulting from physical, social, economic and environmental factors, which determine the likelihood and scale of damage from the impact of a given hazard."	UNDP-BCPR (2004)	United Nations
Vulnerability	"Vulnerability is expressed as the degree of expected damage (i.e., the cost of repair divided by the cost of replacement) given on a scale of 0 to 1, as a function of hazard intensity (or magnitude, depending on the convention used)."	UNDRO (1991) p. 79	United Nations
Vulnerability	"Represents the interface between exposure to the physical threats to human well-being and the capacity of people and communities to cope with those threats."	UNEP (2002)	United Nations

483

Term	Source	Discipline	Definition
Vulnerability	Vrijling, van Hengel and Houben (1995) p. 218	Engineering	"The vulnerability increases with the number of people affected by the impact of a natural hazard, given by the formula: $v_{ij} = 10^{-23} \cdot n_j^2$, for $n \geq 10$ casualties", where v_{ij} is the vulnerability of an individual i at location j.
Vulnerability (urban)	Mileti (1999)	Science (multidisciplinary)	"Urban vulnerability to natural hazards such as earthquakes is a function of human behaviour. It describes the degree to which socioeconomic systems and physical assets in urban areas are either susceptible or resilient to the impact of natural hazards. Vulnerability is independent from any particular magnitude from a specific natural event but dependent on the context in which it occurs. The characteristic of the urban community that can be assessed through a combination of ecological factors associated with the physical conditions of the population in that place. The physical and social conditions are inextricably bound together in many disaster situations that we can use the former as indicative of the latter. V is continuously modified by human actions and therefore it varies over space and time. V cannot be assessed in absolute terms; the performance of the urban place should be assessed with reference to specific spatial and temporal scales. The adaptive and coping capacities that determine the extent to which a society can tolerate damage from extreme events without significant outside assistance."

Conclusions

This comparative glossary demonstrates just how widely the definitions of a single term can range. Many terms are tightly interwoven and are even used interchangeably. The list informs the reader about the multiple definitions in use across various disciplines and sectors, which is an important stepping stone to dispelling the often lamented misunderstandings that arise in discussions of disaster reduction. What the above listing fails to offer is a harmonised concept of core terms that is precise enough to delineate the terms from each other, yet flexible and broad enough so as to be applicable across sectors, disciplines and scales on which disaster reduction operates.

Terms such as "vulnerability" and "risk" are envelopes for complex and interconnected parameters and processes. A paradigm shift has taken place that puts more and more emphasis on non-natural science issues. These are harder to conceptualise since they are often not tangible, or of qualitative nature; they include factors such as coping capacity, resilience, institutional frameworks, and cultural and social aspects.

Terms of such complexity are not easily defined in an exhaustive way. This does not matter, as it is more important to agree on their key characteristics. In that way, it is possible to create a conceptual frame whose content will vary with context, geographic scale and time scale. In the next section, the characteristics of some central terms are described in such a way that they all fit into a logically coherent framework. Once the basic framework is established, each term can always be defined more precisely to fit the specific context, use and scale.

Hazard

Every disaster starts with a hazard, known or unknown. There are many ways to characterise hazards: natural, technical, man-made, nuclear, ecological and so on. The categories are probably as diverse as the disciplines and sectors involved. But they all have in common the potential to cause the severe adverse effects that lie at the bottom of every emergency, disaster and catastrophe.

A hazard can be as general as a "flood" or "storm" and, as such, stand for groups of potentially harmful events of variable severity. In other words, the hazard "storm" refers to all potential wind speeds that can be expected in a given region. A hazard can also be formulated more specifically as a magnitude 7.2 earthquake in Los Angeles or a category 5 hurricane hitting the Philippines. In that case we are dealing with a specific hazard scenario. One important feature of hazard is that it has the notion of probability or a likelihood of occurring. A hazard is a threat,

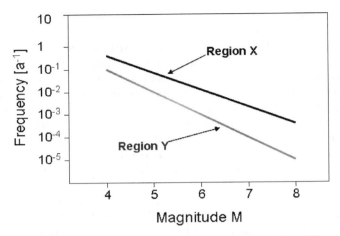

Figure 24.1 For earthquake hazard, the two lines represent the different magnitude-frequency relationships for two different fictitious regions, region x and region y. The two lines are region-specific.
Source: Author.

not the actual event. Any hazard can manifest itself in an actual harmful event. In other words, if it can be measured in terms of real damage or harm it is no longer a hazard but has become an event, disaster or catastrophe.

Every specific hazard magnitude is attached to a site-specific return period, which is usually empirically derived. The return period of a category 5 hurricane is different for New Orleans than for the Philippines. If a hazard is described more broadly – as "epidemic", "drought" or "flood", for example – it is characterised by all possible magnitudes. In order to quantify hazard each magnitude is tied to a specific return period or to its inverse, frequency. The latter ensemble is the magnitude–frequency relationship of a particular hazard and it is always an inherent characteristic of a specific locality or region.

Vulnerability

Another prerequisite for a disaster besides hazard is vulnerability. Vulnerability is a dynamic, intrinsic feature of any community (or household, region, State, infrastructure or any other element at risk) that comprises a multitude of components. The extent to which it is revealed is determined by the severity of the event.

Vulnerability indicates damage potential and is a forward-looking variable. Or as Cannon, Twigg and Rowell (2003) characterised it, "vulnerability (in contrast to poverty, which is a measure of current status) should involve a predictive quality: it is supposedly a way of conceptualizing what may happen to an identifiable population under conditions of particular risk and hazards." Determining vulnerability means asking what would happen if certain events had impacts on particular elements at risk (e.g. a community).

Vulnerability is an intrinsic characteristic of a community that is always there even in quiescent times between events. It is not switched on and off with the coming and going of events; rather, it is a permanent and dynamic feature that is revealed during an event to an extent that depends on the magnitude of the harmful event. This means that vulnerability can often only be measured indirectly and retrospectively, and the dimension normally used for this indirect measure is damage or more general harm.

What is normally seen in the aftermath of a disaster is not the vulnerability per se, but the harm done. Seeing the damage pattern of a community without knowing the magnitude of the event does not allow conclusions regarding the community's vulnerability. In that sense the magnitude–damage relationship reflects the vulnerability of an element at risk (community, household, nation, infrastructure, etc.).

Figure 24.2 illustrates the progression of wind damage. Tornado intensities are marked from F0 to F5 on the Fujita Scale. The full relationship between wind speed and damage characterises the physical vulnerability of a certain building type.

Vulnerability changes continuously over time, and indeed is usually affected by the harmful event itself. It can increase, for example, if pov-

Figure 24.2 Sample residential damage function for the hazard of a tornado.
Source: Doggett, 2003.

erty has been heightened by a disaster, so that the next disaster will have an even more devastating effect on the impoverished community. A small event, however, can raise the awareness of the community and in that way decrease its vulnerability.

Vulnerability is a function of the sensitivity or susceptibility of a system (community, household, building, infrastructure, nation etc.). It is "independent from any particular magnitude from a specific natural event but dependent on the context in which it occurs. Vulnerability cannot be assessed in absolute terms; the performance of the urban place should be assessed with reference to specific spatial and temporal scales" (Rashed and Weeks, 2003).

For practical reasons a vulnerability analysis will often limit itself to a certain scenario – that is, event magnitude – for which an analysis is carried out. This is usually an appropriate approach to assessing vulnerability, but the choice of the event scenario is a subjective one. Which scenario should be chosen: the 100-year event, 200-year event, the largest event that has occurred in the living memory, or the 5-metre flood level?

In earthquake engineering, this susceptibility is often quantified by means of a damage ratio that can vary between no damage (0 per cent) and total destruction (100 per cent). But vulnerability has many dimensions – physical (built environment), social, economic, environmental, institutional and human – and many of them are not easily quantifiable.

The complexity of vulnerability is shown not only by its multiple dimensions but also by the fact that it is site specific and that its parameters change with geographic scale. The parameters that determine vulnerability vary according to the household, community and country level. In the economic dimension of the household level, parameters such as the amount and diversity of income of single persons are relevant, whereas on a country level, inflation rate and GDP are more appropriate.

The limitations of vulnerability theory in addressing complex and dynamic reality are noted in Duryog Nivaran's book, *Understanding Vulnerability*:

Vulnerability is too complicated to be captured by models and frameworks. There are so many dimensions to it: economic, demographic, political, and psychological. There are so many factors making people vulnerable: not just a range of immediate causes but – if one analyses the subject fully – a host of root causes too. Investigations of vulnerability are investigations into the workings of human society, and human societies are complex – so complex and diverse that they easily break out of any attempts to confine them within the neatly drawn frameworks, categories, and definitions. They are also dynamic, in a state of constant change, and, because they are complex and diverse, all the elements within societies are moving, so that these changes occur in different parts of society, in different ways and at different times. (Quoted in Twigg, 1998)

On a more optimistic note, every vulnerability analysis has to be adapted to its specific objectives and scales. Professionals in the field must be aware that there are many answers to the question of vulnerability. One potential answer to the question of vulnerability is given by Birkmann (this volume), who defines vulnerability in a more encompassing way so that it includes exposure and the coping capacities of a community.

Exposure

Together with vulnerability and hazard, exposure is another prerequisite of risk and disaster. Here, exposure is understood as the number of people and/or other elements at risk that can be affected by a particular event. In an uninhabited area the human exposure is zero. No matter how many hurricanes affect an uninhabited island, the human exposure, and hence the risk of human loss, remains zero. While the vulnerability determines the severity of the impact an event will have on the elements at risk, it is the exposure that drives the final tally of damage or harm. So in its economic dimension, vulnerability is depicted by the projection that in a given event a family will probably lose 50 per cent of its assets. The number of families that will be affected and lose 50 per cent of their assets is related to the exposure. In an overly simplified example, the poverty of a community will determine the degree to which it will be affected by an event of a certain magnitude ($\rightarrow$ susceptibility) and the number of community members represents the exposure. In that sense a densely populated area is at a higher risk than a sparsely populated one, all other conditions being equal.

Coping capacity and resilience

In real life the harm done depends not only on hazard, vulnerability and exposure, but also on the coping capacity and the resilience of the element at risk. In the literature most definitions show a large overlap between coping capacity and resilience, which are often used as synonyms. These two dimensions of a harmful event are not easily separated from each other.

Here, coping capacity encompasses those strategies and measures that act directly upon damage during the event by alleviating or containing the impact or by bringing about efficient relief, as well as those adaptive strategies that modify behaviour or activities in order to circumvent or avoid damaging effects.

Resilience is all of these things, plus the capability to remain functional

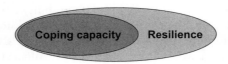

Figure 24.3 Coping capacity and resilience are hard to delineate. Resilience is understood to be the more encompassing term.
Source: Author.

during an event and to completely recover from it. So resilience includes coping capacity but at the same time goes beyond it.

The difficult question that arises from this definition is: does vulnerability already account for coping capacity and resilience or are they separate and counteracting parameters? The answer depends on how we define the damage or harm caused. If the extent of the damage or harm is defined also by the duration of the adverse effects and by its repercussions on people's poverty, economy or awareness, then vulnerability has to include coping capacity and resilience. This conclusion follows from the postulation that vulnerability describes susceptibility to damage or harm.

Risk

Vulnerability is measured in terms of expected harm or damage and so is risk. How can those terms be delineated from each other?

Risk always involves the notion of probability of occurrence. So information on "when" or on "how often" indicates that we are talking about risk. This could be captured in a continuous damage–frequency relationship or just as the definition of the return period for a particular event scenario. While vulnerability informs about the consequences of possible adverse events, risk also provides information on how often or with what probability those scenarios have to be expected.

For example, information on expected losses for an event during which the water level rises 5 metres above normal refers to hazard and vulnerability. Information on expected losses for a 200-year event during which the water level increases by 5 metres above normal refers to risk. In another context, projecting the consequences of a 15-metre tsunami is important, but in order to make informed disaster management decisions it is necessary to know how often such an event can be expected. Disaster management decisions are based on risk and not only on hazard.

Despite all the known shortcomings of databases of historic events, they do provide some means to create a magnitude–frequency relationship over a range of event magnitudes. This magnitude–frequency relationship can be an important tool for supporting the decision-making

Figure 24.4 Risk seen as a function of hazard, vulnerability, exposure and resilience, while the mathematical relationship between the variables is unknown. Source: Author.

process with respect to the level of acceptable risk. Responsible disaster managers have to decide for what type of event a community should be prepared. To prepare for the biggest possible event would be the safest way to go, but this is rarely economically feasible; such high levels of protection are simply unaffordable and the benefits would not justify the costs. In addition, maintenance and alertness would be unmanageable over such long periods of time because the largest events can only be expected to occur after many years of quiescence.

To summarise, risk is understood as a function of hazard, vulnerability, exposure and

resilience (see also Figure 24.4 above):

$$\text{Risk} = f(\text{hazard, vulnerability, exposure, resilience})$$

The frequency or return period of adverse effects allows the individual or official decision maker to define a level of acceptable consequences. This is only possible if the decision maker understands what events to expect over time. Decisions will be different for a 10-year event than for a 5,000-year event. For decision-making, information on the probability of occurrence is crucial.

Often the historical record is too short to provide reliable magnitude–frequency relationships for particular hazards and regions. In addition, climate change has started to alter those relationships. This can be seen in Germany where the return period of the 100-year event for the Rhine and the Danube had to be revised as a 20-year or even a 10-year event (Alt, 2002). It is seen also in the United States, where the Missouri River has had six 100-year floods since 1946 (Albright Seed Company, 1998). Is this a fluke of nature or a real trend? It is hard to decide. But many scientists agree that the trend is strongly supported by data. In situations of uncertainty it would be most appropriate to heed the precautionary principle. After all, if we are not even prepared to deal with the current risk situation, how will we cope with and adapt to a deteriorating situation due to climate change?

REFERENCES

Adger, N., M. Kelly and G. Bentham (2001) "New Indicators of Vulnerability and Adaptive Capacity", paper presented at the International Workshop on Vulnerability and Global Environmental Change, Stockholm, 17–19 May 2001.

Albright Seed Company (1998) *The 100-year Flood: Why It Comes Every 10 Years*, available at http://www.albrightseed.com/flood100.htm.

Allen, M.R. and R. Lord (2004) "The Blame Game", *Nature* 432: 551–552.

Alt, F. (2002) *Jahrhunderthochwasser Wird es Öfter Geben*, available at http://www.sonnenseite.com/.

Alwang, J., P.B. Siegel and S.L. Jorgensen (2001) "Vulnerability: A View from Different Disciplines", Social Protection Discussion Paper No. 115, The World Bank, Social Protection Unit, Human Development Network, available at http://www1.worldbank.org/sp/.

Annan, K. (2005) *Towards a Culture of Peace: Letters to Future Generations*, available at http://www.unesco.org/opi2/lettres/TextAnglais/AnnanE.html.

Blaikie, P., T. Cannon, I. Davis and B. Wisner (1994) *At Risk: Natural Hazards, People's Vulnerability and Disasters*, London: Routledge.

Britton, N.R. (2002) "EdM Framework for Urban Vulnerability Reduction", Proceedings of the Second Annual IIASA-DPRI Meeting "Integrated Disaster Risk Management: Megacity Vulnerability and Resilience", Laxenburg, Austria, 29–31 July 2002, available at http://www.iiasa.ac.at/Research/RMS/dpri2002/Papers/britton.pdf.

Buckle, P. (1998) "Re-defining Community and Vulnerability in the Context of Emergency Management", *Australian Journal of Emergency Management*, Summer 1998/99: 21–29, available at http://online.northumbria.ac.uk/geography_research/radix/resources/buckle-community-vulnerability.pdf.

Buckle, P., G. Marsh and S. Smale (2000) "New Approaches to Assessing Vulnerability and Resilience", *Australian Journal of Emergency Management*, Winter 2000: 8–15, available at http://online.northumbria.ac.uk/geography_research/radix/resources/buckle-marsh.pdf.

Buckle, P., G. Marsh and S. Smale (2001) *Assessment of Personal and Community Resilience and Vulnerability*, Report: EMA Project 15/2000, available at http://online.northumbria.ac.uk/geography_research/radix/resources/assessment-of-personal-and-community-resilience.pdf.

Cannon, T. (1994) "Vulnerability Analysis and Natural Disasters", in A. Varley, ed., *Disasters, Development and Environment*, West Sussex: Wiley.

Cannon, T., J. Twigg and J. Rowell (2003) *Social Vulnerability: Sustainable Livelihoods and Disasters, Report to DFID Conflict and Humanitarian Assistance*

Department (CHAD) and Sustainable Livelihoods Support Office, available at http://www.benfieldhrc.org/disaster_studies/projects/soc_vuln_sust_live.pdf.

Cardona, O.D. (2003) *The Notion of Disaster Risk: Conceptual Framework for Integrated Risk Management*, IADB/IDEA Program on Indicators for Disaster Risk Management, Universidad Nacional de Colombia, Manizales, available at http://idea.manizales.unal.edu.co/ProyectosEspeciales/adminIDEA/CentroDocumentacion/DocDigitales/documentos/01%20Conceptual%20Framework%20IADB-IDEA%20Phase%20I.pdf.

Chambers, R. (1989) "Editorial Introduction: Vulnerability, Coping and Policy", in *Vulnerability: How the Poor Cope*, IDS Bulletin 20(2): 1–7.

Chapmann, D. (1994) *Natural Hazards*, Oxford: Oxford University Press.

Chauvin, B. and D. Hermand (2002) "From Disaster Risk Assessment to Risk Management in Megacities", Proceedings of the Second Annual IIASA-DPRI Meeting "Integrated Disaster Risk Management: Megacity Vulnerability and Resilience", Laxenburg, Austria, 29–31 July 2002, available at http://www.iiasa.ac.at/Research/RMS/dpri2002/Papers/chauvin.pdf.

Commission on Human Security (CHS)(2003) *Human Security Now*, New York, available at http://www.humansecurity-chs.org/finalreport/FinalReport.pdf.

Corell, R., W. Cramer and H.-J. Schellnhuber (2001) Potsdam Sustainability Days. Symposium on "Methods and Models of Vulnerability Research, Analysis and Assessment".

Correia, F., M. Santos and R. Rodrigues (1987) "Engineering Risk in Regional Drought Studies", in L. Duckstein and E.J. Plate, eds, *Engineering Reliability and Risk in Water Resources*, Dordrecht/Bosten: Martinus Nijhoff, pp. 61–86.

Crichton, D. (1999) "The Risk Triangle", in J. Ingleton, ed., *Natural Disaster Management*, London: Tudor Rose, pp. 102–103.

Department of Human Services (2000) *Assessing Resilience and Vulnerability in the Context of Emergencies: Guidelines*, Melbourne: Victorian Government Publishing Service.

Department of Humanitarian Affairs (DHA, now OCHA) (1994) *Disaster Mitigation. Disaster Management Training Programme*, 2nd edn, available at http://www.proventionconsortium.org/files/undp/DisasterMitigation.pdf.

Department for International Development (DFID) (2000) *Sustainable Livelihoods Guidance Sheets*, London: DFID, available at http://www.livelihoods.org/info/info_guidancesheets.html.

DKKV (German Committee for Disaster Reduction) (2002) *Journalist's Manual on Disaster Management 2002*, 7th revised and supplemented edition, Bonn: DKKV.

Doggett, T. (2003) *The Impact of Severe Thunderstorms: The Forgotten Peril*, available at http://www.air-worldwide.com/_public/images/pdf/The_Impact_of_Severe_Thunderstorms.pdf.

Einstein, H.H. (1988) "Landslide Risk Assessment Procedure", in *Proceedings of the 5th International Symposium on Landslides*, Lausanne, pp. 1075–1090.

European Spatial Planning Observation Network (2003) *Glossary of Terms*, available at http://www.gsf.fi/projects/espon/glossary.htm.

Feldbrügge, T. and J. von Braun (2002) "Is the World Becoming a More Risky

Place? Trends in Disasters and Vulnerability to Them", ZEF – Discussion Papers on Development Policy No. 46, Bonn: Center for Development Research.

Hammerstad, A. (2000) "Whose Security? UNHCR, Refugee Protection and State Security After the Cold War", *Security Dialogue* 31(4): 395.

Handmer, J. (2002) *We are All Vulnerable*, available at http://online. northumbria.ac.uk/geography_research/radix/resources/vulmeeting-pbmelbourne11.doc/.

Handmer, J. and B. Wisner (1998) "Conference Report: Hazards, Globalization and Sustainability", *Development in Practice* 9(3): 342–346.

Hori, T., J. Zhang, H. Tatano, N. Okada and S. Iikebuchi (2002) "Microzonation-based Flood Risk Assessment in Urbanized Floodplain", Proceedings of the Second Annual IIASA-DPRI Meeting "Integrated Disaster Risk Management: Megacity Vulnerability and Resilience", Laxenburg, Austria, 29–31 July 2002, available at http://www.iiasa.ac.at/Research/RMS/dpri2002/Papers/hori.pdf.

Horsch, K. (2004) *Indicators: Definition and Use in a Results-Based Accountability System*, Harvard Family Research Project, available at http://www.gse. harvard.edu/hfrp/pubs/onlinepubs/rrb/indicators.html.

International Federation of Red Cross and Red Crescent Societies (IFRC) (1993) *World Disaster Report 1993*, Geneva: IFRC.

IFRC (1999) *Vulnerability and Capacity Assessment: An International Federation Guide*, Geneva: IFRC.

Investor Dictionary.com (2004) *Economic Indicators*, available at http://www. investordictionary.com/definition/Economic+indicators.aspx/.

Intergovernmental Panel for Climate Change (IPCC) (2001) *Climate Change 2001, Synthesis Report: A Contribution of Working Groups I, II, and III to the Third Assessment Report of the Intergovernmental Panel on Climate Change*, R.T. Watson, et al., eds, Cambridge/New York: Cambridge University Press.

Journal of Prehospital and Disaster Medicine (2004) "Glossary of Terms", available at http://pdm.medicine.wisc.edu/vocab.htm.

King County Indicators Initiative Partners, *Communities Count*, available at http://www.communitiescount.org/indicator_descrip.htm.

Klein, R. (2003) *Environmental Vulnerability Assessment*, available at http:// www.pik-potsdam.de/~richardk/eva/.

Lavell, A. (2003) "International Agency Concepts and Guidelines for Disaster Risk Management", in A. Lavell, ed., *Information and Indicators Program for Disaster Risk Management*, IADB/IDEA/ECLAC, Manizales: National University of Colombia, pp. 1–7, available at http://idea.manizales.unal.edu.co/ ProyectosEspeciales/adminIDEA/CentroDocumentacion/DocDigitales/ documentos/AllanLavellEMBarcelonaJuly2003.pdf.

Lonergan, S., K. Gustavson, and B. Carter (2000) "The Index of Human Security", *AVISO Bulletin* 6, available at http://www.gechs.org/aviso/AvisoEnglish/ six/six.shtml.

Luhmann, N. (1993) "Die Moral des Risikos und das Risiko der Moral", in Bechmann, G., Hrsg., *Risiko und Gesellschaft*, Opladen: Westdeutscher Verlag, pp. 327–338.

Mileti, D.S. (1999) *Disasters By Design: A Reassessment of Natural Hazards in the United States*, Brookfield: Rothstein Associates.

MunichRe (2002) *MunichRe Group Annual Report 2002*, Munich: MunichRe Group, available at http://www.munichre.com/publications/302-03661_en.pdf.

Petak, W.J. (2002) "Earthquake Resilience Through Mitigation: A System Approach", Proceedings of the Second Annual IIASA-DPRI Meeting "Integrated Disaster Risk Management: Megacity Vulnerability and Resilience", Laxenburg, Austria, 29–31 July 2002, available at http://www.iiasa.ac.at/Research/RMS/dpri2002/Papers/petak.pdf.

Plate, E.J. (2002) "Risk Management for Hydraulic Systems under Hydrological Loads", in J.J. Bogardi and Z.W. Kundzewicz, eds, *Risk, Reliability, Uncertainty, and Robustness of Water Resources Systems*, UNESCO International Hydrology Series, Cambridge: Cambridge University Press, pp. 209–220.

Quarantelli, E.L. (1998) "Epilogue: Where We Have Been and Where We Might Go", in E.L. Quarantelli, ed., *What is a Disaster? Perspectives on the Question*, London: Routledge, pp. 234–273.

Rahn, P.H. (1996) *Engineering Geology: An Environmental Approach*, 2nd edn, Upper Saddler River: PTR Prentice Hall, 657 pp.

Rashed, T. and J. Weeks (2003) "Assessing Vulnerability to Earthquake Hazards through Spatial Multicriteria Analysis of Urban Areas", *International Journal of Geographical Information Science* 17(6): 547–576.

Reveau, P. (2004) "Intérêts et Limites des Études de Vulnérabilité", *Risques Naturelles* 36: 36–42, September/October.

Schneiderbauer, S. and D. Ehrlich (2004) *Risk, Hazard and People's Vulnerability to Natural Hazards: A Review of Definitions, Concepts and Data*, Brussels: European Commission-Joint Research Centre (EC-JRC).

Shamir, U. (2002) "Risk and Reliability in Water Resources Management: Theory and Practice", in J.J. Bogardi and Z.W. Kundzewicz, eds, *Risk, Reliability, Uncertainty, and Robustness of Water Resources Systems*, UNESCO International Hydrology Series, Cambridge: Cambridge University Press, pp. 162–168.

Shrestha, B.P. (2002) "Uncertainty in Risk Analysis of Water Resources Systems Under Climate Change", in J.J. Bogardi and Z.W. Kundzewicz, eds, *Risk, Reliability, Uncertainty, and Robustness of Water Resources Systems*, UNESCO International Hydrology Series, Cambridge: Cambridge University Press, pp. 153–160.

Sinha, R. and A. Goyal (2004) "Seismic Risk Scenario for Mumbai", in D. Malzahn and T. Plapp (eds) *Disasters and Society: From Hazard Assessment to Risk Reduction*, Logos Verlag Berlin, pp. 107–114.

Suarez, P. (2002) "Urbanization, Climate Change and Flood Risk: Addressing the Fractal Nature of Differential Vulnerability", Proceedings of the Second Annual IIASA-DPRI Meeting "Integrated Disaster Risk Management: Megacity Vulnerability and Resilience", Laxenburg, Austria, 29–31 July 2002, available at http://www.iiasa.ac.at/Research/RMS/dpri2002/Papers/suarez.pdf.

Susman, P., P. O'Keefe and B. Wisner (1983) "Global Disasters: A Radical Interpretation", in K. Hewitt, ed., *Interpretations of Calamity*, Boston: Allen and Unwin, pp. 263–283.

Swiss Re (2005) *Online Glossary*, available at http://www.swissre.com/.

Tiedemann, H. (1992) *Earthquakes and Volcanic Eruptions: A Handbook on Risk Assessment*, Geneva: SwissRe.

Twigg, J. (2001) "Sustainable Livelihoods and Vulnerability to Disasters", Disaster Management Working Paper 2/2001, Benfield Greig Hazard Research Centre, available at http://www.benfieldhrc.org/disaster_studies/working_papers/workingpaper2.pdf.

UN/ISDR (International Strategy for Disaster Reduction) (2004) *Living with Risk: A Global Review of Disaster Reduction Initiatives*, Geneva: UN Publications.

United Nations Development Programme-Bureau for Crisis Prevention and Recovery (UNDP-BCPR) (2004) *Reducing Disaster Risk: A Challenge for Development*. A Global Report, New York: UNDP Publications.

UNDRO (Office of the United Nations Disaster Relief Co-Ordinator) (1991) *Mitigating Natural Disasters: Phenomena, Effects and Options. A Manual for Policy Makers and Planners*, Geneva: UNDRO/MND.

United Nations Environment Programme (UNEP) (2002) *Global Environment Outlook 3: Past, Present and Future Perspectives*, London: Earthscan.

van Ginkel, H. and E. Newman (2000) "In Quest of Human Security", *Japan Review of International Affairs* 14(1): 79.

Vrijling, J.K., W. van Hengel and R.J. Houben (1995) "A Framework for Risk Evaluation", *Journal of Hazardous Materials* 43: 245–261; quoted in J.J. Bogardi and Z.W. Kundzewicz, eds, *Risk, Reliability, Uncertainty, and Robustness of Water Resources Systems*, UNESCO International Hydrology Series, Cambridge: Cambridge University Press, p. 218.

Wahlström, R., S. Tyagunov, G. Grünthal, L. Stempniewski, J. Zschau and M. Müller (2004) "Seismic Risk Analysis for Germany: Methodology and Preliminary Results", in D. Malzahn and T. Plapp, eds, *Disasters and Society: From Hazard Assessment to Risk Reduction*, Berlin: Logos Verlag, pp. 83–90.

Wong, C. (2001) *Key Phrase: Sustainability Indicators*, European Spatial Planning Research and Information Database, available at http://www.esprid.org/keyphrases%5C29.pdf.

Yodmani, S. (2001) "Disaster Risk Management and Vulnerability Reduction: Protecting the Poor", paper presented at the Asia Pacific Forum on Poverty, Asian Development Bank, available at http://unpan1.un.org/intradoc/groups/public/documents/APCITY/UNPAN009672.pdf.

World Bank (2005) *Hazard Risk Management*, available at http://web.worldbank.org/WBSITE/EXTERNAL/TOPICS/EXTURBANDEVELOPMENT/EXTDISMGMT/0,,menuPK:341021~pagePK:149018~piPK:149093~theSitePK:341015,00.html.

List of Contributors

Masaru Arakida is a Senior Researcher at the Asian Disaster Reduction Center (ADRC), where he is coordinating the GLIDE programme, in which a universal coding system for natural hazard databases has been developed in order to enable the linking of national and global datasets on natural hazards. Other recent foci of his work are post-tsunami assessments in Indonesia as well as, for example, field surveys dealing with the impacts of the 2001 earthquake in West India.
Email: arakida@adrc.or.jp

Peter Billing is responsible for the Civil Protection Monitoring and Information Centre in the Directorate-General for the Environment of the European Commission. He has a PhD in political science from the University of Heidelberg. Milestones in his professional career include research experience at the University of Tübingen (Germany), holding positions at the Federal Ministry for the Environment, the International Relations Department, Bonn (Germany) and the European Commission (Directorate General for the Environment, Brussels, and the European Commission Humanitarian Aid Office, ECHO). His main fields of professional interest include international relations, conflict management, humanitarian aid and civil protection.
Email: peter.billing@cec.eu.int

Jörn Birkmann is an Academic Officer at UNU-EHS and holds a PhD in spatial planning from the Dortmund University as well as a licence as an urban planner. He has broad experience of working in the field of vulnerability, sustainable development and environmental assessment, having specialised in the area concerning the development of assessment methodologies and

497

indicators to estimate and evaluate different socio-economic trends at sub-national, local and household scale. His work experience also encompasses research activities and lectures around the globe, for example in London and Mexico. Besides research and training activities at UNU-EHS, Birkmann also teaches at the Institute for Geography at the University of Bonn. His current field of activities is focused on vulnerability assessment in coastal communities in Sri Lanka, Indonesia, Spain and Egypt. He also coordinates the development and testing of indicators to measure vulnerability to floods in Russia, Hungary, Serbia and Rumania.
Email: birkmann@ehs.unu.edu

Janos J. Bogardi is Director of the United Nations University Institute for Environment and Human Security (UNU-EHS) in Bonn, Germany. He is also co-opted professor of the University of Bonn. Following studies of water resources engineering at the Technical University of Budapest he earned a doctoral degree at the University of Karlsruhe in Germany in 1979. He holds doctor honoris causa degrees from universities in Warsaw and Budapest. Prior to his present work at UNU he was professor at the Wageningen Agricultural University in the Netherlands and served as Chief of Section at the Division of Water Sciences of UNESCO in Paris.
Email: bogardi@ehs.unu.edu

Christina Bollin has been Programme Manager for Disaster Risk Management at the German Technical Cooperation (GTZ) until 2005. She is now working as an independent consultant. Bollin has a PhD in political science. She has gained regional experience in development cooperation and disaster risk management in Latin America, Mozambique and South East Asia. Bollin's special focus is on linking disaster risk management with development, strengthening institutional and governance structures, capacity development especially for local actors, and impact monitoring.
Email: c.bollin@t-online.de

Omar D. Cardona is Professor of Integrated Disaster Risk Management at the Institute of Environmental Studies (IDEA) of the National University of Colombia (UNC). Previously he was a Professor at the University of Los Andes in Bogotá. Cardona took the lead in several disaster-related institutions. He is the former President of the Colombian Association for Earthquake Engineering. Between 1992 and 1995 he was the Director General of the National Directorate of Disaster Prevention and Attention of Colombia. He is also a founding member of the Latin-American Network of Social Studies on Disaster Prevention (LA RED). In 2004 Cardona was the winner of the UN Sasakawa Prize for Disaster Prevention. He studied civil engineering at UNC and has a PhD in earthquake engineering and structural dynamics from the Technical University of Catalonia.
Email: ocardona@uniandes.edu.co

Daniele Ehrlich is Leader of the Information Support for Effective and Rapid Action Project at the

Institute for the Protection and Security of the Citizen of the Joint Research Centre of the European Commission. The project addresses special needs in response strategies to natural and man-made disasters. It developed an information infrastructure providing statistical and geographical information to the European Commission services about EU aid, development and assistance policies. Previous employment included a post-doctorate position at UNEP/GIRD in Sioux Falls, South Dakota, 1992, and at the Joint Research Centre of the European Commission. Ehrlich has a PhD in geography from UC Santa Barbara, USA.
Email: Daniele.Ehrlich@jrc.it

Maxx Dilley is a Senior Policy Advisor at the United Nations Development Programme (UNDP). He works in the Bureau for Crisis Prevention and Recovery Disaster Reduction Unit where he manages a global natural disaster risk identification programme and oversees work in the area of climate risk management. Prior to joining UNDP, from 2002 through June 2005, he worked at the International Research Institute for Climate Prediction (IRI) at Columbia University in New York as a Research Scientist in disaster and risk management. Dilley directed the IRI's Africa programme and co-directed climate impacts research. Before that he worked for two years at the World Bank Disaster Management Facility and for seven years at the US Agency for International Development's Office of US Foreign Disaster Assistance. He has PhD and MSc degrees from

the Pennsylvania State University and a BA from the University of Delaware, all in geography. Dilley specialises in disaster and risk assessment, climate risk management and food security. He has designed and managed disaster risk management programmes in Africa, Latin America and Asia. Recent work includes co-authoring the report, *Natural Disaster Hotspots: A Global Risk Assessment*. Email: maxx.dilley@undp.org

Nishara Fernando is a Senior Lecturer at the Department of Sociology of the University of Colombo, Sri Lanka, and a Research Associate in Social Policy Analysis at the Research Centre of the University of Colombo. He has also worked as a Research Officer at the Centre for the Study of Human Rights in the Faculty of Law at the University of Colombo. Nishara Fernando holds a Graduate Diploma in sociology from La Trobe University in Melbourne, Australia, and an MPhil from the University of Colombo, Sri Lanka. His MPhil thesis was on "Vulnerability to Chronic Poverty and Livelihood Strategies of Urban Resettled Families". His specialisation lies in the fields of applied research methodology, urban livelihood analysis, globalisation, education and youth in Sri Lanka. Moreover, he has broad work experience as a visiting lecturer for institutions like the National Science Foundation, the Prison Training and Rehabilitation Centre and the Sri Lanka Foundation Institute. His research and consulting experience includes projects for UNDP, the German Development Cooperation (GTZ) and the Colombo Municipal

Council, to only name a few.
Email: nishara2000@yahoo.com

Stefan Greiving has been a Researcher at the University of Dortmund since 1995 and a Lecturer since 2001. Currently he is the Project Manager of the ARMONIA project, the "Applied Multi Risk Mapping of Natural Hazards for Impact Assessment". He is also Chair (temporary head) of Planning and Environmental Law at the Faculty of Spatial Planning of the University of Dortmund and a full member of the German Academy for Spatial Planning. Greiving has a Diploma in spatial planning and a PhD in urban land use planning. His post-doctorate thesis deals with linkages on spatial planning and administration. His special field of professional interest includes spatially oriented risk management, risk governance, institutional vulnerability, planning and environmental law, and comparative planning studies.
Email: stefan.greiving@uni-dortmund.de

Hans van Ginkel is the Rector of the United Nations University, situated in Tokyo since September 1997. He has a PhD cum laude from Utrecht University and honorary doctorates from Universitatea Babes-Bolyai, Cluj, Romania, the State University of California (Sacramento), and the University of Ghana. He was elected President of the International Association of Universities (IAU, Paris), serving until July 2004. He is Vice Chair of the Board of Trustees of the Asian Institute of Technology (AIT, Bangkok), a member of the Academia Europaea; Honorary

Fellow of the Institute for Aerospace Survey and Earth Sciences (ITC, Enschede) and a former Rector of Utrecht University in the Netherlands. He serves as a member and officer of several professional associations and organisations. His fields of professional interest include urban and regional development, population and housing studies, science policy, internationalisation and university management. He has published widely on these areas, having contributed extensively to the work of various international educational organisations.
Email: rector@hq.unu.edu

Siri Hettige is a Professor at the Department of Sociology of the University of Colombo, Sri Lanka. He holds a PhD in social anthropology from Monash University, Australia. Moreover, he was awarded the status of Senior Fulbright Visiting Scholar at the University of Pennsylvania, USA, in 1994–1995 as well as having four other international visiting professor and fellowship titles at the Universities of Edinburgh (UK), Adelaide (Australia), Bonn (Germany), and Kuopio (Finland) in the following years. Siri Hettige is currently working in a joint project with UNU-EHS to assess the impact of the 2004 tsunami on coastal communities in Sri Lanka. Since 2003, he has been President of the Sociological Association of Sri Lanka. He is also a member of the Sri Lanka Association for the Advancement of Science as well as of the National Science Foundation. Since the beginning of the 1990s, he has published a vast number of

monographs and research articles dealing particularly with the topics of the sociology of youth, globalisation and education, labour migration and the sociology of ethnic conflicts.
Email: sthetti@webmail.cmb.ac.lk

Ria Hidajat works for the German Development Cooperation (GTZ) within the Advisory Project on Disaster Risk Management in Development Cooperation. Prior to taking up this position, she acted as a Scientific Advisor to the Centre for Natural Risks and Development at the University Bonn (ZENEB) and the German Committee on Disaster Reduction (DKKV). Furthermore, she has been a seconded expert to the Federal Institute of Geosciences and Natural Resources (BGR) in Indonesia within a GTZ/BGR Cooperation project "Mitigation of Geohazards". She has an MSc in geography. Her fields of special professional interest include disaster risk management and development, risk perception/ risk awareness and community-based disaster risk management.
Email: ria.hidajat@gtz.de

Stefan Hochrainer is associated with the International Institute for Applied Systems Analysis (IIASA, www.iiasa.ac.at). He is currently working on his PhD about optimal financial strategies against natural disasters in developing countries in the Risk and Vulnerability Program. Hochrainer's main fields of interest include disaster risk management and international solution strategies for catastrophic events, insurance and disaster mitigation, extreme value theory and queuing systems. He has studied statistics at the University of Vienna where he received his Master's degree in 2002. He also received a Master's degree in sociology in 2004 at the University of Vienna.
Email: hochrain@iiasa.ac.at

Jill Jäger is a Senior Researcher at the Sustainable European Research Institute (SERI) in Vienna, Austria. Her projects deal with a range of activities related to scientist–practitioner dialogues on sustainability, as well as on integrated sustainability assessment. Jäger received her BSc degree in environmental sciences from the University of East Anglia (UK) in 1971. The University of Colorado (USA) awarded her a PhD in geography (climatology) in 1974. In 1991 Jäger became Director of the Climate Policy Division of the Wuppertal Institute for Climate, Environment and Energy in Germany. In 1994 she joined the International Institute for Applied Systems Analysis (IIASA) as Deputy Director for Programs, where she was responsible for the implementation and coordination of the research programme. From 1996 until 1998 she was Deputy Director of IIASA and Executive Director of the International Human Dimensions Programme on Global Environmental Change (IHDP) until 2002. Jäger's main fields of interest are the linkages between science and policy in the development of responses to global environmental issues.
Email jill.jaeger@seri.at

Robert Benjamin Kiunsi is a Senior Lecturer in Environmental Planning, Natural Resources Assessment, Environmental Impact Assessment

and Disaster Management at the University College of Lands and Architectural Studies (UCLAS), Tanzania. Furthermore, since 2001 he acts as a Director of Postgraduate Studies at UCLAS. Robert Kiunsi is responsible for the recently conducted Vulnerability Assessment in Tanzania. He has an MSc in rural land ecology from the International Institute for Geo-information, Science and Earth Observation in the Netherlands. He received his PhD in the context of desertification control in 2002 at the University of Cape Town, South Africa.
Email: kiunsi@uclas.ac.tz

Marcel T.J. Kok works in the Climate and Global Sustainability Programme of the Netherlands Environmental Assessment Agency. He studied public policies and planning and environmental science at Utrecht University in the Netherlands. He worked for five years as a Programme Officer at the Netherlands' National Research Programme on Climate Change. Research interests include global environmental governance and the science–practice interface.
Email: Marcel.Kok@mnp.nl

Vladimir Kotov is a Professor and Doctor of Economics in the Russian Academy of Sciences and the President of the EcoPolicy Research and Consulting. He has been a Guest Professor at the University of Augsburg, Germany, at the Institute for Applied Systems Analysis (IIASA), Austria, and the Institute for Global Environmental Strategies (IGES), Japan. He has also been a Vice Chairman of the Department of Education at the Union of

Russia's Entrepreneurs. His current activities include research on global environmental change and sustainable development policies in Russia during the transition to a market economy, on natural resource economics, and national and international natural disaster reduction policies. Kotov is currently involved in a number of international research projects such as the collaborative project on institutional capacity in natural disasters risk reduction in Asia (APN), the project on institutional coordination among stakeholders in environmental risk management in large river basins (EC/INCO), and the assessment of implications of global climate change for various sectors in Russia. Earlier he participated in numerous international research projects. He also has been involved in consultancy for the UNDP and OECD on various aspects of Russian environmental policies.
Email: vl-kotov@mtu-net.ru

Elisabeth Krausmann is a Scientific Officer at the Institute for the Protection and Security of the Citizen (IPSC) of the Joint Research Centre (JRC) to the European Comission in Ispra, Italy. In 1999 she earned her PhD in physics at the Institute for Nuclear Physics at the University of Technology in Vienna, Austria. Her main task at the IPSC is to contribute to the Natural and Environmental Disaster Information Exchange System (NEDIES) project and to assist in its scientific management. Within the NEDIES project she is in charge of the analysis of natural-hazard-triggered technological accidents

(NATECHs), and the development of tools to assess NATECH risk. Krausmann's fields of special professional interest include the interrelationship between natural and technological hazards, NATECH accidents, lesson learning, accident and disaster management and consequence modelling.
E-mail: elisabeth.krausmann@jrc.it

Louis Lebel is Director of the Unit for Social and Environmental Research (USER), which is a research organisation affiliated with the Faculty of Social Science of the Chiang Mai University in Thailand. He received a PhD from the Department of Zoology at the University of Western Australia in Perth in 1993. Currently, he acts as a Science Coordinator for the Global Environmental Change Programme in South East Asia. He also helps coordinates collaborative research networks on sustainable production–consumption systems, urbanisation in Asia, and water governance in the Mekong Region.
Email: louis@sea.user.org

Joanne Linnerooth-Bayer is the Leader of the Programme on Risk and Vulnerability at the International Institute for Applied Systems Analysis (IIASA) in Laxenburg, Austria. She received her PhD in economics from the University of Maryland. At IIASA, she has worked within interdisciplinary teams exploring social and economic issues related to environmental and technological risks, including risk estimation, risk–benefit analysis, risk perception, culturally determined risk construction and risk burden sharing. She collaborated with the World Bank and the Inter-American Development Bank on activities concerning the improvement of the financial capacity of disaster-prone countries to cope with extreme events. Together with the Kyoto University, she is currently involved in the organisation of an annual conference on Integrated Disaster Risk Management. Dr Linnerooth-Bayer is on the editorial board of three international journals dealing with risk.
Email: bayer@iiasa.ac.at

Peter Lauwe works at the Centre for Critical Infrastructure Protection of the Federal Office for Civil Protection and Disaster Assistance. His responsibilities include the implementation of risk analyses methodologies in the context of critical infrastructures. Part of his work is also the development of safety/security concepts for water supply, food supply, health care services and operators handling hazardous material. Lauwe has a Diploma in civil engineering from the Darmstadt Technical University in Germany. As a civil engineer he worked for several institutions including Björnsen Consulting Engineers (BCE) in Koblenz, Germany, and CAU (Gesellschaft für Consulting und Analytik im Umweltbereich mbH) in Dreieich, Germany.

Peter Lauwe's special professional interests are hazard, vulnerability and risk assessment and safety/security concepts.
Email: peter.lauwe@bbk.bund.de

Ulrike Madengruber is a Desk Officer for Pakistan and the Balkans with an Austrian humanitarian NGO. She studied economics and

languages at the University of Vienna, followed by a Master's course of European studies at the same university. She has also been associated with other institutions such as the Ludwig Boltzmann Institute of Human Rights, the Austrian Federal Chamber of Commerce and the Austrian Foreign Ministry.
Email: ullima@yahoo.com

Jesse Bacamante Manuta is an Associate Professor at the College of Arts and Sciences of Ateneo de Davao University in the Philippines and a Visiting Research Associate at the Unit for Social and Environmental Research of the Faculty of Social Sciences at Chiang Mai University, Thailand. He has a PhD in urban affairs and public policy with a focus on energy and environmental policy from the Centre for Energy and Environmental Policy at the University of Delaware in the United States.

Another important milestone in Manuta's CV has been his position as a Research Fellow at the Advanced Institute on Vulnerability to Global Environmental Change Programme funded by the Systems for Analysis, Research and Training (START).

Further fields of professional interests include vulnerability and disaster risk governance, sustainable and empowered livelihoods and environmental justice.
Email: jbmanuta@addu.edu.ph

Reinhard Mechler first joined the International Institute for Applied Systems Analysis (IIASA) in September 1999. He is currently a Research Scholar in the Risk and Vulnerability Programme. He studied economics, mathematics and English. He holds a Diploma in economics from the University of Heidelberg and a PhD in economics from the University of Karlsruhe in Germany. Mechler has been analysing the impacts and costs of natural disasters in developing countries, as well as strategies to reduce these costs, in particular those related to risk financing. Furthermore, he has worked on costs, impacts and benefits analysis for the reduction of the effects of air pollution and climate change. He has been a Consultant to the World Bank, the Inter-American Development Bank and the German Technical Cooperation (GTZ).
Email: Mechler@iiasa.ac.at

Manoris Victor Meshack is an Associate Professor at the Department of Urban and Regional Planning and Deputy for Principal Academic Affairs at the University College of Lands and Architectural Studies (UCLAS) in Dar es Salaam, Tanzania. He holds an MA(DS) with a focus on technology, planning, and political economy from the Teachers' College in Dar es Salaam. Meshack was also awarded a PhD in regional planning by the University of Dortmund, Germany, in 1989. Other qualifications include certificates in disaster management, conflict analysis and peace building from the United Nations staff college at the University of Wisconsin, Madison, USA.
Email: Meshack@uclas.ac.ta

Fesil Mushtaq works at the Institute for the Protection and Security of the Citizen (IPSC) of the European Commission Joint Research Centre,

Ispra in Italy. He is carrying out research in the area of risk assessment and industrial safety at the Major Accident Hazards Bureau (MAHB) in Ispra. One of his main activities is the continuous development, operation and management of the Major Accident Reporting System (MARS) and the Seveso Plants Information Retrieval System (SPIRS), which are essential components of the Seveso II Directive. Mushtaq has an MEng in Chemical Engineering which includes a Diploma in Industrial Studies, and a PhD ("Safe Design of Computer-Controlled Pipeless Batch Plants") from the Department of Chemical Engineering at Loughborough University. His fields of special professional interest include industrial risk, process safety, chemical engineering, accident reporting and analysis. Email: fesil.mushtaq@jrc.it

Vishal Narain is an Associate Professor at the School of Public Policy and Governance at the Management Development Institute, Gurgaon, India. His research and teaching interests lie primarily in the field of analysis of public policy and institutions, governance and management of natural resources, local governance, irrigation management reform, water rights and legal pluralism. Vishal holds a PhD from Wageningen University, the Netherlands. He is the author of a book titled *Institutions, Technology and Water Control* published by Orient Longman. His research has been published in international refereed journals such as *Water Policy*. Email: vishalnarain@mdi.ac.in

Elena Nikitina is Director of the EcoPolicy Research and Consulting Institute. She has a PhD from the Institute of World Economy and International Relations of the Russian Academy of Sciences. Her interests include the analysis of environmental governance systems and institutional designs for natural disaster risk reduction and the implementation and effectiveness of international environmental agreements such as the climate change policy. She has been working at the Russian Academy of Sciences and the International Institute for Applied Systems Analysis (IIASA). Moreover, she has acquired broad international work experience, having participated in numerous projects, including the "Volga Vision" of UNESCO and the "Sustainable Water Management Systems in NIS: Problems of Transfer and Adaptation" of the European Commission. Elena Nikitina contributed to the UN/ISDR publication "Living with Risk" and is a lead author of the WGIII Third and Fourth Assessment Report of the IPCC. She is also a member of the scientific committee for the IHDP Global Environment Change and Human Security (GECHS) programme. Email: enikitina@mtu-net.ru

Pascal Peduzzi has been Head of the Early Warning Unit of the UNEP Global Resource Information Unit (GRID) since 1998, where he is responsible for different projects such as reporting of forest fires, the project on Madagascar Fires Scars Mapping. His main tasks at GRID include remote sensing and GIS

analysis, cartography and web design. He previously worked as a teacher in earth sciences and scientific observation for secondary schools in Geneva, as a university tutor in remote sensing and as a scientific collaborator in Australia. Peduzzi has an MSc in remote sensing from James Cook University in Australia.
Email: Pascal.peduzzi@grid.unep.eh

Mark Pelling is a Researcher and Lecturer at the Department of Geography at Kings College, University of London. To date his research focuses on issues of risk governance at a range of scales, including studies of grassroots leadership, linkages between community-based organisations (CBOs) and international development agencies, and the influence of contrasting national governance regimes with respect to risk reduction. Pelling has undertaken substantial research for UNDP and DFID. In 2003 he worked as the editor of the UNDP publication "Disaster Risk: A Challenge for Development". In this context he is involved in the final analysis of the Disaster Risk Index, a global-scale tool developed to assess patterns of vulnerability to disaster risk at the national scale. With DFID, Pelling co-authored a scoping study into reducing disaster risk in 2004, which examined the institutional barriers preventing the mainstreaming of disaster risk reduction into international development agencies' agendas, and opportunities for such integration set against the Millennium Development Goals. Mark is Chair of the RGS-IBG Climate Change Research Group.
Email: mark.pelling @kcl.ac.uk

Georg Pflug is associated with the International Institute for Applied Systems Analysis (IIASS), where he joined the Adaptation and Optimization Project in 1990 to work on the optimisation of stochastic systems and computationally intensive methods in simulation and optimisation. He is currently working in the "Risk and Vulnerability" Programme. Pflug received his PhD in mathematics and statistics and habilitated thereafter in 1981. He was a Professor of Probability and Statistics at the University of Giessen in Germany and then moved to the Department of Statistics and Computer Science at the University of Vienna. Pflug subsequently held guest professorships at the University of California in the United States, at the Université de Rennes, France, at the Technion-Israel (Institute of Technology) in Haifa, Israel, and at Princeton University in the United States.
Email: pflug@iiasa.ac.at

Erich J. Plate is Emeritus Professor of Hydrology and Water Resources Planning at the University of Karlsruhe in Germany. Before he became a Professor at the University of Karlsruhe in 1970, he was Professor of Civil Engineering at Colorado State University, USA. He chaired many committees associated with water both at the national level, for example, the Committee for Water Research of the German Science Foundation and the Scientific Advisory Board of the German National Committee on

Disaster Reduction, and at the international level as Chairman of the Committee for Water Research (COWAR) and the International Council of Scientific Unions. Plate has also been a member of the Scientific and Technological Committee for the IDNDR. Much of his extensive work in water resources has been on flood modelling, risk analysis and risk management.
Email: plate@iwk.uka.de

Angela Queste is Head of the Centre for Critical Infrastructure Protection of the Federal Office for Civil Protection and Disaster Response in Germany. Previously she held positions at the Institute of Public Health in North-Rhine Westphalia (NRW), Germany, and at the WHO Collaborating Centre for Health Related Water Management and Risk Communication. She has a Diploma in geography and a Master's in public health. Her fields of special professional interest include the protection of critical infrastructures, natural disasters, public health and epidemiology.
Email: angela.queste@bbk.bund.de

Fabrice Renaud is an Academic Officer at the Institute for Environment and Human Security of the United Nations University (UNU-EHS). He does research on environmental aspects of vulnerability, especially on the post-tsunami impact on groundwater resources and agriculture in Sri Lanka. Before joining UNU-EHS, Renaud was a Research Officer and Lecturer at Cranfield University in Silsoe, UK, where he focused on the fate of pesticides in the environment through modelling, laboratory and

lysimeter studies. Other previous fields of activities include multiple aspects of land use and land degradation in Thailand and Namibia. Renaud has an MSc in agricultural engineering from Cranfield University. He did his PhD in soil physics (Agronomy) at the University of Arkansas, USA. Further scientific interests today are land degradation (including its socio-economic dimensions), and ground and surface water quality.
Email: renaud@ehs.unu.edu

Stefan Schneiderbauer is a Researcher at the Institute for the Protection and Security of the Citizen (IPSC) of the European Commission Joint Research Centre in Ispra, Italy. He is a scientific collaborator and expert for population density and vulnerability estimations within the project Information Support for Effective and Rapid Action. After taking his MSc in geography at the University of Cologne in Germany, he was scientific employee for digital image analysis and GIS applications at the University for Applied Sciences in Berlin. He also acted as an international consultant and trainer, focusing on the application of remote sensing and GIS technology for natural resource management, in particular in Africa and Europe. Schneiderbauer's fields of special professional interest include the analysis of geo-spatial data (remotely sensed and others) for population density estimations and vulnerability assessments.
Email: stefan.schneiderbauer@jrc.it

Katharina Thywissen is an Academic Officer at the United Nations University, Institute for Environment and Human Security

(UNU-EHS). Her focal area of research is interdisciplinary disaster mitigation, where she looks closely at factors leading to risk. She is also involved in post-tsunami research on economic vulnerability in Sri Lanka.

Thywissen has a Diploma in marine geology (seismic stratigraphy) and a PhD in seismology/geophysics (damage pattern of the Northridge earthquake).

Previously she worked for the Swiss Reinsurance Company, New York, where she dealt with disaster management including hazard modelling, risk calculation, pricing and worldwide exposure control. At UNEP in Nairobi she was engaged in the area of early warning for natural disasters.
Email: thywissen@ehs.unu.edu

Juan Carlos Villagrán de León is an Academic Officer at the Institute for Environment and Human Security of the United Nations University (UNU-EHS) in Bonn, Germany. Currently he is coordinating a project sponsored by the Platform for the Promotion of Early Warning of the United Nations International Strategy for Disaster Reduction (UN/ISDR), focusing on the strengthening of early warning capacities in Sri Lanka. He is also involved in the German–Indonesian Tsunami Early Warning System Project. Contributions from UNU-EHS spanned the coordination of post-doctoral and PhD programmes linking German and Indonesian universities and research centres. Villagrán de León provided various consulting services to GTZ, US-AID, UNDP and Regional Disaster Risk Management Agencies in Central America.

He has a PhD in experimental condensed matter physics from the University of Texas at Austin. His fields of special professional interest include disaster-risk management spanning research, policy development and practical implementation.
Email: villagran@ehs.unu.edu

Ben Wisner is a researcher at the London School of Economics and professor at Oberlin College, Cleveland, USA. He began work in Tanzania in 1966, doing his PhD on coping capacity of households dealing with drought in Kenya (1978). During the period from 1966 to 1995 he worked mostly in Africa on issues concerning wood fuel and rural energy, water and sanitation, drought, flood, community health and food security. In 1996 his work took an urban turn, and he has coordinated research for the United Nations University on urban social vulnerability to disaster. He was Vice Chair of the Earthquakes and Megacities Initiative from 1997–2002, and is currently Vice Chair of the Commission on Risk and Hazards of the International Geographical Union (IGU) and of IGU's Task Force on Megacities. He served as senior technical editor for the UNDP's report, "Reducing Disaster Risk: A Challenge for Development" (Geneva: UNDP, 2004). He is lead author of the second edition of *At Risk* (London: Routledge, 2004), a textbook on vulnerability to natural hazards.
Email: bwisner@igc.org

Steven Wonink works for the Netherlands Environmental Assessment Agency. Currently he is a contributing author for Chapter 8 and was involved in scenario modelling for Chapter 9 of the UNEP's Global Environment Outlook.

He was also a co-author of the Energy for Development paper, prepared for the Energy for Development conference in the Netherlands in December 2004. Wonink has studied physical geography (MSc) at Utrecht University in the Netherlands. His fields of special professional interest include vulnerability research and human security.

Email: steven.wonink@mnp.nl

Index